THE ATMOSPHERE

THe aTM

an introduction to meteorology

Frederick K. Lutgens
Illinois Central College

Edward J. Tarbuck
Illinois Central College

OSPHERE

Prentice-Hall, Inc., Englewood Cliffs, New Jersey 07632

Library of Congress Cataloging in Publication Data

Lutgens, Frederick K.
    The atmosphere.

    Includes bibliographical references and index.
    1. Atmosphere.   2. Meteorology.   3. Weather
    I. Tarbuck, Edward J., joint author. II. Title.
QC861.2.L87   551.5   78-24541
ISBN 0-13-050104-2

THE ATMOSPHERE
an introduction to meteorology

Frederick K. Lutgens / Edward J. Tarbuck

Printed in the United States of America

10 9 8 7 6 5 4 3

PRENTICE-HALL INTERNATIONAL, INC., *London*
PRENTICE-HALL OF AUSTRALIA PTY. LIMITED, *Sydney*
PRENTICE-HALL OF CANADA, LTD., *Toronto*
PRENTICE-HALL OF INDIA PRIVATE LIMITED, *New Delhi*
PRENTICE-HALL OF JAPAN, INC., *Tokyo*
PRENTICE-HALL OF SOUTHEAST ASIA PTE. LTD., *Singapore*
WHITEHALL BOOKS LIMITED, *Wellington, New Zealand*

to our parents—our first and best teachers;
and to our wives and children
for their encouragement and patience.

Contents

Chapter Thirteen **312**
world climates

Appendix A   **353**
metric units

Appendix B   **357**
explanation and decoding
of the daily weather map

Preface

There are few aspects of the physical environment that influence our daily lives more than the phenomena we collectively call weather. With increasing frequency, newspapers, magazines, and television stations are reporting a wider range of weather events as major news stories; an obvious reflection of people's growing interest and curiosity about the atmosphere. Since a meteorology course can take advantage of this increased interest and curiosity about the weather, it is not surprising that introductory courses dealing with the atmosphere have become very popular on many college campuses.

The Atmosphere: An Introduction to Meteorology is designed to meet the needs of students who enroll in such a course. The book is a current and comprehensive, yet not encyclopedic, survey of the atmosphere intended for an introductory audience. Although a portion of modern meteorology is highly quantitative, it was our intent to produce a nontechnical and nonmathematical text. It is our firm belief that weather phenomena can be understood and appreciated by people who do not have a strong background in mathematics.

Since student use of the text was one of our primary concerns, we made a concerted effort to write a book that is highly usable—a tool for learning the basic concepts of meteorology. To achieve this goal, the book is written concisely and is well illustrated. In addition, a number of useful learning aids have been included. For example, important terms are italicized within the body of the text and also appear in a vocabulary review at the end of each chapter. A glossary also serves as a useful vocabulary reference. Review questions and problems that highlight major concepts and aid student review conclude each chapter.

Inasmuch as the aim of this book is to bring about a basic understanding of the atmosphere, the first ten chapters are devoted to a presentation of the major elements and concepts of meteorology. Where appropriate, topics such as air pollution and intentional weather modification are integrated into these chapters. Next, Chapter 11 on weather analysis serves to reinforce and apply many of the ideas from the first ten chapters again. In addition to introducing the reader to the nature of modern weather forecasting, the chapter provides a case history of a storm. The several different perspectives and views of the storm as it passes across the United States are designed to increase the reader's insight and understanding of our often complex middle-latitude weather. The text concludes with two chapters on climate. Chapter 12, "The Changing Climate," explores a topic that is the focus of ever-increasing interest and concern, not only to scientists, but to governments and individuals as well. How is man modifying the climate of cities? Is global climate changing and, if so, in what ways? Is man causing or contributing to any of these changes? These are some of the questions Chapter 12 investigates. Finally, Chapter 13 examines climates in a more traditional manner. Here, the reader will investigate the broad diversity of climatic types around the world and the controls that have created them.

The authors wish to express their thanks to the many individuals, institutions, government agencies, and businesses that provided photographs and illustrations for use in the text. Further, we would also like to acknowledge the aid of our students. Their comments on the preliminary drafts of many chapters helped us to maintain our focus on readability and understanding. Finally, we feel very fortunate to have had the following individuals review and evaluate the manuscript:

Professor Robert B. Batchelder
Boston University

Professor Eric S. Johnson
Illinois State University

Professor Robert C. King
San Jose City College

Professor David W. Marczely
Southern Connecticut State College

Professor David H. Miller
University of Wisconsin, Milwaukee

Professor Robert A. Muller
Louisiana State University

Professor Peter J. Robinson
University of North Carolina

Mr. Malcolm P. Sillars
Commercial Weather Services, Inc.

Professor Phillip J. Smith
Purdue University

Professor John Trischka
Syracuse University

All of their comments, both critical and positive, were welcome and very useful. In many cases their insights and constructive comments were incorporated into our writing. The authors, of course, accept full responsibility for any remaining errors.

Frederick K. Lutgens
Edward J. Tarbuck

Chapter One

INTRODUCTION TO THE ATMOSPHERE

A view of the earth from space affords us a unique perspective of our planet (Figure 1-1). At first it may strike us that the earth is a fragile appearing sphere surrounded by the blackness of space. It is, in fact, just a speck of matter in an infinite universe. As we look more closely, it becomes apparent that the earth is much more than rock and soil. Indeed, the most conspicuous features are not the continents, but the swirling clouds suspended above the surface and the vast global ocean. From such a vantage point we can appreciate why the earth's physical environment is traditionally divided into three major parts: the solid earth or *lithosphere*; the water portion of our planet, the *hydrosphere*; and the earth's life-giving gaseous envelope, the *atmosphere*. It is this latter portion of our physical environment, the atmosphere, that will be the primary focus of this text. However, it should be emphasized that our environment is highly integrated and is not dominated by rock, water, or air alone. Rather, it is characterized by continuous interaction as air comes in contact with rock, rock with water, and water with air. Moreover, the *biosphere*, the totality of life forms on our planet, extends into each of the three physical realms and is an equally integral part of the earth.

People's interest in the atmosphere is probably as old as the history of mankind. Certainly there are few other aspects of our physical environment that affect our daily lives more than the phenomena we collectively call the weather. The clothes we wear and the activities we engage in are strongly influenced by the weather. However, in addition to the day-to-day personal decisions we must make about the weather, there have been and will continue

1

to be an increasing number of political decisions to make involving the atmosphere. Answers to questions regarding air pollution and its control, the effect of supersonic transports and propellants from aerosol spray cans on the upper atmosphere, the possibility of climatic change and its effect on world food production, and the positive or adverse effects of intentional weather modification are all very important and in some cases perhaps even vital. Hence, there is a need for increased awareness and understanding of our atmosphere and its behavior.

weather and climate

Acted upon by the combined effects of the earth's motions and energy from the sun, our planet's formless and invisible envelope of air reacts by producing an infinite variety of weather, which, in turn, creates the basic pattern of global climates. Although they are not identical, weather and climate have much in common. *Weather* is constantly changing, sometimes from hour to hour and at other times from day to day. It is a term that denotes the state of the atmosphere at a given time and place. While changes in the weather are constant and sometimes seemingly erratic, it is nevertheless possible to arrive at a generalization of these variations. Such a description of aggregate weather conditions is termed *climate*. Climate is often defined simply as "average weather," but this is an inaccurate definition. In order to truly portray the character of an area, variations and extremes must also be included. Thus, climate is the sum of all statistical weather information that helps describe a place or region.

The nature of both weather and climate is expressed in terms of the same basic *elements*, those quantities or properties that are measured regularly. The most important of these are: (*a*) the temperature of the air, (*b*) the humidity of the air, (*c*) the type and amount of cloudiness, (*d*) the type and amount of precipitation, (*e*) the pressure exerted by the air, and (*f*) the speed and direction of the wind. These elements constitute the variables from which weather patterns and climatic types are deciphered. Although we shall study these elements separately at first, keep in mind that they are very much interrelated. A change in one of the elements often brings about changes in the others.

Before you study the elements of weather and climate in detail, which begins in Chapter 2, you should read the remainder of this chapter as a general introduction to the physical and chemical nature of the atmosphere.

composition of the atmosphere

In the days of Aristotle air was believed to be one of four fundamental substances that could not be further subdivided into constituent components. The other three substances were fire, earth (soil), and water. Even today the term *air* is sometimes used as if it were a specific gas, which, of course, it is not. The envelope of air that surrounds our planet is a mixture of many discrete gases, each with its own physical properties, in which varying quantities of tiny solid and liquid particles are suspended.

The composition of air is not constant; it varies from time to time and from place to place. However, if the suspended particles, water vapor, and other variable gases were removed from the atmosphere, we would find that its makeup is, in fact, very stable all over the earth up to an altitude of about 80 kilometers. As can be seen in Table 1-1, two elements, nitrogen and oxygen, make up 99 percent of the volume of clean, dry air; most of the remaining 1 percent is accounted for by the inert gaseous element, argon. Although these elements are the most plentiful components of the atmosphere, they are of little or no importance in affecting weather phenomena. Carbon dioxide, however, is present in only minute quantities, but it is nevertheless a meteorologically important constituent of air.

Since carbon dioxide is an efficient absorber of radiant energy emitted by the earth, it is of great interest to meteorologists because it influences the flow of energy through the atmosphere. Although the proportion of carbon dioxide is relatively uniform in the air, its percentage has been rising steadily for more than a century. This increase is attributed to the burning of fossil fuels such as coal, oil, and natural gas, which from 1890 to 1970 increased

3

Table 1-1

Principal Gases of Dry Air

Constituent	Percent by Volume
Nitrogen (N_2)	78.084
Oxygen (O_2)	20.946
Argon (A)	0.934
Carbon dioxide (CO_2)	0.0325
Neon (Ne)	0.00182
Helium (He)	0.000524
Methane (CH_4)	0.00015
Krypton (Kr)	0.000114
Hydrogen (H_2)	0.00005

more than ten times. Much of this additional carbon dioxide is absorbed by the waters of the ocean or is used by plants, but some (about 50 percent) remains in the air. Estimates project that by the year 2000 the atmosphere's carbon dioxide content will rise by approximately 18 percent over the 1970 level. Although the impact of the increased carbon dioxide is difficult to predict, some scientists are concerned that it may effect a warming of the lower atmosphere and hence produce global climatic change. The role of carbon dioxide in the atmosphere and its possible effect on climate are discussed in more detail in Chapter 12.

Like carbon dioxide, some of the atmosphere's variable components, notably water vapor, ozone, and dust particles, are also present in small amounts, but they are nonetheless very important. Water vapor is one of the most variable gases in our atmosphere. In the warm and wet tropics it may account for up to 4 percent, by volume, of the atmosphere, while in the air of deserts and polar regions, it may comprise but a tiny fraction of 1 percent. The significance of atmospheric moisture to all life is universally recognized and does not require further emphasis here. However, the roles that it plays in the physical processes of the atmosphere are less well-known and are extremely important meteorologically.

In addition to the fact that water vapor is the source of all clouds and precipitation, it has the ability to absorb not only radiant energy given off by the earth but also some solar energy. Therefore, along with carbon dioxide, it is a controlling influence upon energy transfer through the atmosphere. Further, water is the only substance that can exist in all three states of matter (solid, liquid, and gas) at the temperatures that normally exist on earth. As water changes from one state to another, it absorbs or releases heat energy (termed latent heat). In this manner, heat absorbed at one location is transported by winds to other locales and released. The released latent heat, in turn, supplies the energy that drives many storms. Chapters 4 and 5 examine the many important roles of water in the air in detail.

Another important component of the atmosphere is *ozone*, the triatomic form of oxygen (O_3). Ozone is not the same as the oxygen we breathe, which

4

Significance of ozone ✳ ↓

has two atoms per molecule (O_2). There is very little of this gas in the atmosphere (less than 0.00005 percent by volume) and its distribution is not uniform. In the lowest portion of the atmosphere it represents less than one part in 100 million. It is concentrated well above the surface between 10 to 50 kilometers with a peak near the altitude of 25 kilometers. In this altitude range molecules of oxygen (O_2) are split into single atoms of oxygen after absorbing shortwave solar energy. Ozone is then formed when an atom of oxygen collides with a molecule of oxygen in the presence of a third, neutral molecule that allows the reaction to occur without being consumed in the process. Ozone is probably concentrated where it is because of a balance between two factors: (1) the availability of shortwave ultraviolet energy from the sun that is necessary to produce atomic oxygen and (2) an atmosphere dense enough to bring about the required collisions between molecular oxygen and atomic oxygen.

The presence of the ozone layer in our atmosphere is of vital importance to those of us on earth. The reason lies in the capability of ozone to <u>absorb the burning ultraviolet radiation from the sun</u>. If ozone did not act to filter a great deal of the ultraviolet radiation and if the rays were allowed to reach the surface of the earth, our planet would likely be uninhabitable for most life as we know it. Part of the controversy surrounding the development of supersonic transports is linked to the fear that these high-flying aircraft might disturb the ozone layer. The fear, which has yet to be verified, rests upon the fact that the engines of these planes emit nitrogen oxides that would react with and reduce the amount of protective ozone. It is believed that a decrease of as little as 10 percent in the ozone layer would produce a 20-percent increase in the incidence of skin cancer. Research is also being conducted to determine the effect of fluorocarbons on the ozone layer. Many scientists believe that fluorocarbons, which are used as propellants in some aerosol sprays, drift upward and release chlorine, which, in turn, reacts with and destroys ozone.

The movements of the atmosphere are sufficient to keep a large quantity of solid particles suspended within it. Although visible dust sometimes clouds the sky, these relatively large particles are too heavy to remain for very long in the air. However, many particles are microscopic and remain suspended for considerable periods of time. They may originate from many sources, both natural and man-made and include: sea salts from breaking waves, fine soil blown into the air, smoke and soot from fires, pollen and microorganisms lifted by the wind, ash and dust from volcanic eruptions, and more. Such particles are most numerous in the lower part of the atmosphere near their primary source, the earth's surface. Nevertheless, the upper atmosphere is not free of them, because some dust is carried to great heights by rising currents of air and other particles are contributed by meteors that disintegrate as they pass through the earth's envelope of air. From a meteorological standpoint these tiny, often invisible particles can be significant. First, many are water absorbent and consequently act as surfaces upon which water vapor may condense, a very important function in the formation of clouds and fog. Second, dust may intercept and reflect incoming solar radiation. Thus, when dust loading of the atmosphere is great, as it may be following an explosive volcanic eruption, the amount of sunlight reaching the earth can be measurably reduced. Third, dust in the air contributes to an optical phenomenon we have all observed—the varied hues of red and orange at sunrise and sunset.

origin of the atmosphere

The atmosphere has not always consisted of the same relatively stable mixture of gases that we breathe today. On the contrary, the present mixture of gases that makes up our atmosphere is the result of very gradual change, a slow evolutionary process that began soon after the earth came into being 4.5 to 5 billion years ago.

Initially our planet was a molten spheroidal mass with surface temperatures in excess of 8000°C. Because of the excessively high temperatures, the gases making up the newly formed planet's first atmosphere readily escaped the earth's gravity and were lost to space. As the earth cooled, a solid crust formed and the gases that had been dissolved in the molten rock were gradually released, a process called *degassing*. Thus, an atmosphere believed to be made up of gases similar to those that are currently released during volcanic eruptions came into being. The principal components of this "new" atmosphere were probably water vapor, carbon dioxide, and nitrogen.

As the earth continued to cool, clouds formed and the greatest rains the earth has ever experienced commenced. At first, the water evaporated before reaching the surface or was quickly boiled away. This helped to speed the cooling of the earth's surface. When the earth had cooled sufficiently, the torrential rains continued for thousands of years, filling the ocean basins. This event not only diminished the amount of water vapor in the air, but it carried away much of the carbon dioxide as well.

We are now faced with an interesting dilemma. If the earth's primitive atmosphere resulted from volcanic degassing, it could not have contained free oxygen. How then did our present oxygen-rich atmosphere come into existence? Scientists have proposed two probable sources of the free oxygen in the atmosphere. It is known that water vapor that is carried into the upper atmosphere is dissociated into hydrogen and oxygen by the action of the sun's ultraviolet radiation. Hydrogen, being a very light gas, escapes the atmosphere while the heavier oxygen atoms remain and combine to form molecular oxygen (O_2). Although there is little doubt that some of the atmosphere's free oxygen was created in this manner, this very slow process is not adequate to account for the present percentage of oxygen in our atmosphere.

A second more important source of oxygen is believed to have been (and, indeed, continues to be) green plants. By the process of *photosynthesis* plant life utilizes sunlight to generate oxygen by changing water and atmospheric carbon dioxide into organic matter. This method of oxygen production obviously implies the need for life on earth prior to the time that oxygen was present in the atmosphere. Scientists believe that the first life forms, probably

6

bacteria, carried out their metabolic processes without oxygen. Even today many of these anaerobic bacteria still exist. Next, primitive green plants evolved, which, in turn, supplied most of the free oxygen to support higher forms of life. Slowly the amount of oxygen in the atmosphere increased. By the beginning of the Paleozoic era, some 600 million years ago, the fossil record reveals that organisms that require oxygen became abundant in the sea. Two hundred million years later, during the Devonian period, land plants became widespread. Hence, since the makeup of our atmosphere is related to life, its composition must have evolved through time, just as the abundance and variety of life forms changed through the ages.

In summary, the basic steps in the evolution of our atmosphere include the following:

1. Because of the very high surface temperatures the earth's first atmosphere escaped to space.

2. A new atmosphere was created by the gradual release of gases that had been dissolved in molten rock. This atmosphere, however, contained no free oxygen.

3. As the earth cooled, torrential rains filled the ocean basins, thus reducing the amount of water vapor in the air as well as "washing out" much of the carbon dioxide.

4. The free oxygen of the atmosphere was slowly added primarily by green plants carrying on the process of photosynthesis.

exploring the atmosphere

For more than 200 years people have been attempting to explore that part of the atmosphere that seemed out of reach. In 1752 Benjamin Franklin, using a kite, made his famous discovery that lightning is an electrical charge. Not many years later kites were being used to observe temperatures above the earth's surface. In the late eighteenth century manned balloons were used in an attempt to investigate the composition and properties of the "upper" atmosphere. Several manned ascents were attempted over the years, but they were dangerous, sometimes fatal undertakings and seldom exceeded the 5 to 8 kilometer level. Unmanned balloons, on the other hand, could rise to greater heights, but there was no assurance that the instruments that were carried aloft would be recovered. Notwithstanding the difficulties and dangers, considerable data were gathered on the nature of the air above.

Today, improved unmanned balloons continue to play a significant role in the systematic investigation of the atmosphere. Since the late 1920s balloons have carried aloft *radiosondes*, lightweight packages of instruments fitted with radio transmitters (Figure 1-2). Since the radiosonde sends back data on temperature, pressure, and relative humidity, the recovery of the equipment is no longer essential. When the radiosonde is tracked by radar in order to obtain data on wind speed and direction at various levels, the system is known as *rawinsonde*.

Although the lower atmosphere was, and still is explored by kite- and balloon-borne instruments, the direct exploration of the upper atmosphere is a relatively recent accomplishment. Immediately after World War II the United States acquired a vehicle that revolutionized the exploration of the atmosphere—the rocket. Over the years scientists devised several methods of gathering data on the upper atmosphere by using rockets, several of which are briefly discussed below:

1. A sphere, ejected from a rocket, falls toward the earth. As it falls, air density increases and friction slows the sphere. Through the use of ground-based radar, velocities and accelerations are determined and the air's density at different levels is deduced. If the density is known, the air temperature can be calculated.

2. Instrument packages are ejected. As they slowly descend by parachute, data is radioed to the ground.

3. Information on the nature of winds in the upper atmosphere is sometimes gathered by using radar to track the parachutes just mentioned or small metal-coated strips that are ejected by rockets high up in the atmosphere.

4. Sometimes rockets carry grenades that are fired at intervals as the vehicle speeds through the upper atmosphere. Microphones at the earth's

(a)

Figure 1-2

(a) A radiosonde sends back data on conditions aloft. (b) Radiosonde stations in the United States. (Courtesy of NOAA.)

(b)

9

surface then detect the arrival time and angle of the sound waves and through a series of complicated calculations scientists translate this information into wind and temperature data for various layers above.

Since 1957, the year the Soviets launched Sputnik I, satellites, too, have played an important part in gathering information on the upper atmosphere. In addition, these "eyes in the sky" transmit a wealth of data on the lower atmosphere, especially on the nature of clouds and the atmospheric circulation patterns they reveal.

Prior to the time of rockets and satellites knowledge of the atmosphere beyond about 30 kilometers had come solely from indirect, ground-based measurements. Some of these techniques are still in use today. In pre-rocket days much was learned about the temperature of the upper atmosphere by studying the behavior of meteor trails and examining the refraction (bending) of sound waves by the atmosphere. Still other data were collected on the electrical nature of the upper atmosphere by examining the behavior of radio waves. Moreover, by analyzing the light emitted by the gases of the upper atmosphere through spectrographic means, the composition of the air at great heights was determined. Thus, although direct measurements by rockets and satellites are vital to a complete understanding of the upper atmosphere, much had been deciphered in pre-rocket days by using many ingenious ground-based observational techniques.

vertical structure of the atmosphere

To say that the atmosphere begins at the earth's land–sea surface and extends upward is rather obvious. But, where does the atmosphere end and outer space begin? In order to understand the vertical extent of the atmosphere, let us examine the changes in atmospheric pressure with height. The pressure of the atmosphere is simply the weight of the air above. At sea level the average pressure is slightly more than 1000 millibars, a pressure that corresponds to a weight of 1 kilogram of air above every square centimeter of earth. Obviously, the pressure at higher altitudes is less. This was borne out in a dramatic way by some nineteenth-century balloonists attempting to study the upper atmosphere. Several were known to have passed out upon reaching heights in excess of 6 kilometers and some even perished in the rarified upper air.

By examining Table 1-2 you can see that one-half of the atmosphere lies below an altitude of 5.6 kilometers. At about 16 kilometers 90 percent of the atmosphere has been traversed, and above 100 kilometers only 0.00003 percent of all the gases making up the atmosphere remains. At this latter altitude the atmosphere is so tenuous that the density of air is less than could be found in the most perfect man-made vacuum at the earth's surface. Moreover, the graphic portrayal of pressure data in Figure 1-3 shows that the rate of pressure decrease is not constant. Rather, pressure decreases at a decreasing rate with an increase in altitude until, beyond an altitude of about 50 kilometers, the decrease is very slight. Put another way, the data

pressure at level (handwritten annotation)

Table 1-2

Percent of Sea-Level Pressure Encountered
at Selected Altitudes

Altitude (km)	Percent of Sea-Level Pressure
0	100
.5.6	50
16.2	10
31.2	1
48.1	0.1
65.1	0.01
79.2	0.001
100	0.00003

illustrate that air is highly compressible, that is, it expands with decreasing pressure and becomes compressed with increasing pressure. Consequently, traces of our atmosphere extend for thousands of kilometers beyond the earth's surface. Thus, to say where the atmosphere ends and outer space begins is very arbitrary and to a large extent depends upon what phenomena one is studying. It is apparent that there is no sharp boundary.

In summary, the data on vertical pressure changes reveal that the vast bulk of the gases making up the atmosphere is very near the earth's surface and

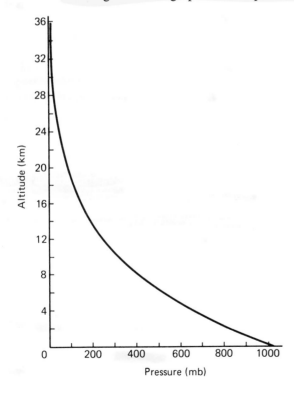

Figure 1-3

Pressure variations with altitude.

that the gases gradually merge with the emptiness of space. When compared to the size of the earth, with its radius of about 6400 kilometers, the envelope of air surrounding our planet is indeed very shallow.

temperature changes in the vertical

By the early part of the twentieth century much had been learned about the lower atmosphere and through the use of indirect methods there was some knowledge of the upper air. Data from balloons and kites had revealed that the air temperature dropped with increasing height above the earth's surface. Although measurements had not been taken above a height of about 10 kilometers, scientists believed that the temperature continued to decline with height to a value of absolute zero ($-273°C$) at the outer edge of the atmosphere. However, in 1902 the French scientist, Leon Philippe Teisserenc de Bort, refuted the notion that temperature decreases continuously with an increase in altitude. In studying the results of more than 200 balloon flights Teisserenc de Bort found that at an altitude between 8 and 12 kilometers the temperature stopped decreasing and leveled off. This surprising discovery was at first doubted, but subsequent data gathering confirmed his findings. Later, through the use of radiosondes and rocket-sounding techniques, the temperature structure of the atmosphere up to great heights became clear.

Today the atmosphere is divided vertically into four layers on the basis of temperature (Figure 1-4). The bottom layer, where temperature decreases with an increase in altitude, is known as the *troposphere*. The term was coined in 1908 by Teisserenc de Bort and literally means the region where air "turns over," a reference to the appreciable vertical mixing of air in this lowermost zone. Although the temperature decrease in the troposphere is variable, it averages $6.5°C$ per kilometer, a figure known as the *normal lapse rate*. The lapse rate continues to an average height of approximately 12 kilometers. However, the thickness of the troposphere is not everywhere the same. It reaches to heights in excess of 16 kilometers in the tropics, but in polar regions it is more subdued, extending to 9 kilometers or less (Figure 1-5). Warm temperatures and highly developed thermal mixing are responsible for the greater vertical extent of the troposphere near the equator. As a result, the lapse rate extends to great heights, and in spite of the relatively high temperatures below, the lowest tropospheric temperatures are found aloft in the tropics and not at the poles.

The troposphere is the chief focus of meteorologists, for it is in this layer that essentially all of the phenomena that we collectively refer to as weather occur. Almost all clouds and certainly all of the precipitation as well as all of our violent storms are born in this lowermost layer of the atmosphere. There should be little wonder why the troposphere is often called the "weather sphere."

Beyond the troposphere lies the *stratosphere*; the boundary between the troposphere and the stratosphere is known as the *tropopause*. In the strato-

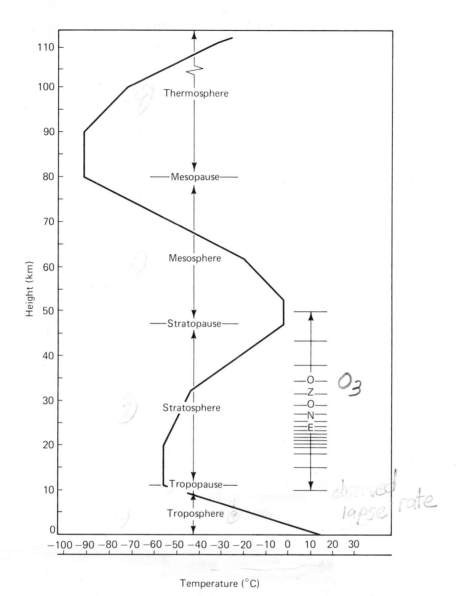

Figure 1-4

Thermal structure of the atmosphere to a height of about 110 kilometers.

sphere the temperature at first remains constant to a height of about 20 kilometers before it begins a gradual increase that continues until the *stratopause* is encountered at a height of about 50 kilometers above the earth's surface. Below the tropopause atmospheric properties are readily transferred by large-scale turbulence and mixing, but above it, in the stratosphere, they are not. The reason for the increased temperatures in the stratosphere is the fact that it is in this layer that the atmosphere's ozone is concentrated. Recall

13

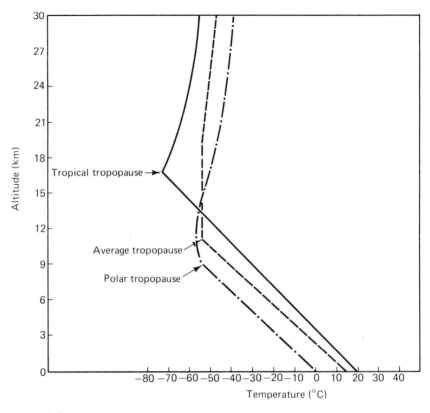

Figure 1-5

Differences in the height of the tropopause. (After Weather For Aircrews, AFM 51-12.)

that ozone absorbs ultraviolet radiation from the sun. As a consequence, the stratosphere is heated.

In the *mesosphere* temperatures again decrease with height until, at the *mesopause*, some 80 kilometers above the surface, the temperature approaches −100°C. Extending upward from the mesopause and having no well-defined upper limit is the *thermosphere*, a layer that accounts for but a minute fraction of the atmosphere's mass. In the extremely rarified air of this outermost layer temperatures again increase as a result of the absorption of very short-wave solar energy by atoms of oxygen and nitrogen. Although temperatures rise to extremely high values of more than 1000°C, such temperatures are not strictly comparable with those experienced near the earth's surface. Temperature is defined in terms of the average speed at which molecules are moving. Since the gases of the thermosphere are moving at very high speeds, the temperature is obviously very high. However, the gases are so sparse that only a very few of these fast-moving air molecules would collide with a foreign body and hence only an insignificant quantity of energy would be transferred. Thus, the temperature of a satellite orbiting the earth in the thermosphere is determined chiefly by the amount of solar radiation it absorbs

14

and not by the temperature of the surrounding air, and if an astronaut inside were to expose his hand, it would not feel hot.

vertical variations in composition

In addition to the layers defined by varying temperatures in the vertical, other layers or zones are also recognized in the atmosphere. Based upon composition, the atmosphere is often divided into two layers, the homosphere and the heterosphere. From the earth's surface to an altitude of about 80 kilometers the makeup of the air is uniform in terms of the proportions of its component gases. That is, the composition is the same as that shown earlier in Table 1-1. This lower uniform layer is termed the *homosphere*; the zone of homogeneous composition. By contrast, the rather tenuous atmosphere above 80 kilometers is not uniform. Since it has a heterogeneous composition, the term *heterosphere* is used. Here the gases are arranged into four roughly spherical shells, each with a distinctive composition. The lowermost layer is dominated by molecular nitrogen (N_2); next, a layer of atomic oxygen (O) is encountered, followed by a layer dominated by helium (He) atoms, and finally a region consisting of hydrogen (H) atoms. The stratified nature of the gases making up the heterosphere is in accordance with their weights. Since molecular nitrogen is the heaviest, it is lowest. The lightest gas, hydrogen, is outermost.

Located in the altitude range between 80 to 400 kilometers, and thus coinciding with the lower portions of the thermosphere and heterosphere, is an electrically charged layer known as the *ionosphere*. Here molecules of nitrogen and atoms of oxygen are readily ionized as they absorb high-energy, short wavelength solar energy. In this process each affected molecule or atom loses one or more electrons and becomes a positively charged ion. Although ionization occurs at heights as great as 1000 kilometers and extends as low as perhaps 50 kilometers, positively charged ions and negative electrons are most dense in the 80- to 400-kilometer height range. The concentration of ions is not great below this zone because much of the needed shortwave radiation has already been depleted. In addition, greater atmospheric density results in a large percentage of the free electrons that are produced being captured by positively charged ions. Beyond the 400-kilometer upward limit of the ionosphere the concentration of ions is low because of the extremely low density of the air. Because so few molecules and atoms are present, only a relatively few ions and free electrons can be produced. The electrical structure of the ionosphere is not uniform. It consists of five distinct layers of varying ion density. Furthermore, because the ionization process requires direct solar radiation, the layers are well developed on the illuminated side of the earth, but under nighttime conditions the layers tend to weaken and all but one or two disappear.

The primary significance of this electrically charged zone is its importance to long-distance radio communication. Radio waves transmitted from the

15

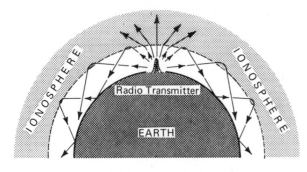

Figure 1-6

Certain layers in the ionosphere are important to long-distance communications because they reflect radio waves back to earth.

earth are reflected back to the earth by the ionosphere (Figure 1-6). It was this fact which in 1902 led two men, Kennelly and Heaviside (working independently), to conclude that a strongly ionized layer must indeed exist. In their honor the name *Kennelly–Heaviside layer* was given to the lower portion of the ionosphere where most of the significant reflection of long radio waves takes place. Unlike some of the other zones within the ionosphere, this reflecting layer exists both day and night. Communications, however, are often better at night because during sunlit hours an absorbing layer shields the Kennelly–Heaviside layer from below.

Figure 1-7

Aurora borealis (northern lights). (Courtesy of NOAA.)

The auroras, certainly one of nature's most interesting spectacles, take place in the ionosphere (Figure 1-7). The *aurora borealis* (northern lights) and its Southern Hemisphere counterpart, the *aurora australis* (southern lights), appear in a wide variety of forms. Sometimes the displays consist of vertical streamers in which there can be considerable movement. At other times the auroras appear as a series of luminous expanding arcs or as a quiet glow that has an almost foglike quality.

The occurrence of auroral displays is closely correlated in time with solar flare activity and in geographic location with the earth's magnetic poles. Solar flares are massive magnetic storms on the sun that emit enormous amounts of energy and great quantities of fast-moving atomic particles. As the clouds of protons and electrons from the solar storm approach the earth, they are captured by the earth's magnetic field which, in turn, guides them toward the magnetic poles. Then as the ions impinge upon the ionosphere, they energize the atoms of oxygen and molecules of nitrogen and cause them to emit light—the glow of the auroras. Since the occurrence of solar flares is closely correlated with sunspot activity, auroral displays increase conspicuously at times when sunspots are most numerous.

REVIEW
1. List and briefly define the four "spheres" that constitute our environment.
2. Define the terms weather and climate.
3. The statements below refer to either weather or climate. On the basis of the definitions you wrote for question 2, determine which statements refer to weather and which refer to climate. (NOTE: One statement includes aspects of both weather and climate.)
   a. The baseball game was rained out today.
   b. January is Peoria's coldest month.
   c. North Africa is a desert.
   d. The high this afternoon was 25°C.
   e. Last evening a tornado ripped through Canton.
   f. I am moving to southern Arizona because it is warm and sunny.
   g. The highest temperature ever recorded at this station is 43°C.
   h. Thursday's low of −20°C is the coldest temperature ever recorded for that city.
   i. It is partly cloudy.
4. What are the basic elements of weather and climate?
5. What are the major components of clean dry air?
6. What is responsible for the increasing carbon dioxide content of the air? What is one possible effect of increased carbon dioxide in the atmosphere?
7. Why are water vapor, ozone, and dust particles important constituents of our atmosphere?
8. Outline the stages in the formation of our atmosphere.
9. What is a radiosonde? What is a rawinsonde?
10. Rockets have become important tools in atmospheric research. Briefly describe some methods of gathering data by using rockets.
11. Recall that average sea-level pressure is about 1000 millibars which is equivalent to a weight of 1 kilogram per square centimeter. What is the pressure (in millibars) and weight of the atmosphere (in kilograms per square centimeter) above the following heights: 5.6 kilometers, 16 kilometers, 100 kilomters?

12. With an increase in altitude, pressure decreases at a(n) (constant, increasing, decreasing) rate. Underline the correct answer.

13. The atmosphere is divided vertically into four layers on the basis of temperature. List the names of these layers and their boundaries in order (from lowest to highest) and list as many characteristics of each as you can.

14. If the temperature at sea level were 23°C, under average conditions what would the air temperature be at a height of 2 kilometers?

15. a. On a spring day a middle-latitude city (about 40° N latitude) has a surface (sea-level) temperature of 10°C. If vertical soundings reveal a nearly constant lapse rate of 6.5°C per kilometer and a temperature at the tropopause of −55°C, what is the height of the tropopause?

    b. On the same spring day a station near the equator has a surface temperature of 25°C, 15°C higher than the middle-latitude city mentioned above. Vertical soundings reveal a lapse rate of 6.5°C per kilometer and indicate that the tropopause is encountered at 16 kilometers. What is the air temperature at the tropopause?

16. Why does the temperature increase in the stratosphere?

17. Why are temperatures in the thermosphere not strictly comparable to those experienced near the earth's surface?

18. Distinguish between the homosphere and heterosphere.

19. What is the basis for the stratified nature of the heterosphere?

20. What is ionization?

21. Why is the concentration of ions greatest in the 80- to 400-kilometer height range?

22. What is the primary significance of the ionosphere?

23. What is the primary cause of auroral displays?

VOCABULARY REVIEW

Review your understanding of important terms in this chapter by defining and explaining the importance of each term listed below. Terms are listed in order of occurrence in the chapter.

lithosphere	normal lapse rate
hydrosphere	stratosphere
atmosphere	tropopause
biosphere	stratopause
weather	mesosphere
climate	mesopause
elements (of weather and climate)	thermosphere
air	homosphere
ozone	heterosphere
degassing	ionosphere
photosynthesis	Kennelly–Heaviside layer
radiosonde	aurora borealis
rawinsonde	aurora australis
troposphere	

Chapter Two

solar radiation

From our experiences we know that the sun's rays feel hotter on a clear day than on an overcast day. After taking a barefoot walk on a sunny day, we realize that city pavement gets much hotter than does a grassy boulevard. From a picture of a snow-capped mountain we conclude that temperature decreases with altitude. And we know that the fury of winter is always replaced by the newness of spring. However, you may not know that these occurrences are manifestations of the same phenomenon that causes the blue color of the sky and the red color of a brilliant sunset. All of these common occurrences are the result of the interaction of solar energy with the earth's atmosphere and its land–sea surface. That is the essence of this chapter.

earth–sun relationships

The earth intercepts only a minute percentage of the energy given off by the sun—less than one two-billionth. This may seem to be a rather insignificant amount until we realize that it is several hundred thousand times the electrical generating capacity of the United States. Energy from the sun is undoubtedly the most important control of our weather and climate. Solar radiation represents more than 99.9 percent of the energy that heats the earth, drives the oceans, and creates winds. Therefore, in order to have a basic understanding of atmospheric processes, we must understand what causes the temporal and spatial variations in the amount of solar energy reaching the earth.

motions of the earth

The earth has two principal motions—rotation and revolution. *Rotation* is the spinning of the earth about its axis, which is an imaginary line running through the poles. Our planet rotates once in approximately 24 hours. During any 24-hour period one-half of the earth is experiencing daylight and one-half is experiencing darkness. The line separating the dark half of the earth from the lighted half is called the *circle of illumination*.

The other motion of the earth, *revolution*, refers to the movement of the earth in its orbit around the sun. Hundreds of years ago most people believed that the earth was stationary in space. The reasoning was that if the earth were moving, people would feel the movement of the wind rushing past them. However, today we know that the earth is traveling at nearly 113,000 kilometers per hour in an elliptical orbit about the sun. Why do we not feel the air rushing past us? The answer is that the atmosphere, bound by gravity to the earth, is carried along at the same speed as the earth.

The distance between the earth and sun averages about 150 million kilometers. However, since the earth's orbit is not perfectly circular, the distance varies during the course of a year. Each year, on about January 3, our planet is 147 million kilometers from the sun, closer than at any other time. This position is called *perihelion*. About 6 months later, on July 4, the earth is 152 million kilometers from the sun, farther away than at any other time. This position is called *aphelion*. The variations in the amount of solar radiation received by the earth as the result of its slightly elliptical orbit, however, are slight and of little consequence when explaining major seasonal temperature variations. By way of illustration, consider that the earth is closest to the sun during the Northern Hemisphere winter.

the seasons

We know that it is colder in the winter than it is in the summer, but if the variations in solar distance do not cause this, what does? All of us have made adjustments for the continuous change in the length of daylight that occurs throughout the year by planning our outdoor activities accordingly. The gradual but significant change in length of daylight certainly accounts for some of the difference we notice between summer and winter. Further, a gradual change in the altitude (angle above the horizon) of the noon sun during the course of a year's time is evident to most people. At mid-summer the sun is seen high above the horizon as it makes its daily journey across the sky. But as summer gives way to autumn, the noon sun appears lower and lower in the sky and sunset occurs earlier each evening.

The seasonal variation in the altitude of the sun affects the amount of energy received at the earth's surface in two ways. First, when the sun is directly overhead (90-degree angle), the solar rays are most concentrated. The lower the angle, the more spread out and less intense is the solar radiation that reaches the surface. This idea is illustrated in Figure 2-1. Second, and of

20

Why don't we feel the wind from the earth's revolution?

Earth's changing distance from the sun.

reasons for diff. between seasons?

use in lecture

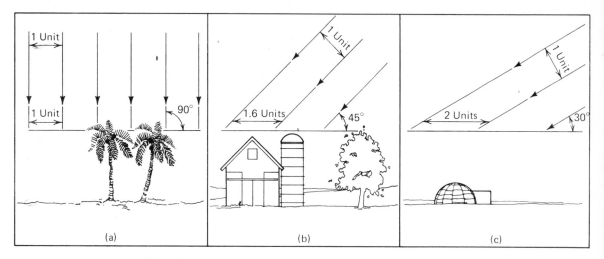

Figure 2-1

Changes in the sun angle cause variations in the amount of solar energy reaching the earth's surface. The higher the angle, the more intense the solar radiation.

lesser importance, the angle of the sun determines the amount of atmosphere the rays must traverse (Figure 2-2). When the sun is directly overhead, the rays only pass through a thickness of 1 atmosphere, while rays entering at a 30-degree angle travel through twice this amount and 5-degree rays travel through a thickness roughly equal to 11 atmospheres (Table 2-1). The longer

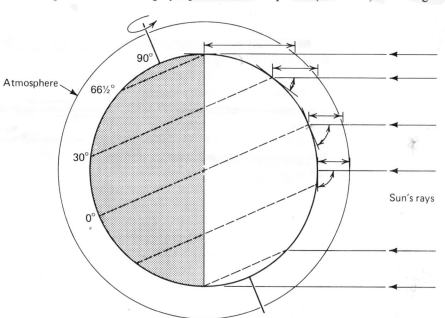

Figure 2-2

Rays striking the earth at a low angle must traverse more of the atmosphere than rays striking at a high angle and thus are subject to greater depletion by reflection and absorption.

Table 2-1

Distance Radiation Must Travel Through Atmosphere

Altitude of Sun	Number of Atmospheres
90°	1.00
80°	1.02
70°	1.06
60°	1.15
50°	1.31
40°	1.56
30°	2.00
20°	2.92
10°	5.70
5°	10.80
0°	45.00

the path, the greater the chance for absorption, reflection, and scattering by the atmosphere which reduces the intensity at the surface. These same effects account for the fact that the midday sun can be literally blinding and that the setting sun can be a sight to behold.

It is important to remember that the earth has a spherical shape. Hence, on any given day only places located at a particular latitude will receive vertical (90-degree) rays from the sun. As we move either north or south of this location, the sun's rays strike at an ever decreasing angle. Thus, the nearer a place is situated to the latitude receiving the vertical rays of the sun, the higher will be its noon sun.

In summary, the most important causes for the variation in the amount of solar energy reaching the earth are the seasonal changes in the angle at which the sun's rays strike the surface and in the length of day.

What causes the yearly fluctuations in the sun angle and length of day? They occur because the earth's orientation to the sun continually changes. The earth's axis is not perpendicular to the plane of its orbit around the sun; instead, it is tilted $23\frac{1}{2}$ degrees from the perpendicular. This is termed the *inclination of the axis* and as we shall see, if the axis were not inclined, we would have no seasonal changes. In addition, because the axis remains pointed in the same direction (toward the North Star) as the earth journeys around the sun, the orientation of the earth's axis to the sun's rays is always changing (Figure 2-3). On one day each year the axis is such that the Northern Hemisphere is "leaning" $23\frac{1}{2}$ degrees toward the sun. Six months later, when the earth has moved to the opposite side of its orbit, the Northern Hemisphere leans $23\frac{1}{2}$ degrees away from the sun. On days between these extremes the earth's axis is leaning at amounts less than $23\frac{1}{2}$ degrees to the rays of the sun. This change in orientation causes the vertical rays of the sun to make a yearly migration from $23\frac{1}{2}$ degrees north of the equator to $23\frac{1}{2}$ degrees south of the equator. In turn, this migration causes the altitude of the noon sun to vary by as much as 47 degrees ($23\frac{1}{2}$ plus $23\frac{1}{2}$) for many locations during the course

22

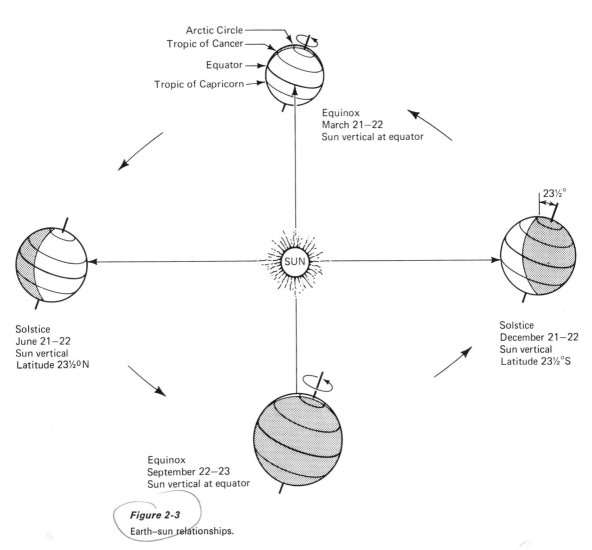

Arctic Circle
Tropic of Cancer
Equator
Tropic of Capricorn

Equinox
March 21–22
Sun vertical at equator

23½°

SUN

Solstice
June 21–22
Sun vertical
Latitude 23½°N

Solstice
December 21–22
Sun vertical
Latitude 23½°S

Equinox
September 22–23
Sun vertical at equator

Figure 2-3
Earth–sun relationships.

of a year. For example, a midlatitude city like New York has a maximum noon sun angle of 73 degrees when the sun's vertical rays have reached their farthest northward location and a minimum noon sun angle of 26 degrees 6 months later (Figure 2-4).

Historically, 4 days a year have been given special significance based on the annual migration of the direct rays of the sun and its importance to the yearly cycle of weather. On June 21 or 22 the earth is in a position where the axis in the Northern Hemisphere is tilted $23\frac{1}{2}$ degrees toward the sun (Figure 2-3). At this time the vertical rays of the sun are striking $23\frac{1}{2}$ degrees north latitude ($23\frac{1}{2}$ degrees north of the equator), a line of latitude known as the *Tropic of Cancer*. For people living in the Northern Hemisphere, June 21 or 22 is known as the *summer solstice*. Six months later, on about December 21 or 22, the earth is in an opposite position, where the sun's vertical rays are

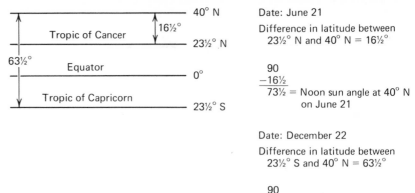

Date: June 21

Difference in latitude between
23½° N and 40° N = 16½°

90
−16½
73½ = Noon sun angle at 40° N
on June 21

Date: December 22

Difference in latitude between
23½° S and 40° N = 63½°

90
−63½
26½ = Noon sun angle at 40° N
on December 22

Figure 2-4

Calculating the noon sun angle. The noon sun angle at 23½ degrees north latitude on June 21 is 90 degrees. A place located 1 degree away receives an 89-
degree sun angle; a place 2 degrees away an 88-degree angle, and so forth.
To calculate the noon sun angle, simply find the number of degrees of latitude
separating the location receiving the vertical rays of the sun and the location
in question; then subtract that answer from 90 degrees. In this figure we see
the large variation in sun angles experienced at 40 degrees north latitude for
the summer and winter solstices (the extremes).

striking at 23½ degrees south latitude. This line is known as the *Tropic of
Capricorn*. For those of us in the Northern Hemisphere, December 21 or 22
is the *winter solstice*. However, at the same time in the Southern Hemisphere,
people are experiencing just the opposite—the summer solstice.

The equinoxes occur midway between the solstices. September 22 or 23 is
the date of the *autumnal equinox* in the Northern Hemisphere, and March 21
or 22 is the date of the *vernal* or *spring equinox*. On these dates the vertical
rays of the sun are striking at the equator (0 degrees latitude), for the earth is
in such a position in its orbit that the axis is tilted neither toward nor away
from the sun.

Further, the length of daylight versus darkness is also determined by the
position of the earth in its orbit. The length of daylight on June 21, the summer
solstice in the Northern Hemisphere, is greater than the length of night. This
fact can be established from Figure 2-5 by comparing the fraction of a given
latitude that is on the "day" side of the circle of illumination to the fraction
on the "night" side. The opposite is true for the winter solstice when the
nights are longer than the days (Table 2-2). Again for comparison let us
consider New York City. It has 15 hours of daylight on June 21 and only 9
hours on December 21. Also note from Table 2-2 that on June 21 the farther
you are north of the equator the longer the period of daylight until the Arctic
Circle is reached where the length of daylight is 24 hours. During an equinox
(meaning "equal night") the length of daylight is 12 hours everywhere on
earth, for the circle of illumination passes directly through the poles, thus
dividing the latitudes in half.

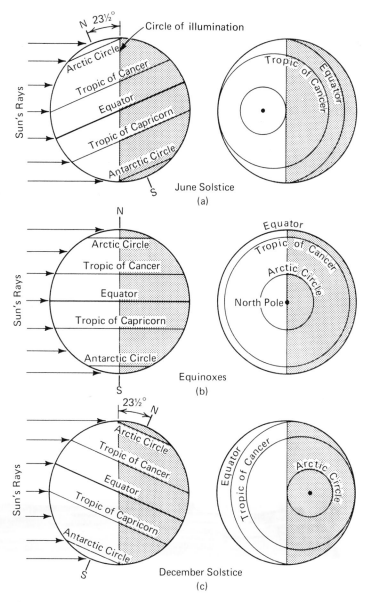

Figure 2-5

Characteristics of the summer and winter solstices and equinoxes.

As a review of the characteristics of the summer solstice for the Northern Hemisphere, examining Figure 2-5 and Table 2-2 and consider the following facts:

1. The date of occurrence is June 21 or 22.

2. The vertical rays of the sun are striking the Tropic of Cancer ($23\frac{1}{2}$ degrees north latitude).

Table 2-2

Length of Day

Latitude (degrees)	Summer Solstice	Winter Solstice	Equinoxes
0	12 hr	12 hr	12 hr
10	12 hr 35 min	11 hr 25 min	12
20	13 12	10 48	12
30	13 56	10 04	12
40	14 52	9 08	12
50	16 18	7 42	12
60	18 27	5 33	12
70	2 mo	0 00	12
80	4 mo	0 00	12
90	6 mo	0 00	12

3. Locations in the Northern Hemisphere are experiencing their longest days and highest sun angles (opposite for the Southern Hemisphere).

4. The farther you are north of the equator, the longer the period of daylight until the Arctic Circle is reached where the day is 24 hours long (opposite for the Southern Hemisphere).

The facts about the winter solstice will be just the opposite. It should now be apparent why a midlatitude location is warmest in the summer. It is then when the days are longest and the altitude of the sun is the highest.

All places situated at the same latitude have identical sun angles and lengths of days. If the earth–sun relationships just described were the only controls of temperature, we would expect these places to have identical temperatures as well. Obviously, this is not the case. Although the altitude of the sun is the main control of temperature, it is not the only control as we shall see in the following chapter.

In summary, seasonal fluctuations in the amount of solar energy reaching places on the earth's surface are caused by the migrating vertical rays of the sun and the resulting variations in sun angle and length of day.

radiation

From our everyday experience we know that the sun emits light and heat as well as the rays that give us a suntan. Although these forms of energy comprise a major portion of the total energy that radiates from the sun, they are only a part of a large array of energy called *radiation* or *electromagnetic radiation*. This array or spectrum of electromagnetic energy is shown in Figure 2-6. All radiation, whether it be X-rays, radio, or heat waves, is capable of transmitting energy through the vacuum of space at 300,000 kilometers per second and only slightly slower through air. Nineteenth-century physicists were so puzzled by the seemingly impossible task of energy traveling without the aid of an intervening medium to transmit it that they assumed that a material, which they named ether, existed between the sun and the earth. This medium

26

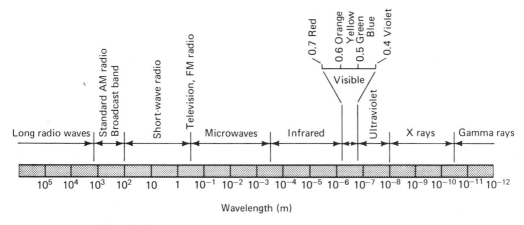

Figure 2-6

The electromagnetic spectrum.

was thought to transmit radiant energy in much the same way that air transmits sound waves produced by the vibration of one person's vocal cords to another person's eardrums. Today we know that, like gravity, radiation requires no material to transmit it. Yet, physicists still do not know why this is possible.

In some respects, the transmission of radiant energy parallels the motion of the gentle swells one sees in the open ocean. Not unlike ocean swells these electromagnetic waves, as they are called, come in various sizes. For our purpose, the most important difference between electromagnetic waves is their wavelength or distance from one crest to the next (Figure 2-7). Radio waves have the longest wavelengths, ranging to tens of kilometers. Gamma waves are the very shortest, being less than a billionth of a centimeter long. *Visible light*, as the name implies, is the only portion of the spectrum we can see. We often refer to visible light as white light since it appears "white" in color. However, it is easy to show that white light is really an array of colors, each color corresponding to a specific wavelength. By using a prism, white

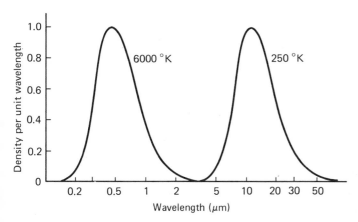

Figure 2-7

Radiation spectrums for the emissions from a blackbody at 6000°K (temperature of the sun's surface) and 250°K (approximate temperature of the earth's surface). Note that these two curves are separated into two spectral ranges, greater and less than 4 micrometers. Thus, we call the sun's radiation shortwave and the earth's radiation long-wave. (From R. G. Fleagle and J. A. Businger, *An Introduction to Atmospheric Physics.* © 1963 by Academic Press, reprinted by permission of the publisher.)

light can be divided into the colors of the rainbow, violet having the shortest wavelength, 0.4 micrometer (1 micrometer is 0.0001 centimeter) and red having the longest, 0.7 micrometer. Located adjacent to red, and having a longer wavelength, is *infrared* radiation that we cannot see, but it is detected as heat. The closest invisible waves to violet are called *ultraviolet* rays and are responsible for the sunburn one can obtain after an intense exposure to the sun. Although we divide radiant energy into groups based on our ability to perceive them, all forms of radiation are basically the same. When any form of radiant energy is absorbed by an object, the result is to increase molecular motion, which causes a corresponding increase in temperature.

Most of the radiant energy from the sun is concentrated in the visible and near-visible parts of the spectrum. The narrow band of visible light, between 0.4 and 0.7 micrometer, represents over 43 percent of the total emitted. The bulk of the remainder lies in the near-infrared section (49 percent) and ultraviolet section (7 percent). Less than 1 percent of solar radiation is emitted as X-rays, gamma rays, and radio waves.

To obtain a better appreciation of how the sun's radiant energy interacts with the earth's atmosphere and surface, it is necessary to have a general understanding of the basic laws governing radiation. The principles that follow were set forth by physicists during the late 1800s and early 1900s. Although the mathematical implications of these laws are beyond the scope of this book, the concepts themselves are well within your grasp.

1. All objects, at whatever temperature, emit radiant energy. Hence, not only hot objects like the sun, but the earth, including its polar ice caps, continually emit energy.

2. Hotter objects radiate more total energy per unit area than colder objects. The sun, which has a surface temperature of 6000°K, emits hundreds of thousands of times more energy than the earth, which has an average surface temperature of 288°K.

3. The hotter the radiating body, the shorter the wavelength of maximum radiation. This law can be easily visualized if we examine something from our everyday experiences. For example, a very hot metal rod will emit visible radiation and produce a white glow. Upon cooling, it will emit more of its energy in longer wavelengths and will glow a reddish color. Eventually, no light will be given off, but if you place your hand near the rod, the still longer infrared radiation will be detectable as heat. The sun, which has a surface temperature of 6000°K, radiates maximum energy at 0.5 micrometer which is in the visible range (Figure 2-7). The maximum radiation for the earth occurs at a wavelength of 10 micrometers, well within the infrared (heat) range. Because the maximum earth radiation is roughly 20 times longer than the maximum solar radiation, terrestrial radiation is often referred to as long-wave radiation and solar radiation is called shortwave radiation.

4. Objects that are good absorbers of radiation are also good emitters. The perfect absorber (emitter) is a theoretical object called a *blackbody*. The term "black" is misleading since the object need not be black in color.

However, a dull black surface is nearly a blackbody since it absorbs roughly 90 percent of the radiation striking it.

Technically, a blackbody is any object that radiates, for every wavelength, the maximum intensity of radiation possible for that temperature. The earth's surface and the sun approach being blackbodies (perfect radiators) since they absorb and radiate with nearly 100 percent efficiency for their respective temperatures. On the other hand, gases are selective absorbers and radiators. Thus, the atmosphere which is nearly transparent (does not absorb) to certain wavelengths of radiation is nearly opaque (good absorber) to others. Our experience tells us that the atmosphere is transparent to visible light; hence, it readily reaches the earth's surface. This is not the case for long-wave terrestrial radiation, as we shall see later. Freshly fallen snow is another example of a selective absorber. It is a poor absorber of visible light (reflects 90 percent), but at the same time it is a very good absorber (and emitter) of infrared heat. Consequently, the air directly above a snow-covered surface is colder than it would otherwise be had the snow surface been barren ground, since much of the incoming radiation is reflected away. However, a blanket of snow has the opposite effect on the ground below it. As the ground radiates heat (infrared) upward, the lower layer of the snow absorbs this energy and reradiates most of it downward. Thus, the depth at which a winter's frost can penetrate into the ground is much less when the ground has a snow cover than in an equally cold region without snow. The statement, "The ground is blanketed with snow" can be taken literally.

mechanisms of heat transfer

Three mechanisms of heat transfer are recognized: radiation, conduction, and convection. Since radiation is the only one of these that can travel through the relative emptiness of space, the vast majority of energy coming to and leaving the earth must be in this form. Radiation also plays an important role in transferring heat from the earth's land–sea surface to the atmosphere and vice versa.

Conduction is familiar to most of us through our everyday experiences. Anyone who has attempted to pick up a metal spoon that was left in a hot pan is sure to realize that heat was conducted through the spoon. Conduction is the transfer of heat through matter by molecular activity. The energy of molecules is transferred through collisions from one molecule to another, with the heat flowing from the higher temperature to the lower temperature. The ability of substances to conduct heat varies considerably. Metals are good conductors, as those of us who have touched a hot spoon have quickly learned. Air, on the other hand, is a very poor conductor of heat. Consequently, conduction is only important between the earth's surface and the air directly in contact with the surface. As a means of heat transfer for the atmosphere as a whole, conduction is the least significant and can be disregarded when considering most meteorological phenomena.

Heat gained by the lowest layer of the atmosphere from radiation or

29

Figure 2-8
Conduction, convection, and radiation.

conduction is most often transferred by *convection*. Convection is the transfer of heat by the movement of a mass or substance from one place to another. It can only take place in liquids and gases. Convective motions in the atmosphere are responsible for the redistribution of heat from equatorial regions to the poles and from the surface upward. The term *advection* is usually reserved for horizontal convective motions such as winds; the term convection is used to describe vertical motions in the atmosphere.

Figure 2-8 summarizes the various mechanisms of heat transfer. The heat produced by the campfire passes through the pan by conduction and warms the water at the bottom of the pan. Convection currents carry the warmed water throughout the pan and heat the remaining water. Meanwhile, the camper is warmed by radiation from the fire and the pan. Furthermore, since metals are good conductors, the camper's hand is likely to be burned if he does not use a potholder. In most situations involving heat transfer, conduction, convection, and radiation occur simultaneously.

incoming solar radiation

Although the atmosphere is very transparent to incoming solar radiation, less than 25 percent penetrates directly to the earth's surface without some sort of interference on the part of the atmosphere. The remainder is either absorbed by the atmosphere, scattered about until it reaches the earth's surface or returns to space, or is reflected back to space. What determines whether radiation will be absorbed, scattered, or reflected outward? As we shall see, it depends greatly upon the wavelength of the energy being transmitted, as well as on the size and nature of the intervening material.

30

scattering

When light is scattered by very small particles, primarily gas molecules, it is distributed in all directions, forward as well as backward. Some of the light that is back-scattered is lost to space, but the remainder continues downward where it interacts with other molecules which further scatter it by changing the direction of the light beam, but not its wavelength.

The light that reaches the earth's surface after having its direction changed is called *diffused daylight*. It is the diffused light that keeps a shaded area or a room lit when direct sunlight is absent.

The brightness of the daytime sky away from the direct rays of the sun is also attributed to the scattering process. Bodies like the moon and Mercury, which are without atmospheres, have dark skies even during "daylight" hours. Scattering also produces the blue color of the sky. Air molecules are more effective scatterers of the shorter wavelength (blue and violet) portion of "white" sunlight than the longer wavelength (red and orange) portion. Thus, when we look in a region of the sky away from the direct solar rays, we see predominantly blue light, which was more readily scattered.

On the other hand, the sun appears to have a yellowish to reddish tint when viewed near the horizon. When the sun is in this position, the solar beam must travel through a great deal of atmosphere before it reaches your eye. Hence, most of the blues and violets will be scattered out, leaving a beam of light that consists mostly of reds and yellows. This phenomenon is especially obvious on a day when fine dust or smoke particles are present. The reddish appearance of clouds during sunrise and sunset also results because the clouds are illuminated by light in which the blue color has been subtracted by scattering. The most spectacular sunsets occur when large quantities of dust penetrate into the stratosphere. For 3 years after the great volcanic eruption of Krakatoa in 1883 beautiful sunsets occurred worldwide. The European summer that followed this colossal explosion was cooler than normal, a fact which has been attributed to the added loss of radiation caused by back-scattering.

Large particles associated with haze, fog, or smog scatter light more equally in all wavelengths. Since no color is predominant over any other, the sky appears white on days when numerous large particles are abundant. We can conclude, then, that the color of the sky gives an indication of the number of large particles present. The bluer the sky, the cleaner the air.

reflection

About 35 percent of the solar energy reaching the outer atmosphere is reflected back to space. Included in this figure is the amount sent skyward by back-scattering. This energy is lost to the earth and does not play a role in heating the atmosphere.

The fraction of the total radiation encountered that is reflected by a surface is called its *albedo*. Thus, the albedo for the earth as a whole (planetary albedo) is 35 percent. However, the albedo from place to place as well as from time to time in the same locale varies considerably, depending upon the amount of cloud cover and particulate matter in the air, as well as upon the angle of the sun's rays and the nature of the surface. A lower angle means that more atmosphere must be penetrated, thus making the "obstacle course" longer and the loss of solar radiation greater.

Table 2-3 gives the albedo for various surfaces and clouds. Note that the angle at which the sun's rays strike a water surface greatly affects its albedo. The amount of light reflected from the earth's land–sea surface represents only about 4 percent of the total planetary albedo of 35 percent. Clouds are responsible for most of the earth's "brightness" as seen from space. The high reflectivity of clouds should not surprise anyone who has tried to drive with his bright lights on a foggy night. By comparison, the moon, which is without clouds or an atmosphere, has an average albedo of only 7 percent. Even

Table 2-3
Albedo of Various Surfaces

Surface	Percent
Fresh snow	80–85
Old snow	50–60
Sand	20–30
Grass	20–25
Dry earth	15–25
Wet earth	10
Forest	5–10
Water (sun near horizon)	50–80
Water (sun near zenith)	3–5
Thick cloud	70–80
Thin cloud	25–50
Earth and atmosphere	35

though a full moon gives us a good bit of light on a clear night, the much brighter earth would provide an explorer of the moon with even more light for an "earth-lit" walk at night.

absorption within the atmosphere

As stated earlier, gases are selective absorbers, meaning that they absorb strongly in some wavelengths, moderately in others, and only slightly in still others. When a gas molecule absorbs light waves, this energy is transformed into internal molecular motion, which is detectable as a rise in temperature. Thus, it is the gases which are good absorbers of the available radiation that play the primary role in heating the atmosphere as a whole.

Figure 2-9 gives the absorptivity of the principal atmospheric gases in various wavelengths. Nitrogen, the most abundant constituent in the atmo-

Wavelength (μm)

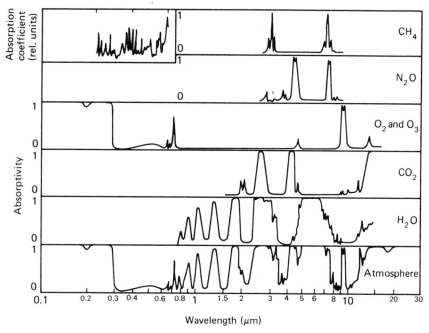

Figure 2-9

The absorptivity of selected gases of the atmosphere and the atmosphere as a whole. (From R. G. Fleagle and J. A. Businger, *An Introduction to Atmospheric Physics* © 1963 by Academic Press, reprinted by permission of the publisher.)

sphere, is a rather poor absorber of incoming solar radiation, the vast bulk of which comes in wavelengths between 0.2 and 2 micrometers. Oxygen (O_2) and ozone (O_3) are efficient absorbers of ultraviolet radiation shorter than 0.29 micrometer. Oxygen removes most of the shorter ultraviolet radiation high in the atmosphere, and ozone absorbs longer wavelength ultraviolet rays in the stratosphere between 10 and 50 kilometers. The absorption of ultraviolet radiation in the stratosphere accounts for the high temperatures experienced there. A discussion of the importance of the removal of harmful ultraviolet radiation by ozone is found in Chapter 1. The only other significant absorber of incoming solar radiation is water vapor, which along with oxygen and ozone accounts for most of the 15 percent of the total solar radiation that is absorbed within the atmosphere.

From Figure 2-9 we see that for the atmosphere as a whole, none of the gases is an effective absorber of radiation that has wavelengths between 0.3 and 0.7 micrometer; hence, a large "gap" exists. This region of the spectrum corresponds to the visible range, to which a large fraction of solar radiation belongs. This explains why most visible radiation reaches the ground and we say that the atmosphere is transparent to incoming solar radiation. Thus, direct solar energy is a rather ineffective "heater" of the earth's atmosphere. The fact that the atmosphere does not acquire the bulk of its energy directly from the sun but is heated by reradiation from the earth's surface is of the utmost importance to the dynamics of the weather machine.

terrestrial radiation

Approximately 50 percent of the solar energy that strikes the top of the atmosphere reaches the earth's surface directly or indirectly (scattered) and is absorbed. Most of this energy is then reradiated skyward. Since the earth has a much lower surface temperature than the sun, terrestrial radiation is emitted in longer wavelengths than is solar radiation. The bulk of terrestrial radiation has wavelengths between 1 and 30 micrometers, placing it well within the infrared range. Referring to Figure 2-9 we see that the atmosphere as a whole is a rather efficient absorber of radiation between 1 and 30 micrometers (terrestrial radiation). Water vapor and carbon dioxide are the principal absorbing gases in that range. Water vapor absorbs roughly five times more terrestrial radiation than do all the other gases combined and it accounts for the warm temperatures found in the lower troposphere where it is most highly concentrated. Because the atmosphere is very transparent to solar (shortwave) radiation and more absorptive to terrestrial (long-wave) radiation, the atmosphere is heated from the ground up instead of vice versa. This explains the general drop in temperature with increased altitude experienced in the troposphere. The farther from the "radiator," the colder it gets. On the average, the temperature drops 6.5°C for each kilometer increase in altitude, a figure known as the normal lapse rate.

When the gases in the atmosphere absorb terrestrial radiation, they warm, but they eventually radiate this energy away. Some travels upward where it may be reabsorbed by other gas molecules, a possibility that becomes much less likely with increasing height because the concentration of water vapor decreases with altitude. The remainder travels downward and is again absorbed by the earth. Thus, the earth's surface is being continually supplied with heat from the atmosphere as well as from the sun. This energy will again be emitted by the earth's surface and some will be returned to the atmosphere which will, in turn, radiate some earthward and so forth. This very complicated game of "pass the hot potato" that is played between the earth's surface and selected gases in the atmosphere keeps the earth's average temperature 35°C warmer than it would otherwise be. Without these absorptive gases in our atmosphere, the earth would not provide a suitable habitat for man and numerous other life forms.

This very important phenomenon has been termed the *greenhouse effect* because it was once thought that greenhouses were heated in a similar manner (Figure 2-10). The glass in a greenhouse allows shortwave radiation to enter and be absorbed by the objects inside. These objects, in turn, reradiate the heat, but at longer wavelengths, to which glass is nearly opaque. The heat, therefore, is trapped in the greenhouse. However, it has been shown that a more important factor in keeping a greenhouse warm is that it prevents mixing of the air inside with cooler air outside. Nevertheless, the term greenhouse effect remains.

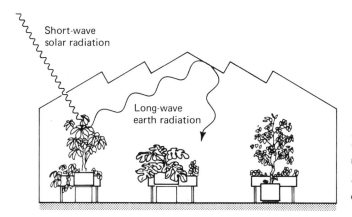

Figure 2-10

The greenhouse effect. Shortwave solar radiation passes through the glass and is absorbed by the objects in the greenhouse. The longer wavelengths radiated by these objects cannot as readily penetrate the glass and thus warm the greenhouse.

The importance of water vapor in keeping the atmosphere warm is well known to those living in mountainous regions. More radiant energy is received on mountaintops than in the valleys below because there is less atmosphere to hinder its arrival. Yet the decrease in water vapor content with altitude allows much of the heat to escape these lofty peaks. This more than compensates for the extra radiation received. Hence, the valleys remain warmer than the adjacent mountains even though they receive less solar radiation.

Clouds, like water vapor, are very good absorbers of infrared (terrestrial) radiation and play a key role in keeping the earth's surface warm, especially at night. A thick cloud cover will absorb most terrestrial radiation and return it to the surface. This explains why on clear, dry nights the surface cools considerably more than on cloudy or humid evenings.

heat budget

Except for a small amount of energy stored as fossil fuels, a balance exists between the amount of incoming solar radiation and the amount of terrestrial radiation returned to space; otherwise the earth would be getting progressively colder or progressively warmer. The proposed balance of incoming and outgoing radiation has been termed the earth's *heat budget*. An examination of the earth's heat budget provides a good review of the processes just discussed. Figure 2-11 quantitatively illustrates this balance by using 100 units to represent the solar radiation intercepted at the outer atmosphere.

Of the total radiation reaching the earth, roughly 35 units are reflected back to space. The remaining 65 units are absorbed, 15 units within the atmosphere and 50 units by the earth's land–sea surface. If all of the energy absorbed at the earth's surface were reradiated directly back to space, the earth's heat budget would be a simple matter. But as previously stated, much of the long-wave terrestrial radiation is absorbed by the atmosphere. Only 5 units pass directly through the atmosphere and are lost to space. Radiation between 8 and 11 micrometers in length escape the troposphere most readily because water vapor and carbon dioxide do not absorb these wavelengths (see Figure 2-9). This zone is called the *atmospheric window* since it is

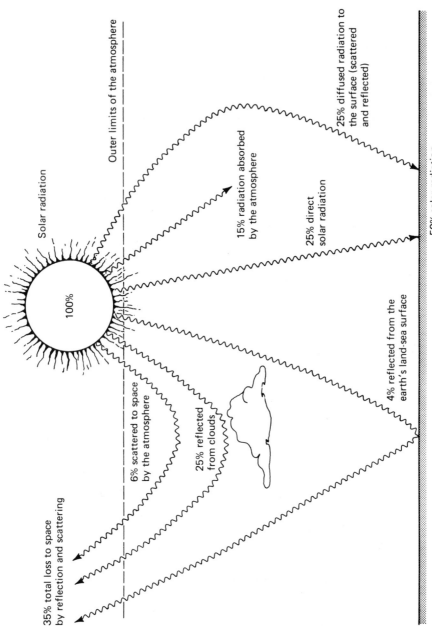

Solar radiation

Outer limits of the atmosphere

100%

35% total loss to space by reflection and scattering

6% scattered to space by the atmosphere

25% reflected from clouds

4% reflected from the earth's land-sea surface

15% radiation absorbed by the atmosphere

25% direct solar radiation

25% diffused radiation to the surface (scattered and reflected)

50% solar radiation absorbed at the surface

Figure 2-11

Distribution of incoming solar radiation by percentage. More solar energy is absorbed by the earth's surface than by the atmosphere. Consequently, the air is not heated directly by the sun but is heated indirectly from the earth's surface.

36

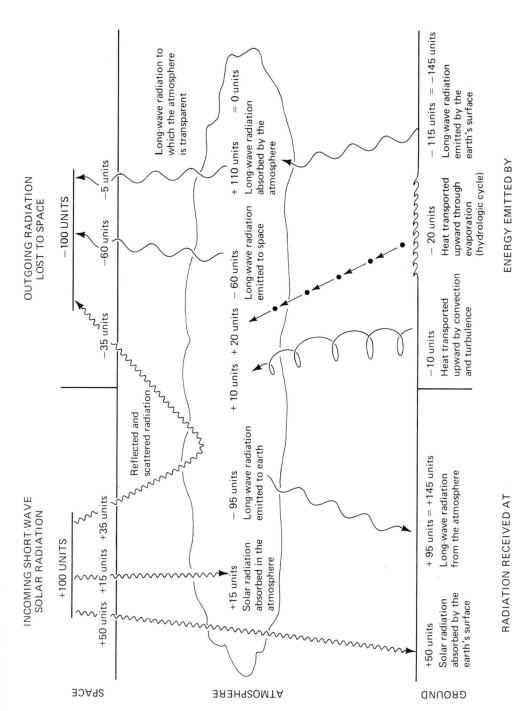

Figure 2-12

Solar radiation budget of the earth and atmosphere.

37

transparent to terrestrial radiation of that length. However, the vast majority of outgoing radiation is absorbed by the atmosphere and reradiated earthward. The atmosphere radiates 96 units of energy earthward and with the 50 units of direct solar radiation, the earth receives a total of 146 units. To maintain a balance, these 146 units must be lost. About 114 units of this energy is emitted as long-wave terrestrial radiation. Heat is also transferred from the earth's land–sea surface upward by water molecules during the process of evaporation (20 units) and by conduction (12 units). A careful examination of Figure 2-12 reveals that the total incoming radiation is balanced by outgoing radiation.

latitudinal heat balance

In the previous discussion it was pointed out that the amount of incoming solar radiation is roughly equal to the amount of outgoing terrestrial radiation for the earth as a whole. Thus, the average temperature of the earth remains rather constant. The balance of incoming and outgoing radiation that holds for the entire earth is not maintained at each latitude, however. At latitudes below 40 degrees more solar radiation is received than is lost to space by the earth. The opposite is true for higher latitudes where more heat is lost through radiation than is received. The conclusion one might draw from this is that the tropics are getting hotter and the poles are getting colder, but this is not happening; instead, the atmosphere and, to a lesser extent, the oceans act as giant thermal engines that transfer heat from the tropics poleward. In effect, it is the imbalance of heat that drives the wind and the ocean currents.

It should be of interest to those who live in the midlatitudes (30 degrees to 50 degrees) that most of the heat transfer takes place across this region. As a consequence, much of the stormy weather we associate with the midlatitudes can be attributed to this unending transfer of heat from the tropics toward the poles. These processes will be discussed in more detail in later chapters.

REVIEW

1. Can distance variations between the earth and sun adequately explain seasonal temperature changes? Explain.

2. Why does the amount of solar energy received at the earth's surface change when the altitude of the sun changes?

3. After referring to Figure 2-4, calculate the noon sun angle on June 21 and December 21 at 50 degrees north latitude, 0 degrees latitude (the equator), and 20 degrees south latitude. Which of these latitudes has the greatest variation in noon sun angle between summer and winter?

4. For the latitudes listed in the previous question, determine the length of daylight and darkness on June 21 and December 21 (refer to Table 2-2). Which of these latitudes has the largest seasonal variation in length of daylight? Which latitude has the smallest variation?

5. How would our seasons be affected if the earth's axis were not inclined $23\frac{1}{2}$ degrees to the plane of its orbit but were instead perpendicular?

6. Most solar radiation is concentrated in what portion of the electromagnetic spectrum?

7. Describe the relationship between the temperature of a radiating body and the wavelengths it emits.

8. Describe the three basic mechanisms of heat transfer. Which of these mechanisms is least important meteorologically?

9. What is the difference between convection and advection?

10. Why does the sky usually appear blue?

11. Although the sky is usually blue during the day, it may appear to have a red or orange tint near sunrise or sunset. Explain.

12. What factors might influence albedo from time to time and from place to place?

13. Explain why the atmosphere is heated chiefly by reradiation from the earth's surface.

14. Which gases are the primary heat absorbers in the lower atmosphere? Which one of these gases is most important?

VOCABULARY
REVIEW

rotation

circle of illumination

revolution

perihelion

aphelion

altitude (of the sun)

inclination of the axis

Tropic of Cancer

summer solstice

Tropic of Capricorn

winter solstice

autumnal equinox

vernal or spring equinox

radiation or electromagnetic radiation

visible light

infrared

ultraviolet

blackbody

conduction

convection

advection

diffused daylight

albedo

greenhouse effect

heat budget

atmospheric window

Chapter Three

temperature

heat and temperature

The concepts of heat and temperature are often confused. The phrase "in the heat of the day" is but one common expression where the word heat is used in a situation in which a scientist would use the term temperature. Essentially, *heat* is a form of energy; it is that which makes things hotter. *Temperature* shows how hot the thing is. Generally, heat refers to the quantity of energy present while temperature refers to intensity, that is, the degree of "hotness." For example, a cup of boiling water is no hotter than a pail of boiling water. The temperature in each case is identical—100°C. However, the pail of boiling water has a greater quantity of heat than the cup of boiling water. Moreover, a cup of boiling water is obviously hotter than a tub full of lukewarm water, but the cup does not contain as much heat. Far more ice would be melted by the tub of lukewarm water than the cup of boiling water. The water temperature in the cup is higher, but the amount of heat is smaller. Thus, the quantity of heat depends upon the mass of material considered; temperature does not. It is now easier to understand why the extremely rarified air of the thermosphere (discussed in Chapter 1), although having a high temperature, contains very little heat.

Although the ideas of temperature and quantity of heat are distinct, they are nevertheless related. Certainly, the addition or subtraction of heat is necessary to raise or lower temperature. In addition, differences in temperature determine the direction of heat flow. Heat moves from a hotter body to a colder body. In fact, a common working definition of temperature is: that property that determines whether heat will flow out of or into an object when in contact with another object.

40

We often use terms like hot, warm, cool, and cold to express temperature without using an instrument. The standard of comparison in such cases is the temperature of our skin. Although convenient and sometimes helpful in gaining a preliminary notion of temperature, the sensation of hot and cold is an extremely unreliable standard. A simple experiment will illustrate the point. Set three containers side by side, one containing cold water, another hot water, and a third lukewarm water. Place your right hand in the cold water and your left hand in the hot water. Then, after a minute or two, place both hands in the lukewarm water. The right hand will feel warm and the left hand will feel cool even though the lukewarm water is the same.

While the foregoing example relates to human perception of water temperature, there are many examples to illustrate that human response to air temperatures is an equally unreliable method of temperature measurement. *Sensible temperature* is the term applied to the sensation of temperature the human body feels in contrast to the actual temperature of the air as recorded by a thermometer. The human body is a heat engine that is continually releasing energy, and anything that influences the rate of heat loss from the body also influences the sensation of temperature the body feels, and thus affects human comfort. Several factors play a part in controlling the thermal comfort of the human body and certainly air temperature is a major factor. However, other environmental conditions, such as relative humidity, wind, and radiation are also significant.

Since evaporation is a cooling process, the evaporation of perspiration from skin is a natural means of regulating body temperature. When the air is very humid, however, heat loss by evaporation is reduced. Hence, a hot and humid day feels warmer and more uncomfortable than a hot and dry day. The *temperature–humidity index (THI)* is a well-known and often used summertime guide to human comfort or discomfort based on the conditions of temperature and humidity. In order to calculate the THI using information commonly given in weather reports, the following formula is used:

$$\text{THI} = T - 0.55\,(1 - RH)\,(T - 14)$$

where T is the temperature of the air (in degrees Celsius) and RH is the relative humidity expressed as a decimal fraction. Values in excess of 25 indicate that most people will feel uncomfortable, while a THI between 15 and 20 is accepted by most as comfortable.

Wind is another significant factor affecting sensible temperature. Because wind increases the rate of evaporation, a cold and windy winter day may feel much colder than the air temperature would seem to indicate. Not only is cooling by evaporation heightened in this situation, but the wind also acts to carry heat away from the body by constantly replacing warmer air with colder air. For example, on a day when the temperature is $-6.5°C$ and the wind speed is 32 kilometers per hour, the sensible temperature would be approximately $-23.5°C$, or $16°C$ less than the actual air temperature. The *wind chill* chart (Table 3-1) illustrates the effects of wind and temperature on the cooling rate of the human body by translating the cooling power of the

41

Table 3-1
Wind-Chill Chart

Wind Speed (km/hr)	Actual Temperature in °C											
	10	4.5	-1	-6.5	-12	-17.5	-23.5	-29	-34.5	-40	-45.5	-51
	Equivalent Temperature (to nearest one-half degree Celsius)											
8	9	3	-3	-9	-14.5	-20.5	-26	-31.5	-38	-44	-49.5	-55.5
16	4.5	-3	-9	-15.5	-23	-29	-36	-43.5	-50	-56.5	-64	-70.5
24	2	-5.5	-13	-20.5	-28	-38	-43	-50	-58	-65	-73	-80
32	0	-8	-15.5	-23.5	-31	-39.5	-47	-55	-63.5	-71	-79	-86.5
40	-1	-9	-18	-26	-33	-42	-50.5	-59	-66.5	-75.5	-83.5	-91.5
48	-2	-10.5	-19	-28	-36	-44.5	-53	-61.5	-70	-78.5	-87	-95.5
56	-3	-11.5	-20	-29	-37	-45	-55	-63.5	-72	-80.5	-89.5	-98
64	-3.5	-12	-21	-29.5	-38.5	-47	-56	-65	-73.5	-82	-91	-100

atmosphere with wind to a temperature under calm conditions. By examining Table 3-1 you well see that it is evident that the cooling power of the wind rises as wind speed increases and as temperature decreases. As a result, wind chill is mainly important in the winter. In contrast to the cold and windy day, a calm and sunny day in winter often feels warmer than the thermometer reading. In this situation, the warm feeling is caused by the absorption of direct solar radiation by the body.

From the preceding discussion it becomes apparent that, because of the sensitivity of the human body to factors other than the actual air temperature, it is necessary to use a more objective means of measuring temperature.

temperature measurement

Thermometers are devices designed to measure temperature accurately. Most substances expand when heated and contract when cooled and the designs of most commonly used thermometers are based upon this property. More precisely, the construction of these instruments is based on the fact that different substances react to temperature changes to varying extents. For example, the common *liquid-in-glass thermometer* functions because, when the liquid (usually mercury or alcohol) is heated, it expands more than the glass and, when cooled, it contracts more than glass. Consequently, because of this relative expansion and contraction, the length of the liquid column in the glass tube is a measure of surrounding temperature.

The highest and lowest temperatures that occur each day are of considerable importance and are often obtained by using specially designed liquid-in-glass thermometers. Mercury is the liquid used in the *maximum thermometer* which has a narrowed passage called a constriction in the bore of the glass tube just above the bulb (Figure 3-1). As the temperature rises, the mercury expands and is forced through the narrow opening. When the temperature falls, the constriction prevents a return of mercury to the bulb. As a result, the top of the mercury column remains at the highest point. The instrument is reset by shaking or by whirling it around a mounting. Once the thermometer is set, it indicates the current air temperature.

In contrast to a maximum thermometer that contains mercury, a *minimum thermometer* contains a liquid of low density such as alcohol. Within the

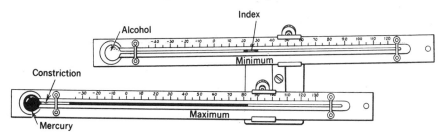

Figure 3-1

Maximum and minimum thermometers. (After Howard J. Critchfield, *General Climatology*, 3rd ed., © 1974 by Prentice-Hall, Inc., reprinted by permission of the publisher.)

alcohol, and resting at the top of the column, is a small dumbbell-shaped index (Figure 3-1). As the air temperature drops, the column shortens and the index is pulled toward the bulb by the effect of surface tension with the top of the liquid. When the temperature subsequently rises, the alcohol flows past the index, leaving it at the lowest temperature reached. To return the index to the top of the alcohol column, the thermometer is simply tilted. Since the index is free to move, it will fall to the bottom if the minimum thermometer is not mounted horizontally.

Another commonly used thermometer that works on the principle of differential expansion is the *bimetal strip*. As the name indicates, this thermometer consists of two thin strips of metal that are welded together and have widely different coefficients of thermal expansion. When the temperature changes, the two metals expand or contract unequally and cause changes in the curvature of the element. These changes are then translated into temperature values on a calibrated dial. The primary meteorological use of the bimetal strip is in the construction of a *thermograph*, an instrument that continuously records temperature. The changes in the curvature of the strip can be used to move a pen arm that records the temperature on a calibrated chart that is attached to a clock-driven, rotating drum (Figure 3-2). Although they are very convenient, thermograph records are generally less accurate than readings obtained from a mercury-in-glass thermometer. To obtain the most reliable values, it is necessary to check and correct the thermograph periodically by comparing it with an accurate, similarly exposed thermometer.

Thermometers have also been devised that do not rely on differential expansion but instead measure temperatures electrically. The most common

Figure 3-2

A common use of the bimetal strip is in the construction of a thermograph, an instrument that continuously records temperature. (Courtesy of WeatherMeasure Corporation.)

Figure 3-3

A standard instrument shelter. A shelter protects instruments from direct sunlight and allows for the free flow of air. (Courtesy of Robert E. White Instruments.)

use of electrical thermometers is in gathering upper-air data. One of two principles is commonly employed in such instruments. *Thermocouples* operate on the principle that differences in temperature between the junction of two unlike metal wires in a circuit will induce a current to flow. *Thermistors* consist of a conductor whose resistance to the flow of current is temperature-dependent. Temperature is therefore indicated as a function of current. This instrument is commonly used in radiosondes.

The accurate determination of air temperature depends not only upon the care with which the thermometer is constructed, but upon its proper exposure as well. To obtain a meaningful temperature reading, thermometers must be shaded from direct sunlight and be shielded from radiating surfaces such as buildings and the ground. Radiation should be prevented from reaching the instruments because thermometers are much more efficient absorbers than the air. It is the air temperature that is desired, not the temperature of the thermometer after absorbing radiation. In addition, good ventilation is essential. Thermometers sheltered from freely moving air will not indicate the true air temperature. Therefore, in order to avoid obtaining inaccurate and misleading data, thermometers are mounted in an instrument shelter (Figure 3-3). The shelter is a white box that has louvered sides that permit the free movement of air through it while shielding the instruments from sunshine and precipitation. Further, the shelter is placed over grass whenever possible and is mounted about 1 meter above the ground and as far from buildings as circumstances permit.

temperature scales

In order to make quantitative measurements of temperature possible, it was necessary to establish scales or units. Such temperature scales are based on the use of reference points, sometimes called *fixed points*. In 1714 Gabriel Daniel *Fahrenheit*, a German physicist who spent much of his life in Holland, devised the temperature scale that bears his name. He constructed a mercury-in-glass thermometer in which the zero point was the lowest temperature he could attain with a mixture of ice, water, and common salt. For his second fixed point he chose human body temperature that he arbitrarily set at 96°. On this scale he determined that the melting point of ice (the *ice point*) was 32° and the boiling point of water (the *steam point*) was 212°. Since Fahrenheit's original reference points were difficult to reproduce accurately, his scale is now defined using the ice point and the steam point. As thermometers improved, human body temperature was later shown to be 98.6°F instead of 96°F.

In 1742, 28 years after Fahrenheit invented his scale, Anders Celsius, a Swedish astronomer, devised a decimal scale on which the melting point of ice was set at 0° and the boiling point of water, 100°.* For many years it was called the centigrade scale, but it is now known as the *Celsius scale* after its inventor.

Since the interval between the melting point of ice and the boiling point of water is 100° on the Celsius scale and 180° on the Fahrenheit scale, a Celsius degree (°C) is larger than a Fahrenheit degree (°F) by a factor of 180/100 or 1.8. Hence, to convert from one system to the other, allowance must be made for the difference in the size of units as well as the fact that the ice point on the Celsius scale is at 0° rather than at 32° as in the Fahrenheit system. This relationship is shown graphically in Figure 3-4 and by the formulas below.

$$°F = (1.8 \times C) + 32 \qquad \text{or} \qquad °C = \frac{F - 32}{1.8}$$

The most awkward of the scales, the Fahrenheit scale, is best known in English-speaking countries where, along with the equally cumbersome British system of weights and measures, its official use is declining. In other parts of the world as well as in the scientific community where the metric system is used, the Celsius temperature scale is also used.

For some scientific purposes a third temperature scale is used, the Kelvin or absolute scale. It is similar to the Celsius scale because its divisions are

---

*The boiling point referred to in the Celsius and Fahrenheit scales pertains to pure water at standard sea-level pressure. It is necessary to remember this fact since the boiling point of water gradually decreases with a decrease in air pressure.

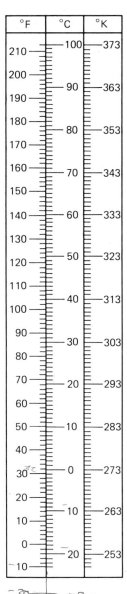

°F	°C	°K
210	100	373
200	90	363
190		
180	80	353
170		
160	70	343
150		
140	60	333
130		
120	50	323
110		
100	40	313
90	30	303
80		
70	20	293
60		
50	10	283
40		
30	0	273
20		
10	−10	263
0		
−10	−20	253

Figure 3-4
Temperature scales.

exactly the same; there are 100° separating the melting point of ice and the boiling point of water. However, on the *Kelvin scale* the ice point is set at 273° and the steam point at 373° (see Figure 3-4). Moreover, the zero point (called *absolute zero*) represents the temperature at which all molecular motion is presumed to cease. Thus, unlike the Celsius and Fahrenheit scales, it is not possible to have a negative value when using the Kelvin scale, for there is no temperature lower than absolute zero. The relationship between the Kelvin and Celsius scales is easily written as follows:

$$°C = °K − 273 \quad \text{or} \quad °K = °C + 273$$

air temperature data

Temperatures collected daily at thousands of stations throughout the world serve as the basis for much of the temperature data compiled by meteorologists and climatologists. Hourly values are either recorded by an observer, or more frequently, are obtained from a thermograph record. At many locations only the maximum and minimum temperatures are obtained. The *daily mean* for a given day may therefore be determined by averaging the 24 hourly readings or, more commonly, by adding the maximum and minimum temperatures for a 24-hour period and dividing by two. From the maximum and minimum, the *daily* or *diurnal range* is also computed by finding the difference between these figures. In addition to the daily mean and range, other data involving longer time periods are also compiled:

1. The *monthly mean* is calculated by adding together the daily means for each day of the month and dividing by the number of days in the month.

2. The *annual mean* is an average of the 12 monthly means.

3. The *annual temperature range* is computed by finding the difference between the warmest and coldest monthly means.

Mean temperatures are especially useful for making comparisons, whether on a daily, monthly, or annual basis. It is common to hear a weatherman report, "Last month was the warmest February on record" or "Today Omaha was 10 degrees warmer than Chicago." Temperature ranges are also useful statistics, because they give an indication of extremes, a necessary part of understanding the weather and climate of a place or an area.

In order to examine the distribution of air temperatures over large areas, isotherms are commonly used. An *isotherm* is a line that connects places that have the same air temperature. Therefore, all points through which an isotherm passes have identical mean temperatures for the time period indicated, either an hour, day, month, or year. Commonly isotherms representing 5° or 10° temperature differences are used, but any interval may be chosen. Figure 3-5 illustrates how isotherms are drawn on a map. Notice that most isotherms do not pass directly through the observing stations because the station readings do not coincide with the values chosen for the isotherms. Since only an occasional station temperature will be exactly the same as the value of the isotherm, it is usually necessary to draw the lines by estimating the proper position between stations.

Isothermal maps are valuable tools because they clearly make the important aspects of temperature distribution visible at a glance. Areas of low and high temperatures and the degree of temperature change per unit of distance (the temperature gradient) are easily seen. On the other hand, a map covered with numbers representing temperatures at tens or hundreds of places would make such patterns difficult to grasp.

controls of temperature

The *controls* of temperature are those factors that cause variations in temperature from place to place. Chapter 2 examined the single greatest cause for temperature variations—differences in the receipt of solar radiation. Since

48

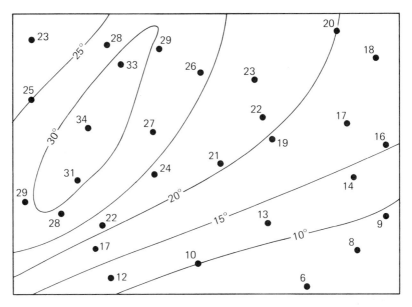

Figure 3-5

Isotherms are lines that connect points of equal temperature. Showing temperature distribution in this way makes patterns easier to see.

variations in sun angle and length of day are a function of latitude, they are responsible for warm temperatures in the tropics and colder temperatures at more poleward locations. However, latitude is not the only control of temperature; if it were, we would expect that all places along the same parallel would have identical temperatures. This is clearly not the case. For example, Eureka, California and New York City are both coastal cities at about the same latitude and both places have an average annual mean temperature of 11°C. However, New York City is 9.4°C warmer than Eureka in July and 9.4°C colder than Eureka in January. Two cities in Ecuador, Quito and Guayaquil, are within a relatively few kilometers of one another, but the mean annual temperatures at these two cities differ by 12.2°C. To explain these situations and countless others, we must realize that factors other than variations in solar radiation also exert a strong influence upon temperatures. In the following pages we will examine these other controls, which include:

1. differential heating of land and water
2. ocean currents
3. altitude
4. geographic position

land and water

The heating of the earth's surface controls the heating of the air above. Therefore, in order to understand variations in air temperatures, we must examine the heating properties of various surfaces. Different land surfaces

49

use in lecture →

reflect and absorb varying amounts of incoming solar energy, which, in turn, cause variations in the temperature of the air above. The greatest contrast, however, is not between different land surfaces but between land and water. Land heats more rapidly and to higher temperatures than water and it cools more rapidly and to lower temperatures than water. Variations in air temperatures, therefore, are much greater over land than over water.

An important reason for surface water temperatures rising and falling much more slowly than surface temperatures on land is the fact that water is highly mobile. As water is heated, turbulence distributes the heat through a considerably larger mass. On a daily basis temperature changes occur to depths of 6 meters or more below the surface, and on a yearly basis oceans and deep lakes are subject to temperature variations through a layer between 200 and 600 meters thick.

By contrast, heat does not penetrate deeply into soil or rock; it remains in a thin surface layer. Obviously, no turbulent mixing occurs on land; heat must be transferred by the slow process of conduction. Consequently, diurnal temperature changes are small below a depth of 10 centimeters, although some change can occur to a depth of perhaps 1 meter. Annual temperature variations usually reach depths of 15 meters or less. Thus, as a result of the mobility of water and the lack of it in the solid earth, a relatively thick layer of water is heated to moderate temperatures during the summer. On land only a thin layer is heated but to much higher temperatures. During winter the shallow layer of rock and soil that was heated in summer cools rapidly. Water bodies, on the other hand, cool slowly as they draw upon the reserves of heat stored within. As the surface of the water cools, vertical motions are established. The chilled surface water sinks and is replaced by warmer water from below. Hence, the entire mass of water must be cooled before the temperature at the surface will drop appreciably.

Other factors also contribute to the differential heating of land and water:

1. Since land surfaces are opaque, heat is absorbed only at the surface. Water, being more transparent, allows some solar radiation to penetrate to a depth of several meters.

2. The *specific heat* (amount of heat needed to raise 1 gram of a substance 1°C) is almost three times greater for water than it is for land. Thus, water requires considerably more heat to raise its temperature the same amount as an equal quantity of land.

3. Evaporation (a cooling process) from water bodies is greater than that from land surfaces.

All of these factors collectively cause water to warm more slowly, store greater quantities of heat energy, and cool more slowly than land.

Temperature data for two cities will demonstrate the moderating influence of a large water body and the extremes associated with land (Table 3-2). San Francisco, California is located along a windward coast, while Springfield, Missouri is in a continental position far from the influence of water. Both cities are at the same latitude and thus experience similar sun angles and lengths of day. Moreover, both places have identical annual means, 13.3°C. Springfield, however, has an average January temperature that is 8.8°C colder than San Francisco's. Conversely, Springfield's July average is 7.7°C higher than San Francisco's. Although their average temperatures for the year are the same, Springfield, which has no water influence, experiences much greater temperature extremes than San Francisco, which does.

Table 3-2

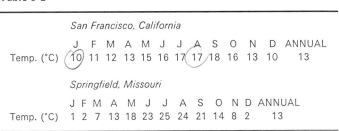

San Francisco, California

	J	F	M	A	M	J	J	A	S	O	N	D	ANNUAL
Temp. (°C)	10	11	12	13	15	16	17	17	18	16	13	10	13

Springfield, Missouri

	J	F	M	A	M	J	J	A	S	O	N	D	ANNUAL
Temp. (°C)	1	2	7	13	18	23	25	24	21	14	8	2	13

On a different scale, the moderating influence of water may also be demonstrated when temperature variations in the Northern and Southern Hemispheres are compared. The two views of earth in Figure 3-6 show the uneven distribution of land and water over the globe. Sixty-one percent of the Northern Hemisphere is covered by water; land represents the remaining 39 percent. However, the figures for the Southern Hemisphere (81 percent

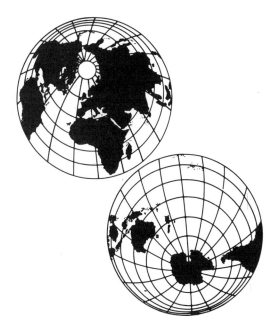

Figure 3-6

Two views of the earth from space showing the uneven distribution of land and water. Most of the land occurs in the Northern Hemisphere; the Southern Hemisphere may properly be termed the water hemisphere. (After M. Grant Gross, *Oceanography*, © 1972 by Prentice-Hall, Inc., reprinted by permission of the publisher.)

Table 3-3

Variation in Mean Annual Temperature Range
with Latitude (°C)

Latitude	Northern Hemisphere	Southern Hemisphere
0	0	0
15	3	4
30	13	7
45	23	6
60	30	11
75	32	26
90	40	31

water, 19 percent land) reveal why it is correctly called the *water hemisphere*. Table 3-3 portrays the considerably smaller annual temperature variations in the water-dominated Southern Hemisphere as compared to the Northern Hemisphere.

ocean currents

The effects of ocean currents upon temperatures of adjacent land areas are variable. The moderating effect of poleward-moving warm ocean currents is wellknown. The North Atlantic Drift, an extension of the warm Gulf Stream (Figure 3-7), keeps wintertime temperatures in Great Britain and much of western Europe warmer than one would expect for their latitudes. Because of the prevailing westerly winds, the moderating effects are carried far inland. For example, Berlin (52 degrees north latitude) has a mean January temperature similar to that experienced at New York City, which lies 12 degrees latitude farther south, while the January mean at London (51 degrees north latitude) is 4.5°C higher than at New York City.

In contrast to warm ocean currents whose effects are felt most during the winter, the influence of cold currents is most pronounced in the tropics or during the summer months in the middle latitudes. Cool currents such as the Benguela Current off the western coast of Southern Africa moderate the tropical heat. Wavis Bay (23 degrees south latitude), a town adjacent to the Benguela Current, is 5°C cooler in summer than Durban which is 6 degrees latitude farther poleward but on the eastern side of South Africa away from the influence of the current. Because of the cold California Current, summer temperatures in subtropical coastal Southern California are lower by 6°C or more when compared to east-coast stations in the subtropical United States.

altitude

The two cities in Ecuador, Quito and Guayaquil, mentioned earlier demonstrate the influence of altitude upon mean temperature. Both cities are near the equator and are relatively close to one another, but the annual mean at

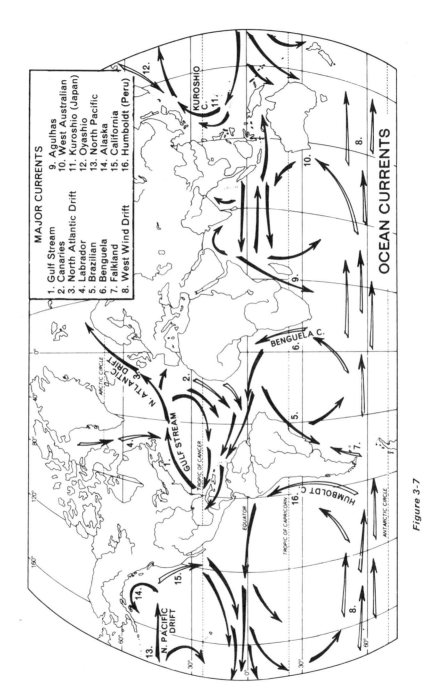

MAJOR CURRENTS

1. Gulf Stream
2. Canaries
3. North Atlantic Drift
4. Labrador
5. Brazilian
6. Benguela
7. Falkland
8. West Wind Drift
9. Agulhas
10. West Australian
11. Kuroshio (Japan)
12. Oyashio
13. North Pacific
14. Alaska
15. California
16. Humboldt (Peru)

OCEAN CURRENTS

Figure 3-7

Major ocean currents. Warm currents are shown by dark arrows and cold currents by open arrows. (After J. B. Hoyt, *Man and the Earth*, 3rd ed., © 1973 by Prentice-Hall, Inc., reprinted by permission of the publisher.)

53

Guayaquil is 25.5°C as compared to Quito's mean of 13.3°C. The difference may be understood when the cities' elevations are noted. Guayaquil is only 12 meters above sea level while Quito is high in the Andes Mountains at 2800 meters. Recall that temperatures drop an average of 6.5°C per kilometer in the troposphere; thus, cooler temperatures are to be expected at greater heights. However, the magnitude of the difference is not totally explained by the normal lapse rate. If this figure were used, we would expect Quito to be about 18.2°C cooler than Guayaquil, but the difference is only 12.2°C. The fact that high-altitude places, such as Quito, are warmer than the value calculated using the normal lapse rate results from the absorption and reradiation of solar energy by the ground surface.

In addition to the effect of altitude upon mean temperatures, the daily temperature range also changes with variations in height. Not only do temperatures drop with an increase in altitude, but atmospheric pressure and density also diminish. Because of the reduced density at high altitudes, the overlying atmosphere absorbs and reflects a smaller portion of the incoming solar radiation. Consequently, with an increase in altitude, the intensity of insolation increases, resulting in rapid and intense daytime heating. Conversely, rapid nighttime cooling is also the rule in high mountain locations. Therefore, stations located high in the mountains generally have a greater diurnal range than stations at lower elevations.

Figure 3-8 summarizes the variations in temperature brought about by changes in altitude. The graph shows data for a 15-day period in July at five

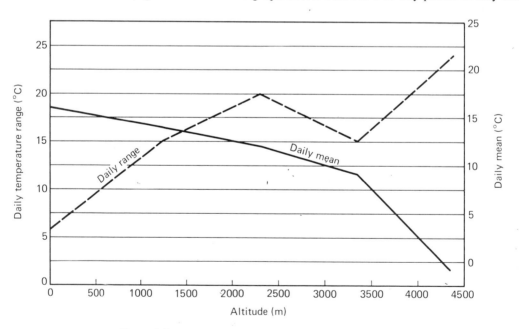

Figure 3-8

Temperature changes associated with changes in altitude. With increasing altitude, the mean temperature drops and the daily temperature range increases.

stations (at approximately 15 degrees south latitude) in Peru whose elevations range from sea level to almost 4400 meters. It can be seen that as altitude increases, the mean temperature drops a total of 16.6°C while the daily range increases from 5.8°C to 24°C.

geographic position

The geographic setting may greatly influence the temperatures experienced at a specific location. For example, a windward coastal location, that is, a place that is subject to prevailing onshore winds, experiences considerably different temperatures than a coastal location where the prevailing winds are directed from the land toward the ocean. In the first situation, the place will experience the full moderating influence of the ocean—cool summers and mild winters as compared to an inland station at the same latitude. However, a leeward coastal situation will have a more continental temperature regime because the winds do not carry the ocean's influence onshore. Eureka, California and New York City, the two cities mentioned in the introduction to the section on temperature controls, illustrate this aspect of geographic position (Table 3-4). The annual temperature range at New York City is 19°C higher than Eureka's.

Table 3-4

	Eureka, California												
	J	F	M	A	M	J	J	A	S	O	N	D	ANNUAL
Temp. (°C)	9	9	9	10	12	13	14	14	14	12	11	9	11
	New York City												
	J	F	M	A	M	J	J	A	S	O	N	D	ANNUAL
Temp. (°C)	−1	−1	3	9	16	21	23	23	21	15	7	2	11

Seattle and Spokane, both in the state of Washington, illustrate a second aspect of geographic position: mountains acting as barriers. Although Spokane is only about 360 kilometers east of Seattle, the towering Cascade Range separates the cities. Consequently, while Seattle's temperatures show a marked marine influence, Spokane's are more typically continental (Table 3-5). Spokane is 7°C cooler than Seattle in January and 4°C warmer than

Table 3-5

	Seattle, Washington												
	J	F	M	A	M	J	J	A	S	O	N	D	ANNUAL
Temp. (°C)	4	5	7	9	12	15	17	17	14	11	8	6	11
	Spokane, Washington												
	J	F	M	A	M	J	J	A	S	O	N	D	ANNUAL
Temp. (°C)	−3	−1	3	9	13	16	21	20	16	9	2	−1	9

Seattle in July. The annual range at Spokane is 11°C greater than at Seattle. The Cascade Range effectively cuts Spokane off from the moderating influence of the Pacific Ocean.

**world
distribution
of
temperatures**

By examining isothermal maps for January and July (Figures 3-9 and 3-10), changing global temperature patterns can be studied and the effects of the controls of temperature, especially latitude, the distribution of land and water, and ocean currents, can be seen. For most places on earth, January and July represent the seasonal extremes of temperature, and for this reason, these months are most often selected for analysis. As is the case for most isothermal maps, all temperatures have been reduced to sea level in order that the complications caused by differences in altitude are eliminated.

On both maps the isotherms generally trend east and west and show a decrease in temperatures poleward from the tropics. They illustrate one of the most fundamental and best-known aspects of the world distribution of temperature. This reflects the fact that the effectiveness of incoming solar radiation in heating the earth's surface and the atmosphere above is largely a function of latitude. Further, there is a latitudinal shifting of temperatures caused by the seasonal migration of the sun's vertical rays.

If the effect of latitudinal variations in the receipt of solar energy were the only control of temperature distribution, our analysis could end at this point, but this, of course, is not the case. The added effect of the differential heating of land and water is also reflected on the January and July temperature maps. The warmest and coldest temperatures are found over land. Hence, because temperatures do not fluctuate as much over water as over land, the north–south migration of isotherms is greater over the continents than over the oceans. In addition, it is clear that the isotherms in the Southern Hemisphere where there is little land and where the oceans predominate are much more regular than in the Northern Hemisphere where they bend sharply northward in July and southward in January over the continents.

Isotherms also reveal the presence of ocean currents. Warm currents cause isotherms to be deflected toward the poles, while cold currents cause an equatorward bending. The horizontal transport of water poleward warms the overlying air and results in air temperatures that are higher than would otherwise be expected for the latitude. Conversely, currents moving toward the equator produce cooler than expected air temperatures.

Since Figures 3-9 and 3-10 show the seasonal extremes of temperature, it is possible to evaluate variations in the annual range of temperature from place to place. A comparison of the two maps shows that a station near the equator will record a very small annual range because it experiences little variation in the length of day and it always has a relatively high sun angle. However, a station in the middle latitudes experiences much wider variations in sun angle and length of day, and thus larger variations in temperature. Thus, we can state that the annual temperature range increases with an increase in latitude. Moreover, land and water also affect seasonal tempera-

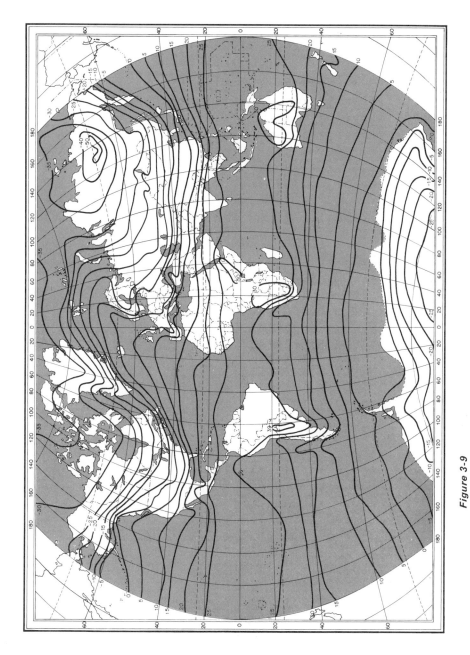

Figure 3-9

World mean sea-level temperatures in January in °C. (After Howard J. Critchfield, *General Climatology*, 3rd ed., © 1974 by Prentice-Hall, Inc., reprinted by permission of the publisher.)

Figure 3-10

World mean sea-level temperatures in July in °C. (After Howard J. Critchfield, *General Climatology*, 3rd ed., © 1974 by Prentice-Hall, Inc., reprinted by permission of the publisher.)

ture variations, especially outside the tropics. A continental location must endure hotter summers and colder winters than a coastal location. Consequently, the annual range will increase with an increase in continentality.

A classic example of the effect of latitude and continentality on annual temperature range is Yatkusk, U.S.S.R. This city is located in Siberia at approximately 60 degrees north latitude and far from the influence of water. As a result, Yatkusk has an average annual temperature range of 62.2°C, among the highest in the world.

cycles of air temperature

A thermograph record such as Figure 3-11 or your own experience reveals that there is a rhythmic rise and fall of air temperature during the day, a phenomenon commonly called the *daily* or *diurnal march of temperature.* After reaching a minimum about sunrise, the temperature curve climbs steadily until midafternoon to late afternoon. Attaining a maximum between 2 P.M. and 5 P.M., the temperature then declines until sunrise the following day. The primary control of this daily cycle of air temperature is probably as obvious as the cycle itself; it is the sun. As the sun angle increases throughout the morning, the intensity of sunlight also rises, reaching a peak at noon and gradually diminishing in the afternoon. Figure 3-12(*a*) is a schematic diagram showing the variation of incoming solar radiation versus outgoing terrestrial radiation during a day for a typical middle-latitude location at the time of an equinox. The time of the minimum temperature follows the radiation curve. During the night the atmosphere and the surface of the earth cool as they radiate heat away that is not replaced by incoming solar energy. The minimum temperature therefore occurs about the time of sunrise, after which the sun again heats the ground which, in turn, heats the air. However, by comparing Figures 3-11 and 3-12(*a*), it is apparent that the time of highest temperature does not coincide with the time of maximum radiation. Although the radiation curve is symmetrical with respect to noon, the daily air temperature curve is not. The delay in the occurrence of the maximum until midafternoon to late afternoon is termed the lag of the maximum. Although

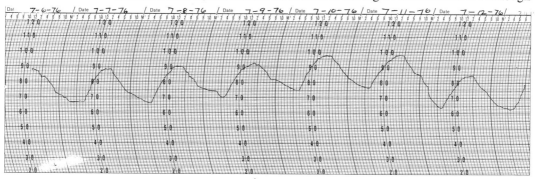

Figure 3-11

This thermograph record for a July week at Peoria, Illinois illustrates the daily rhythm of temperatures typical of the middle latitudes.

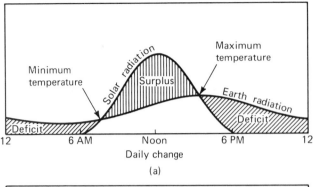

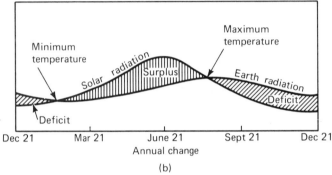

Figure 3-12

The march of incoming solar radiation and outgoing earth radiation for a day at about the time of an equinox (*a*) and annually (*b*).

the greatest intensity of heat is received at noon, during a portion of the afternoon the earth and air near it continue to receive more heat than they lose. Therefore, the temperature continues rising until the amount of heat received from the sun ceases to exceed the amount radiated to space from the earth and atmosphere. Thereafter the temperature drops.

The magnitude of daily temperature changes is variable and may be influenced by locational factors and/or local weather conditions. Three common examples will illustrate this point. The first two relate to location and the third pertains to the influence of clouds.

1. Although variations in sun angle are rather great during the course of a day in the middle and low latitudes, points near the poles experience the same low sun angle all day. Consequently, the diurnal temperature cycle in the high latitudes shows minimal variation.

2. A location on a windward coast is likely to experience small variations in the daily cycle. During a typical 24-hour period the ocean warms less than 1°C. As a result, the air above shows a correspondingly slight change in temperature. Figure 3-13 graphically portrays this oceanic influence on daily temperature range. Eureka, California, a windward coastal station, consistently has a lower daily temperature range than Des Moines, Iowa, an inland city at about the same latitude. On an annual basis the daily range at Des Moines averages 10.9°C compared to 6.1°C at Eureka, a difference of 4.8°C.

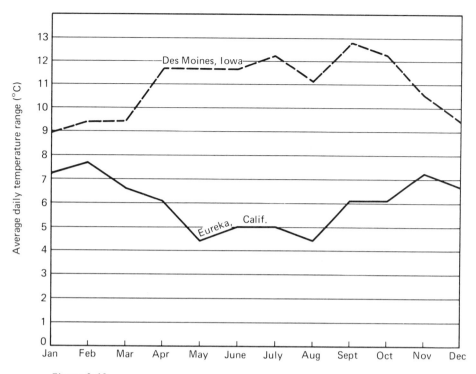

Figure 3-13

Stations having a strong marine influence, like Eureka, California, have smaller daily temperature ranges than stations located far from the influence of water, like Des Moines, Iowa.

3. By comparing the temperature curve for July 25 (a clear day) to that of July 27 (an overcast day), the effect of cloud cover on the daily march of temperature may be seen in Figure 3-14. It is apparent that an overcast day is responsible for a flattened daily temperature curve. By day clouds block incoming solar radiation and hence reduce daytime heating. At night the clouds retard the loss of radiation by the ground and air and reradiate heat earthward. Therefore, nighttime temperatures are not as low as they otherwise would have been.

Although the rise and fall of daily temperature usually reflect the general rise and fall of solar radiation, this is not always the case. If thermograph records for a station were examined for a period of several weeks, non-periodic variations would be seen that are obviously not sun-controlled. Such irregularities are caused primarily by the passage of atmospheric disturbances that are often accompanied by variable cloudiness and winds that bring air having contrasting temperatures. Under these circumstances, the maximum and minimum temperatures may occur at any time of the day or night.

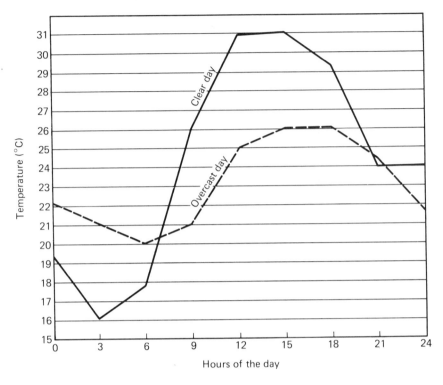

Figure 3-14

The daily march of temperature at Peoria, Illinois on a clear day and on an overcast day. As is typical, the maximum temperature on the clear day was higher and the minimum temperature was lower than for the cloudy day.

Before leaving the topic of temperature cycles, the *annual march of temperature* should also be mentioned. On a yearly basis the dates of the high and low temperatures do not coincide with the times of maximum and minimum solar radiation. Although the greatest intensity of solar radiation occurs at the time of the summer solstice, the months of July and August are generally the warmest of the year in the Northern Hemisphere. Conversely, a minimum of solar energy is received in December at the time of the winter solstice, but January and February are usually colder. The reason for the delay in the time of the warmest and coldest months reflects the balance between incoming solar energy and outgoing radiation from the earth [Figure 3-12(*b*)]. Sometimes the lag of the warmest month, and less often the lag of the coldest month, is greater in areas that have a strong oceanic influence. While at practically all continental stations July is the warmest month and January the coldest, windward coastal cities such as Los Angeles and San Diego demonstrate lags of the maximum. In both instances the months of August and September are warmer than July.

1. Distinguish between the concepts of heat and temperature. Include an example to illustrate the difference.

2. a. What is meant by the term sensible temperature?
   b. What factors influence sensible temperature?

3. If the air temperature is 27°C and the relative humidity is 80 percent, what is the THI? If the relative humidity were 30 percent, what would the THI be?

4. Using Table 3-1, determine equivalent temperatures under the following circumstances:
   a. Temperature = −12°C, wind speed = 24 km/hr.
   b. Temperature = −12°C, wind speed = 48 km/hr.

5. On a calm and sunny winter day the sensible temperature may be higher than the air temperature. Explain.

6. Describe how each of the following thermometers works: liquid-in-glass, maximum, minimum, bimetal strip, thermocouple, thermistor.

7. What is a thermograph?

8. In addition to having an accurate thermometer, what other factors must be considered in order to obtain a meaningful air temperature reading?

9. a. What is meant by the terms steam point and ice point?
   b. What values are given these points on each of the three temperature scales presented in this chapter?

10. Why is it not possible to have a negative value when using the Kelvin temperature scale?

11. How are the following temperature data calculated: daily mean, daily range, monthly mean, annual mean, annual range?

12. What are isotherms and for what purpose are they used?

13. a. State the relationship between the heating and cooling of land versus water.
    b. List and explain the factors that cause the difference between the heating and cooling of land and water.
    c. Since we are interested in the atmosphere, why do we concern ourselves with the heating characteristics at the earth's surface?

14. What is the annual temperature range in the tropics as compared to the annual temperature range in the middle to high latitudes? Explain.

15. Three cities are located at the same latitude (about 45 degrees north latitude). One city is located along a windward coast, another in the center of the continent, and the third along a leeward coast. Compare the annual temperature ranges of these cities.

16. Answer the following questions about world temperature distribution (you may wish to refer to the January and July isotherm maps):
    a. Isotherms generally trend east–west. Why?
    b. Isotherms bend (poleward, equatorward) over continents in summer. Underline the correct answer and explain.
    c. Isotherms shift north and south from season to season. Why?
    d. Where do isotherms shift most, over land or water? Explain.

e. How do isotherms show ocean currents. How can you tell if the current is warm or cold?

f. Why are the isotherms more irregular in the Northern Hemisphere than in the Southern Hemisphere?

17. Although the intensity of incoming solar radiation is greatest at noon, the warmest part of the day is most often midafternoon. Why?

18. How does the daily march of temperature on a completely overcast day compare to a cloudless and sunny day? Explain your answer.

VOCABULARY REVIEW

heat	steam point
temperature	Fahrenheit scale
sensible temperature	Celsius scale
temperature–humidity index (THI)	Kelvin scale
wind chill	absolute zero
thermometer	daily mean
liquid-in-glass thermometer	daily (diurnal) range
maximum thermometer	monthly mean
minimum thermometer	annual mean
bimetal strip	annual temperature range
thermograph	isotherm
thermocouple	controls (of temperature)
thermistor	specific heat
fixed points	water hemisphere
Fahrenheit	daily (diurnal) march of temperature
ice point	annual march of temperature

Chapter Four

HUMIDITY, CONDENSATION, AND ATMOSPHERIC STABILITY

Water vapor constitutes only a small fraction of the atmosphere, varying from almost 0 to 4 percent by volume. However, the importance of water in the air is far greater than this small percentage would indicate. Even as you observe day-to-day weather changes, many questions may come to mind concerning the role of moisture in the atmosphere. What is humidity and how is it measured? Why do clouds form on some occasions but not on others? Why do some clouds look thin and harmless and why do others appear as gray and ominous towers? In the following pages we shall investigate these and other questions involving water in the air.

the hydrologic cycle

An adequate supply of water is vital to life on earth. With increasing demands on this finite resource, science has given a great deal of attention to the continuous exchanges of water between the oceans, the atmosphere, and the continents. This unending circulation of the earth's water supply has come to be called the *hydrologic cycle*. It is a gigantic system powered by energy from the sun in which the atmosphere provides the vital link between the oceans and continents. Water from the oceans, and to a much lesser extent from the continents, is constantly evaporating into the atmosphere. Winds transport the moisture-laden air, often great distances, until the complex processes of cloud formation are set in motion that eventually result in precipitation. The precipitation that falls into the ocean has ended its cycle and is ready to begin another. The water that falls on the continents, however, must still make its way back to the ocean.

65

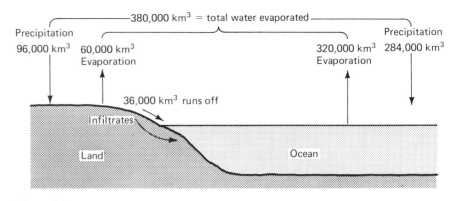

Figure 4-1

The earth's water balance. Huge quantities of water are cycled through the atmosphere each year. Notice that on the continents precipitation exceeds evaporation and that the reverse is true for the oceans. Since the level of the world ocean is not dropping, runoff from the continents must balance the deficit of the oceans.

Once precipitation has fallen on land, a portion of the water soaks into the ground, some of it moving downward, then laterally, finally seeping into lakes and streams or directly into the ocean. When the rate of rainfall is greater than the earth's ability to absorb it, the additional water flows over the surface into streams and lakes. Much of the water that soaks in or runs off eventually finds its way back to the atmosphere. In addition to evaporation from the soil, lakes, and streams, some of the water that infiltrates the ground surface is absorbed by plants which then release it into the atmosphere, a process called *transpiration*.

A diagram of the earth's water balance, a quantitative view of the hydrologic cycle, is shown in Figure 4-1. Although the amount of water vapor in the air at any one time is but a minute fraction of the earth's total water supply, the absolute quantities that are cycled through the atmosphere in one year are immense, some 380,000 cubic kilometers—enough to cover the earth's surface to a depth of about 100 centimeters. Estimates show that over North America almost six times more water is carried within the moving currents of air than is transported by all of the continent's rivers. Since the total amount of water vapor in the atmosphere remains about the same, the average annual precipitation over the earth must be equal to the quantity of water evaporated. However, for all of the continents taken together, precipitation exceeds evaporation. Conversely, over the oceans evaporation exceeds precipitation. Since the level of the world ocean is not dropping, runoff from land areas must balance the deficit of precipitation over the oceans.

In summary, the hydrologic cycle represents the continuous movement of water from the oceans to the atmosphere, from the atmosphere to the land, and from the land back to the sea. The movement of water through the cycle holds the key to the distribution of moisture over the surface of our planet and is intricately related to all atmospheric phenomena.

Water vapor is an odorless, colorless gas that mixes freely with the other gases of the atmosphere. Unlike oxygen and nitrogen—the two most abundant components of the atmosphere—water vapor can change from one state of matter to another with relative ease at the temperatures and pressures experienced near the surface of the earth. It is because of this ability, which allows water to leave the oceans as a gas and return again as a liquid, that the vital hydrologic cycle exists. The processes that involve a change of state require that heat be absorbed or released (Figure 4-2). This heat energy is measured in calories. One *calorie* is the amount of heat required to raise the temperature of 1 gram of water 1°C.

The process of converting a liquid to a vapor is termed *evaporation*. It takes approximately 600 calories of energy to convert 1 gram of water to water vapor. The energy absorbed by the water molecules during evaporation is used solely to give them the motion needed to escape the surface of the liquid and become a gas. Because this energy is subsequently released as heat when the vapor changes back to a liquid, it is generally referred to as *latent heat* (meaning hidden or stored heat).

Since the evaporating molecules require energy to escape, the remaining liquid must be cooled by an equivalent amount; hence, the common expression "evaporation is a cooling process." You have undoubtedly experienced this cooling effect upon stepping dripping wet out of a swimming pool. On a larger scale is the cooling effect of evaporating rainwater during a summer shower.

Condensation denotes the process of water vapor changing to the liquid state. During condensation the water molecules release energy (*latent heat of condensation*) equivalent to that which was absorbed during evaporation. This energy plays an important role in producing violent weather and can act to transfer great quantities of heat energy from tropical oceans to more

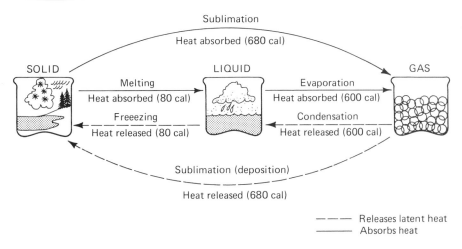

Figure 4-2
Changes of state.

poleward locations. When condensation occurs in the atmosphere, it results in the formation of such phenomena as fog and clouds.

Melting is the process by which a solid is changed to a liquid. It requires absorption of approximately 80 calories of energy per gram of water. *Freezing*, the reverse process, releases these 80 calories per gram as *latent heat of fusion*.

The last of the processes illustrated in Figure 4-2 is *sublimation*. This term is used to describe the conversion of a solid directly to a gas without passing through the liquid state. You may have observed this change as you watched the sublimation of Dry Ice (frozen carbon dioxide). The term sublimation is also used to denote the reverse process, the conversion of a vapor to a solid. This change occurs, for example, during the formation of frost. As shown in Figure 4-2, sublimation involves an amount of energy equal to the total of the other two processes.

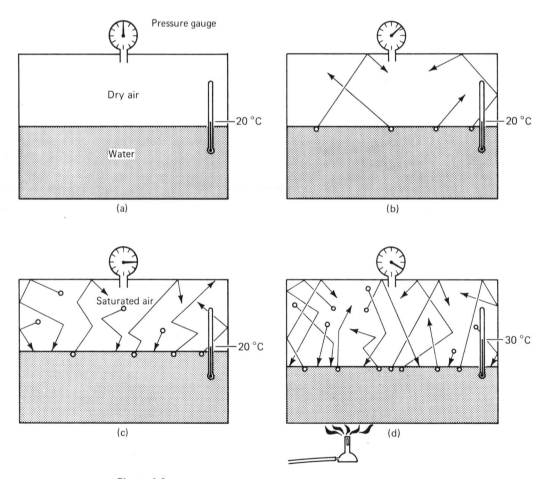

Figure 4-3

Schematic illustration of vapor pressure and saturation.

humidity

Humidity is the general term used to describe the amount of water vapor in the air. Several methods are used to quantitatively express humidity. Among these are (*1*) absolute humidity, (*2*) specific humidity, and (*3*) relative humidity.

Before we consider each of these humidity measures individually, it is important to understand the concept of *saturation*. Imagine a closed container half full of water and overlain with dry air at the same temperature (Figure 4-3). As the water begins to evaporate from the water surface, a small increase in pressure can be detected in the air above. This increase is the result of the motion of the water vapor molecules that were added to the air through evaporation. In the open atmosphere this pressure is termed *vapor pressure* and is defined as that part of the total atmospheric pressure attributable to its water vapor content. As more and more molecules escape from the water surface in the closed container, the steadily increasing vapor pressure in the air above forces more and more of these molecules to return to the liquid. Eventually, the number of vapor molecules returning to the surface will balance the number leaving. At that point the air is said to be saturated or filled to capacity. However, if we increased the temperature of the water and air in the closed container, more water would evaporate before a balance was reached. Consequently, at higher temperatures more moisture is required for saturation. Stated another way, the water vapor capacity of air is temperature-dependent with warm air having a much greater capacity than cold air. The amount of water vapor required for saturation at various temperatures is shown in Table 4-1.

Table 4-1

Water Vapor Capacity
(at average sea-level pressure)

Temperature (°C)	g/kg
−40	0.1
−30	0.3
−20	0.75
−10	2
0	3.5
5	5
10	7
15	10
20	14
25	20
30	26.5
35	35
40	47

Of the methods used to express humidity, absolute and specific humidity are similar in that they both specify the amount of water vapor contained in a unit of air. *Absolute humidity* is stated as the weight of water vapor in a given volume of air (usually as grams per cubic meter). As air moves from one place to another, even without a change in moisture content, changes in pressure and temperature cause changes in volume and consequently the absolute humidity, thus limiting the usefulness of this index. Therefore, meteorologists generally use specific humidity to express the water vapor content of air. *Specific humidity* is expressed as the weight of water vapor per weight of a chosen mass of air, including the water vapor. Since it is measured in units of weight (usually grams per kilogram), specific humidity is not affected by changes in pressure or temperature.

The most familiar and perhaps the most misunderstood term used to describe the moisture content of air is relative humidity. Stated in an admittedly oversimplified manner, *relative humidity* is the ratio of the air's water vapor content to its water vapor capacity at a given temperature. From Table 4-1 we see that at 25°C the capacity of the air is 20 grams per kilogram. If on a 25°C day the air contains 10 grams per kilogram, the relative humidity is expressed as 10/20 or 50 percent. When air is saturated, the relative humidity is 100 percent.

Since relative humidity is based on the air's water vapor content as well as on its capacity, relative humidity can be changed in either of two ways. First, if moisture is added by evaporation, the relative humidity will increase (Figure 4-4). The addition of moisture by this means occurs mainly over the oceans, but plants, soil, and smaller bodies of water do make contributions. The second method, illustrated by Figure 4-5, involves a change in temperature. We can generalize this concept as follows: With the specific humidity at a constant level, a decrease in air temperature will result in an increase in relative humidity, and an increase in temperature will cause a decrease in the relative humidity. In Figure 4-6 the variations in temperature and relative humidity during a typical day demonstrate rather well the relationship just described.

Another important idea related to relative humidity is the dew-point temperature. *Dew point* is the temperature to which a parcel of air would have to

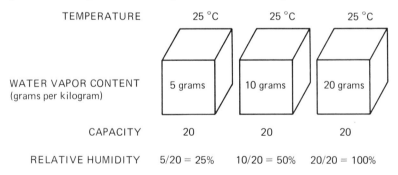

Figure 4-4

At a constant temperature the relative humidity will increase as water vapor is added to the air. Here, the capacity remains constant at 20 grams per kilogram and the relative humidity rises from 25 percent to 100 percent as the water vapor content increases.

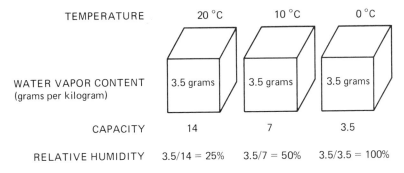

TEMPERATURE   20 °C   10 °C   0 °C

WATER VAPOR CONTENT   3.5 grams   3.5 grams   3.5 grams
(grams per kilogram)

CAPACITY   14   7   3.5

RELATIVE HUMIDITY   3.5/14 = 25%   3.5/7 = 50%   3.5/3.5 = 100%

Figure 4-5

When the water vapor content (specific humidity) remains constant, the relative humidity may be changed by increasing or decreasing the air temperature. In this example the specific humidity remains at 3.5 grams per kilogram. The reduction in temperature from 20°C to 0°C causes a decrease in capacity and thus an increase in the relative humidity.

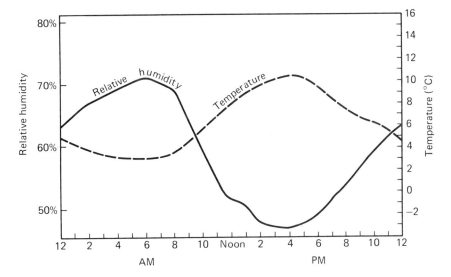

Figure 4-6

Typical daily variations in temperature and relative humidity during a spring day at Washington, D.C.

be cooled in order to reach saturation. Note that in Figure 4-5 unsaturated air at 20°C is cooled to 0°C before saturation occurs. Therefore, 0°C would be the dew-point temperature for this air. If this same parcel of air were cooled further, the air's capacity would be exceeded, and the excess vapor would be forced to condense.

Although relative humidity is used exclusively in our daily weather reports, it does have limitations if improperly interpreted. Recall from our earlier discussion that the capacity of air is temperature-dependent. Table 4-1 illustrates that the capacity of air at 35°C is 35 grams per kilogram and that the capacity of air at 10°C is only 7 grams per kilogram. It should be apparent that when their relative humidities are equal, air at 35°C contains five times as much water vapor as air at 10°C. This explains why cold winter air with a high relative humidity may be described as dry when compared with warm air that has an equally high relative humidity. In summary, relative humidity indicates how near the air is to being saturated, and absolute and specific humidity denote the quantity of water vapor contained in that air.

humidity measurement

Absolute and specific humidity are difficult to measure directly, but if the air temperature and relative humidity are known, they may be readily computed by consulting an appropriate table or graph.

Relative humidity is most commonly measured by using either a psychrometer or a hygrometer. The first of these instruments, the *psychrometer*, consists of two identical thermometers mounted side by side (Figure 4-7). One of the thermometers, called the wet bulb, has a thin muslin wick tied around the end. To use the psychrometer, the cloth sleeve is saturated with

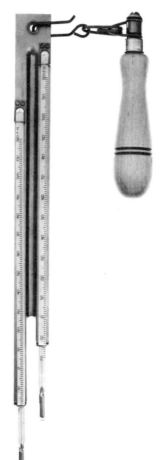

Figure 4-7

Sling psychrometer. This instrument is used to determine relative humidity and dew point. The dry-bulb thermometer gives the current air temperature. The thermometers are spun until the temperature of the wet-bulb thermometer stops declining. Then the thermometers are read and the data are used in conjunction with Tables 4-2 and 4-3. (Courtesy of WeatherMeasure Corporation.)

water and a continuous current of air is passed over the wick, either by swinging the instrument freely in the air or by fanning air past it. As a consequence, water evaporates from the wick and the temperature of the wet bulb drops. The loss of heat that was required to evaporate water from the wet bulb causes the lowering of the thermometer reading.

The amount of cooling that takes place is directly proportional to the dryness of the air. The drier the air, the more the cooling. Therefore, the larger the difference between the thermometers, the lower the relative humidity; the smaller the difference, the higher the relative humidity. If the air is saturated, no evaporation will occur, and the two thermometers will have identical readings.

Tables have been devised for obtaining both the relative humidity and the temperature of the dew point (Tables 4-2 and 4-3). All that is required is to record the air (dry-bulb) temperature and calculate the difference between the wet-and dry-bulb readings. The difference is known as the depression of the wet bulb. For example, assume that the dry-bulb temperature is 20°C and that the wet-bulb reading after swinging or fanning is 15°C. To determine the relative humidity, find the dry-bulb temperature on the left-hand column of Table 4-2 and the depression of the wet bulb across the top. The relative humidity is found where the two meet. In this example the relative humidity is 58 percent. In the same manner (except using Table 4-3) the dew point can be determined. In this case it would be 12°C.

The second commonly used instrument for measuring relative humidity, the *hygrometer*, can be read directly without using tables. The hair hygrometer operates on the principle that hair or certain synthetic fibers change their length in proportion to changes in the relative humidity; the hair lengthens as relative humidity increases and it shrinks as the relative humidity drops. The tension of a bundle of hairs is linked mechanically to an indicator that is calibrated between 0 and 100 percent. Thus, one need only glance at the dial to determine the relative humidity. Unfortunately, the hair hygrometer is much less accurate than the psychrometer. Furthermore, it requires frequent calibration and is slow in responding to changes in humidity, especially at low temperatures.

A different type of hygrometer is used in remote-sensing instrument packages such as radiosondes that transmit upper-air observations back to ground stations. The electric hygrometer contains an electrical conductor coated with a moisture-absorbing chemical. It works on the principle that the passage of current varies as the relative humidity varies.

necessary conditions for condensation

As we learned earlier in this chapter, condensation occurs when water vapor in the air changes to a liquid. The result of this process may be dew, fog, or clouds. Although each of these forms of condensation is very different, they all have two things in common: First, for any form of condensation to occur, the air must be saturated. Saturation occurs either when the air is cooled below the dew point, which most commonly happens, or when water vapor

Table 4-2

Relative Humidity in Percent (1000 mb)

Dry-Bulb Temperature (°C)	Wet-Bulb Depression ($T_d - T_w$)																					
	1	2	3	4	5	6	7	8	9	10	11	12	13	14	15	16	17	18	19	20	21	22
−20	28																					
−18	40																					
−16	48	0																				
−14	55	11																				
−12	61	23																				
−10	66	33	0																			
−8	71	41	13																			
−6	73	48	20	0																		
−4	77	54	32	11																		
−2	79	58	37	20	1																	
0	81	63	45	28	11																	
2	83	67	51	36	20	6																
4	85	70	56	42	27	14																
6	86	72	59	46	35	22	10															
8	87	74	62	51	39	28	17	0														
10	88	76	65	54	43	33	24	13	4													
12	88	78	67	57	48	38	28	19	10	2												
14	89	79	69	60	50	41	33	25	16	8	1											
16	90	80	71	62	54	45	37	29	21	14	7	1										
18	91	81	72	64	56	48	40	33	26	19	12	6	0									
20	91	82	74	66	58	51	44	36	30	23	17	11	5	0								
22	92	83	75	68	60	53	46	40	33	27	21	15	10	4	0							
24	92	84	76	69	62	55	49	42	36	30	25	20	14	9	4	0						
26	92	85	77	70	64	57	51	45	39	34	28	23	18	13	9	5	0					
28	93	86	78	71	65	59	53	47	42	36	31	26	21	17	12	8	4	0				
30	93	86	79	72	66	61	55	49	44	39	34	29	25	20	16	12	8	4	0			
32	93	86	80	73	68	62	56	52	48	43	38	34	28	22	19	14	11	8	4	0		
34	93	86	81	74	69	63	58	52	48	44	40	36	30	26	22	18	14	11	8	5	0	
36	94	87	81	75	69	64	59	54	50	46	42	38	32	28	24	21	17	13	10	7	4	0
38	94	87	82	76	70	66	60	55	51	46	42	38	34	30	26	23	20	16	13	10	7	5
40	94	89	82	76	71	67	61	57	52	48	44	40	36	33	29	25	22	19	16	13	10	7

Table 4-3

Dew-Point Temperature (1000 mb)

Dry-Bulb Temperature (°C)	Saturation Vapor Pressure (mb)	Wet-Bulb Depression $(T_d - T_w)$																					
		1	2	3	4	5	6	7	8	9	10	11	12	13	14	15	16	17	18	19	20	21	22
−20	1.2540	−33																					
−18	1.4877	−28																					
−16	1.7597	−24																					
−14	2.0755	−21	−36																				
−12	2.4409	−18	−28																				
−10	2.8627	−14	−22																				
−8	3.3484	−12	−18	−29																			
−6	3.9061	−10	−14	−22																			
−4	4.5451	−7	−11	−17	−29																		
−2	5.2753	−5	−8	−13	−20																		
0	6.1078	−3	−6	−9	−15	−24																	
2	7.0547	−1	−3	−6	−11	−17																	
4	8.1294	1	−1	−4	−7	−11	−19																
6	9.3465	4	1	−1	−4	−7	−13	−21															
8	10.722	6	3	1	−2	−5	−9	−14															
10	12.272	8	6	4	1	−2	−5	−9	−14	−28													
12	14.017	10	8	6	4	1	−2	−5	−9	−16													
14	15.977	12	11	9	6	4	1	−2	−5	−10	−17												
16	18.173	14	13	11	9	7	4	1	−1	−6	−10	−17											
18	20.630	16	15	13	11	9	7	4	2	−2	−5	−10	−19										
20	23.373	19	17	15	14	12	10	7	4	2	−2	−5	−10	−19									
22	26.430	21	19	17	16	14	12	10	8	5	3	−1	−5	−10	−19								
24	29.831	23	21	20	18	16	14	12	10	8	6	2	−1	−5	−10	−18							
26	33.608	25	23	22	20	18	17	15	13	11	9	6	3	0	−4	−9	−18						
28	37.796	27	25	24	22	21	19	17	16	14	11	9	7	4	1	−3	−9	−16					
30	42.430	29	27	26	24	23	21	19	18	16	14	12	10	8	5	1	−2	−8	−15				
32	47.551	31	29	28	27	25	24	22	21	19	17	15	13	11	8	5	2	−2	−7	−14			
34	53.200	33	31	30	29	27	26	24	23	21	20	18	16	14	12	9	6	3	−1	−5	−12	−29	
36	59.422	35	33	32	31	29	28	27	25	24	22	20	19	17	15	13	10	7	4	0	−4	−10	
38	66.264	37	35	34	33	32	30	29	28	26	25	23	21	19	17	15	13	11	8	5	1	−3	−9
40	73.777	39	37	36	35	34	32	31	30	28	27	25	24	22	20	18	16	14	12	9	6	2	−2

is added to the air. Second, there generally must be a surface on which the water vapor may condense. When dew occurs, objects at or near the ground serve this purpose. When condensation occurs in the air above the ground, tiny bits of particulate matter known as *condensation nuclei* serve as surfaces for the condensation of water vapor. The importance of these nuclei should be noted because if they are absent, a relative humidity of nearly 400 percent is needed to produce clouds. Once condensation occurs under such supersaturated conditions, cloud droplets could grow rapidly and produce downpours of unimaginable magnitude. Fortunately, condensation nuclei such as microscopic dust, smoke, and salt particles are profuse in the lower atmosphere. Because of this abundance of particles, relative humidity rarely exceeds 101 percent. Some particles, like salt from the ocean, are particularly good nuclei because they absorb water. These particles are termed *hygroscopic* ("water-seeking") *nuclei*.

When condensation takes place, the initial growth rate of cloud droplets is rapid, but it diminishes quickly because the excess water vapor is readily consumed by the numerous competing particles. This results in the formation of a cloud consisting of billions of tiny water droplets, all so fine that they remain suspended in air. The slow growth of these cloud droplets by additional condensation and the immense size difference between cloud droplets and raindrops suggest that condensation alone is not responsible for the formation of drops large enough to fall as rain. A discussion of rain-producing mechanisms is found in Chapter 5.

condensation aloft: adiabatic temperature changes

During cloud formation and often in the formation of fog the air is cooled to its dew point. Near the earth's surface heat is readily exchanged between the ground and the air above. This accounts for the cooling involved in the formation of some types of fog. However, because air is a poor conductor of heat, this exchange is virtually nonexistent above a few thousand meters. Thus, some other mechanism must operate during cloud formation. This process is easily visualized if you have ever pumped up a bicycle tire and noticed that the pump barrel became very warm. The heat you felt was the consequence of the work you did on the air in order to compress it. When energy is used to compress air, an equivalent amount of energy is released as heat. Conversely, when air is allowed to escape from a bicycle tire it will cool as it expands. This results because the expanding air pushes (does work on) the surrounding air and must cool by an amount equivalent to the energy expended. The temperature changes just described (no heat was added and no heat was subtracted) are called *adiabatic temperature changes* and result when air is compressed or allowed to expand. In summary, when air is allowed to expand, it cools; when air is compressed, it warms.

Anytime air moves upward it passes through regions of successively lower pressure. As a result, the ascending air expands and cools adiabatically. Unsaturated air cools at the rather constant rate of 1°C for every 100 meters of ascent (10°C per kilometer). Conversely, descending air comes under

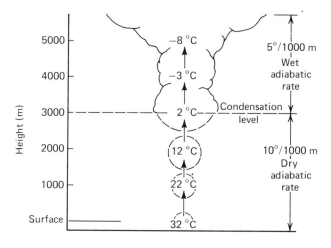

Figure 4-8

Rising air cools at the dry adiabatic rate of 10°C per 1000 meters until the air reaches the dew point and condensation (cloud formation) begins. As air continues to rise, the latent heat released by condensation lowers the rate of cooling. The wet adiabatic rate is therefore always less than the dry adiabatic rate.

increasingly higher pressures, compresses, and is heated 1°C for every 100 meters of descent. This rate of cooling or heating applies only to unsaturated air and is known as the *dry adiabatic rate*. If air rises long enough, it will inevitably cool sufficiently to cause condensation. From this point along its ascent latent heat stored in the water vapor will be liberated. Although the air will continue to cool after condensation begins, the released latent heat works against the adiabatic process, thereby reducing the rate at which the air cools. This slower rate of cooling caused by the addition of latent heat is called the *wet adiabatic rate* of cooling. Since the amount of latent heat released depends upon the quantity of moisture present in the air, the wet adiabatic rate varies from 0.5°C per 100 meters for air with a high moisture content to 0.9°C per 100 meters for dry air. Figure 4-8 illustrates the role of adiabatic cooling in the formation of clouds. Note that from the surface up to the condensation level the air cools at the dry adiabatic rate. The wet adiabatic rate commences at the point of condensation.

stability

It has been pointed out that if air rises, it will cool and eventually produce clouds. Why does air rise on some occasions but not on others? Why do the size of clouds and the amount of precipitation vary so much when air does rise? The answers to these questions are closely related to the stability of the air. Imagine, if you will, a large bubble of air with a thin flexible cover that allows it to expand but prevents it from mixing with the surrounding air. If the imaginary bubble were forced to rise, its temperature would decrease because of expansion. By comparing the bubble's temperature to that of the surrounding air, we could determine its stability. If the bubble's temperature were lower than that of its environment, it would be heavier, and if it were allowed to do so, it would sink to its original position. Air of this type, termed *stable air*, resists vertical displacement.

If, however, our imaginary bubble were warmer, and therefore lighter, than the surrounding air, it would continue to rise until it reached an altitude having the same temperature, much as a hot-air balloon would rise as long as it was lighter than the surrounding air. This type of air is classified as *unstable air.*

determination of stability

In an actual situation the stability of the air is determined by examining the temperature of the atmosphere at various heights. Recall that this measure is called the lapse rate. You must not confuse the lapse rate, which is the temperature of the atmosphere as determined from observations made by balloons and airplanes, with adiabatic temperature changes. The latter measure indicates the change in temperature a parcel of air would experience as it moved vertically through the atmosphere.

For illustration we will examine a situation in which the prevailing lapse rate is 5°C per 1000 meters (Figure 4-9). Under this condition, when the air at the surface has a temperature of 25°C, the air at 1000 meters will be 5°C cooler or 20°C, while the air at 2000 meters will have a temperature of 15°C, and so forth. At first glance it appears that the air at the surface is lighter than the air at 1000 meters since it is 5°C warmer. However, if the air near the surface were unsaturated and were to rise to 1000 meters, it would expand and cool at the dry adiabatic rate of 1°C per 100 meters. Therefore, upon reaching 1000 meters its temperature would have dropped a total of 10°C to 15°C. Being 5°C cooler than its environment, it would be heavier and would tend to sink to its original position. Hence, we say that the air near the surface is potentially cooler than the air aloft and therefore it will not rise. By similar reasoning, if the air at 1000 meters subsided, adiabatic heating would increase its temperature 10°C by the time it reached the surface, making it

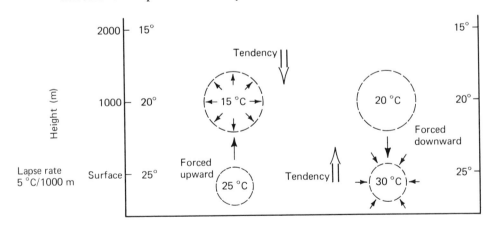

Figure 4-9

Schematic representation of a stable atmosphere. Refer to the parcel of air located on the left side of this drawing and note that the air near the surface is potentially cooler than the air aloft and therefore resists upward motion.

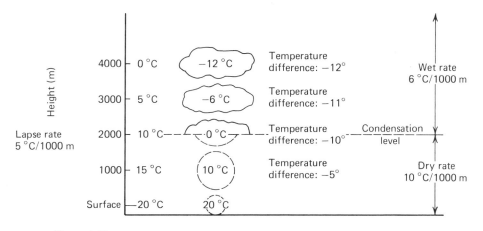

Figure 4-10
Absolute stability prevails when the lapse rate is less than the wet adiabatic rate. The rising parcel of air is therefore always cooler and heavier than the surrounding air.

warmer than the surrounding air; therefore, its buoyancy would cause it to return. The air just described is stable and resists vertical movement.

Stated quantitatively, *absolute stability* prevails when the lapse rate is less than the wet adiabatic rate. Figure 4-10 depicts this situation by using a lapse rate of 5°C per 1000 meters and a wet adiabatic rate of 6°C per 1000 meters. Note that at 1000 meters the temperature of the surrounding air is 15°C and that the rising parcel of air has cooled to 10°C and is therefore the heavier air. Even if this stable air were to be forced above the condensation level, it would remain cooler and heavier than its environment and would have a tendency to return to the surface.

At the other extreme air is said to exhibit *absolute instability* when the lapse rate is greater than the dry adiabatic rate. As shown in Figure 4-11, the ascending parcel of air is always warmer than its environment and will continue to rise because of its own buoyancy.

Another situation existing in the atmosphere is called *conditional instability*. This occurs when moist air has a lapse rate between the dry and wet adiabatic rates (between 0.5°C and 1°C per 100 meters). Notice in Figure 4-12 that for the first 4000 meters the rising parcel of air is cooler than the surrounding air and that it is therefore considered stable. With the addition of latent heat above the condensation level, the parcel eventually becomes warmer than the surrounding air. From this point along its ascent the parcel will continue to rise without an outside force and therefore is considered unstable. Conditionally unstable air can be described as air that begins its ascent as stable air but at some point above the condensation level it becomes unstable. The word conditional is used because only if the air is forced upward initially can it become unstable. Conditional instability is perhaps the most common type of instability.

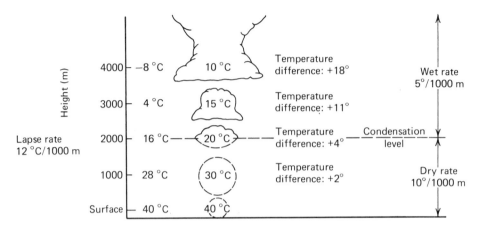

Figure 4-11

Absolute instability illustrated by using a lapse rate of 12°C per 1000 meters. The rising air is always warmer and therefore lighter than the surrounding air.

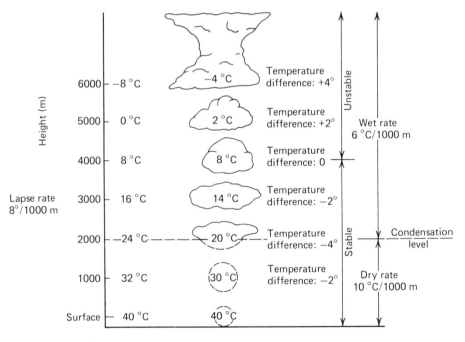

Figure 4-12

Conditional instability illustrated by using a lapse rate of 8°C per 1000 meters that lies between the dry adiabatic rate and the wet adiabatic rate. The rising parcel of air is cooler than the surrounding air below 4000 meters and warmer above 4000 meters.

80

stability and daily weather

From the previous discussion we can conclude that stable air resists vertical movement and that unstable air ascends freely because of its own buoyancy. But how do these facts manifest themselves in our daily weather?

Since stable air resists upward movement, we might conclude that clouds would not form when stable conditions prevail in the atmosphere. Although this seems reasonable, processes do exist that force air aloft. These will be discussed in the following section. When stable air is forced aloft, the clouds that form are widespread and have little vertical thickness in comparison to their horizontal dimension, and precipitation, if any, is light. By contrast, clouds associated with unstable air are towering and are usually accompanied by heavy precipitation. Hence, we can conclude that on a dreary overcast day with light drizzle, stable air was forced aloft. On the other hand, on a day when cauliflower-shaped clouds appear to be growing as if bubbles of hot air were surging upward, we can be relatively certain that the ascending air is unstable.

Instability occurs very often on hot summer afternoons when solar heating is intense. Surface irregularities cause pockets of air to be warmed more than the surrounding air. Consequently, these warmer parcels of air will buoy upward. If they rise above the condensation level, clouds form, which on occasion produce midafternoon rain showers. The height of clouds produced in this fashion is somewhat limited since instability caused solely by surface heating is confined to, at most, the first few kilometers of the atmosphere. Also, the accompanying rains are of short duration because the precipitation readily cools the surface.

The most stable conditions occur during a temperature inversion when temperature increases with height. In this situation the air near the surface is cooler and heavier than the air aloft, and, therefore, little vertical mixing occurs between the layers. Since pollutants are generally added to the air from below, a temperature inversion confines them to the lowermost layer where they continually build in concentration. Widespread fog is another sign of stability. If the layer containing fog were mixing freely with the "dry" layer above, evaporation would quickly eliminate the foggy condition.

In summary, the role of stability in determining our daily weather cannot be overemphasized. The air's stability, or lack of it, determines whether the soot from a smokestack is dispersed widely or is confined to a narrow layer near the earth's surface. Of possibly greater importance, it determines to a large degree whether or not clouds develop and produce precipitation and whether that precipitation will come as a gentle shower or a violent downpour.

changes in stability

Most processes that alter stability occur as a result of the movement of air, although daily temperature changes do play an important role. In general, any factor that causes an increase in the lapse rate renders the air more unstable, while a reduction in the lapse rate increases the air's stability. Recall that the larger the lapse rate, the more rapidly the temperature drops with increased altitude. Therefore, any factor that causes the air near the surface to become warmed in relation to the air aloft increases instability. The opposite is also true; any factor that causes the surface air to be chilled results in the air becoming more stable. Hence, as stated earlier, on a clear day when there is abundant surface heating the lower atmosphere often becomes unstable and causes parcels of air to rise. After the sun sets surface cooling generally renders the air stable again.

Similar changes in stability occur as air moves horizontally, traversing a surface having markedly different temperatures. In the winter warm air from the Gulf of Mexico moves northward over the cold land surface of the Midwest. Since the air is cooled from below, it becomes more stable, often producing widespread fog. The opposite occurs when wintertime polar air moves southward over the open waters of the Great Lakes. The moisture and heat added to the frigid polar air from the water below are enough to make it unstable and generate the clouds that produce heavy snowfalls on the downwind shores of these lakes.

Vertical movements of air also influence stability. When there is a general downward air flow, called *subsidence*, the upper portion of the subsiding layer is heated by compression, more so than the lower portion. Since the air near the surface is generally not involved in the subsidence, its temperature remains unchanged. The net effect is to stabilize the air, since the air aloft is warmed in relation to the surface air. The warming effect of a few hundred meters of subsidence is enough to evaporate the clouds found in any layer of the atmosphere. Thus, one sign of subsiding air is a cloudless sky. Subsidence can also produce a temperature inversion aloft. The most intense and prolonged temperature inversions and associated air pollution episodes are caused by subsidence, a topic discussed more fully at the end of this chapter.

Upward movement of air generally enhances instability. This is especially true when the lower portion of the rising layer has a higher moisture content than the upper portion. As the air moves upward, the lower portion becomes saturated first and cools at the lesser wet adiabatic rate. The net effect of this process is to increase the lapse rate within the rising layer. This process is especially important in producing the instability associated with thunderstorms. In addition, recall that conditionally unstable air can become unstable if it is lifted sufficiently.

Whenever air is flowing together (*convergence*), it results in general upward movement. This occurs because as air converges it occupies a smaller and smaller area, which requires that the height of the air column increase. Consequently, the air within the column must move upward, thus enhancing instability. The Florida peninsula provides an excellent example of the role that convergence plays in initiating instability. On warm days the air flow is off the ocean along both coasts of Florida and causes general convergence over the peninsula (Figure 4-13). This convergence and associated uplift,

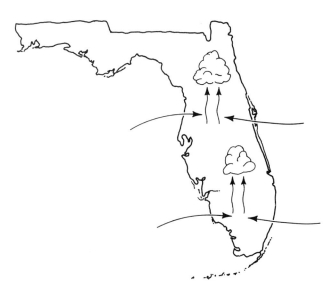

Figure 4-13
Schematic drawing showing convergence over the Florida peninsula.

aided by the intense solar heating, cause midafternoon thunderstorms to frequent this part of the United States more so than any other.

On a smaller scale, the loss of heat by radiation from cloud tops during the evening hours adds to their instability and growth. Unlike air, which is a poor radiator of heat, cloud droplets emit energy to space nearly as well as does the earth's surface. Towering clouds that owe their growth to surface heating lose their source of energy at sunset. However, after sunset radiation cooling at their tops steepens the lapse rate and can initiate additional upward flow of warm parcels from below. This process is believed to be responsible for producing nocturnal thunderstorms from clouds whose growth prematurely ceased at sunset.

A list of factors that modify air's stability provides a summary. Instability is enhanced by:

1. intense solar heating that warms the air from below
2. the heating of an air mass from below as it traverses a warm surface
3. forceful lifting of air, such as over an elevated land surface
4. upward movement of air associated with general convergence
5. radiation cooling from cloud tops

Stability is enhanced by:

1. radiation cooling of the earth's surface after sunset
2. the cooling of an air mass from below as it traverses a cold surface
3. subsidence of an air column.

forceful lifting

Earlier we demonstrated that stable air and conditionally unstable air will not rise on their own; they require some mechanism to trigger the vertical movement. Three such mechanisms are convergence, orographic lifting, and

83

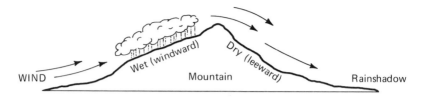

Figure 4-14
Orographic lifting.

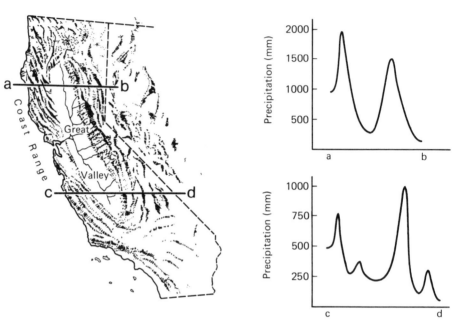

Figure 4-15
Relationship between topography and average annual rainfall from the California coast to the dry plateau of Nevada's Great Basin. As shown in this profile, precipitation maximums occur in the mountainous regions, but precipitation is much lighter in the valleys. (Precipitation data from the U.S. Department of Agriculture.)

frontal wedging. The role of convergence in vertical lifting has already been discussed.

Orographic lifting occurs when sloping terrain such as mountains act as barriers to the flow of air and force the air to ascend (Figure 4-14). Many of the rainiest places in the world are located on windward mountain slopes. A station at Mt. Waialeale, Hawaii, for example, records the highest average annual rainfall in the world, some 1168 centimeters. The station is located on the windward (northeast) coast of the island of Kauai at an elevation of 1523 meters.*

———————

*For other precipitation extremes, see Appendix F.

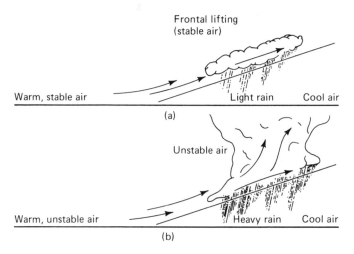

Frontal lifting
(stable air)

Warm, stable air   Light rain   Cool air

(a)

Unstable air

Warm, unstable air   Heavy rain   Cool air

(b)

Figure 4-16

(a) When stable air is lifted, layered clouds usually result. (b) When warm, unstable air is forced to rise over cooler air, "towering" clouds develop.

In addition to providing the lift to render air unstable, mountains further remove more than their share of moisture in other ways. By slowing the horizontal flow of air, they cause convergence as well as retarding the passage of storm systems. Also, the irregular topography of mountains enhance differential heating and surface instability. These combined effects account for the generally higher precipitation we associate with mountainous regions as compared to the surrounding lowlands.

By the time air reaches the leeward side of a mountain much of the moisture has been lost, and if the air descends, it warms, making condensation and precipitation even less likely. As shown in Figure 4-14, the result often is a *rain shadow desert*. The Great Basin Desert of the western United States lies only a few hundred miles from the Pacific Ocean, but it is effectively cut off by the imposing Sierra Nevada (Figure 4-15). The Gobi Desert of Mongolia, the Takla Makan of China, and the Patagonia Desert of Argentina are other examples of deserts found on the leeward sides of mountains.

Frontal wedging occurs when cool air acts as a barrier over which warmer, lighter air rises. This phenomenon is very common throughout the continental United States and is responsible for the bulk of the precipitation in many areas, as we shall see later. Figure 4-16 illustrates frontal wedging of stable and unstable air. As this figure shows, forceful lifting is important in producing clouds. However, the stability of the air determines to a great extent the type of clouds formed and the amount of precipitation that may be expected.

air pollution   Air pollution has become a growing threat to our health and welfare because of the ever-increasing emissions of air contaminants into our never-increasing atmosphere. An average adult male requires about 13.5 kilograms of air each day, as compared with about 1.2 kilograms of food and 2 kilograms of water. The cleanliness of the air, therefore, should certainly be as important to us as the cleanliness of our food and water.

Air is never perfectly clean. There have always been many natural sources of air pollution. Ash from volcanic eruptions, salt particles from breaking waves, pollen and spores released by plants, smoke from forest and brush fires, and windblown dust are all examples of "natural air pollution." Ever since people have been on earth, however, they have added to the frequency and intensity of some of these natural pollutants, especially to the last two (Figure 4-17). With the discovery of fire came an increased number of accidental as well as intentional burnings. Even today, in many parts of the world fire is used to clear land for agricultural purposes (the so-called slash-and-burn method), filling the air with smoke and reducing visibility. When people clear the land of its natural vegetative cover for whatever purpose, soil is exposed and blown into the air. However, when one considers the air in a modern-day industrial city, these man-accentuated forms of pollution, although significant, may seem minor by comparison.

Although some forms of air pollution are relatively recent creations, other forms have been around for centuries. Smoke pollution, for example, plagued London for centuries. Because of the odor and smoke produced by the burning of coal, King Edward I made the following proclamation in 1300: "Be it known to all within the sound of my voice, whosoever shall be found guilty of burning coal shall suffer the loss of his head." Unfortunately, one Londoner did not heed the king's warning and consequently paid the extreme price for his misdeed. As far as is known, however, this is the only case of capital punishment resulting from an air pollution violation!

The ban on the burning of coal led to the use of an alternative fuel—wood. However, because of extensive wood burning, English forests were soon dramatically reduced and coal consumption again increased in spite of royal

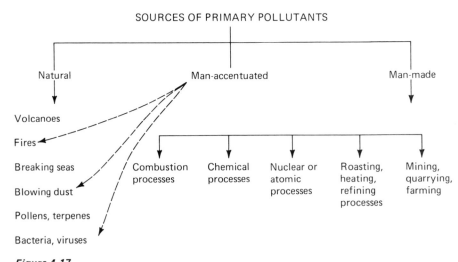

Figure 4-17

Sources of primary pollutants. (After Reid A. Bryson and John E. Kutzbach, *Air Pollution,* Commission on College Geography Resource Paper No. 6, Figure 2, p. 8. Copyright by the Association of American Geographers.)

disapproval. Thus, in 1661 when John Evelyn wrote *Fumifugium, or the Inconvenience of Aer and Smoak of London Dissipated, together with some Remidies Humbly Proposed*, the problem of foul air still plagued Londoners. In his book Evelyn noted that a traveler, although many miles from London, "sooner smells than sees the city to which he repairs." In fact, London continued to have severe air pollution problems well into the twentieth century. It was only after a devastating smog disaster in 1952 that truly decisive action was taken to clean the air.

London, however, has not monopolized the air pollution scene. With the coming of the Industrial Revolution, many cities began to experience "big-time" air pollution. Instead of just simply accelerating natural sources, people found many new ways to pollute the air (Figure 4-17) and many new things to pollute it with—sulfur and nitrogen oxides, carbon monoxide, and hydrocarbons, to name a few. With the rapid growth of the world's population and accelerated industrialization, the quantities of atmospheric pollutants increased drastically.

The recent history of air pollution disasters began in the Meuse valley in Belgium, where the first major episode to be well studied occurred. For 5 days in December, 1930 a blanket of smog hung in the valley, killing 63 people. Since the 1930s many air pollution episodes have demonstrated the devastating effect dirty air can have on life and property. In October, 1948 Donora, Pennsylvania had such an experience. The grime that settled from the air coated houses, streets, and sidewalks so that pedestrians and autos actually left distinct footprints and tire tracks. Almost 6000 of the town's 14,000 inhabitants became ill, and 20 died. The most tragic air pollution episode ever occurred in London in December, 1952. More than 4000 people died as a result of this 5-day ordeal. The people who suffered most were those with respiratory and heart problems, primarily the elderly. Extreme air pollution darkened London again in 1953 and 1962 and affected New York City in 1953, 1963, and 1966. As sensational and tragic as these acute episodes are, health authorities are equally concerned with the slow and subtle effects on our lungs and other organs by air pollution levels that are much lower but that are present every day year after year.

temperature inversions and air pollution

When air pollution episodes occur, they are not generally the result of a drastic increase in the output of pollutants; instead, they occur because of changes in atmospheric stability. Usually, when contaminants are released they are mixed by wind and turbulence with the surrounding air and are diluted. Perhaps you have heard the following well-known phrase: "The solution to pollution is dilution." To some degree, this is true. However, if the air into which the pollutants are released is not dispersed (moved out of the area), the air will become more and more toxic. *Temperature inversions* are among the most significant causes of retarding the dispersion of pollutants. Most of the air pollution episodes cited earlier were linked to the

87

occurrence of these phenomena. Normally, smoke and exhaust fumes are carried upward by air currents and dispersed by winds aloft. However, when an inversion is present, warm air overlying cooler air acts as a lid and prevents further upward movement. The pollutants are thus trapped below. This is illustrated very dramatically by the sequence of photographs of downtown Los Angeles shown in Figure 4-18.

Solar heating can result in high surface temperatures during the late morning and afternoon which steepen the lapse rate and render the lower air unstable. However, during the nighttime hours just the opposite situation may occur; temperature inversions, which result in very stable atmospheric conditions, can develop close to the ground. These surface inversions form because the ground is a more effective radiator than the air above. This being the case, radiation from the ground to a clear night sky causes more rapid cooling at the surface than higher in the atmosphere. Consequently, the coldest air is found next to the ground, yielding a vertical temperature profile resembling the one shown in Figure 4-19(a). Once the sun rises, the ground is heated, and the inversion disappears. Although surface inversions are usually rather shallow, they may be very deep in regions where the land surface is uneven. Because cold air is denser than warm air, the chilled air near the surface gradually drains from the uplands and slopes into adjacent lowlands and valleys (Figure 4-20). As might be expected, this deeper surface inversion will not dissipate as quickly after sunrise. Thus, although valleys are often preferred sites for manufacturing because they afford easy access to water transportation, they are also more likely to experience relatively thick surface inversions which, in turn, will have a negative effect on air quality.

Many extensive and long-lived air pollution episodes are linked to temperature inversions that develop in association with the sinking air that characterizes slow-moving centers of high air pressure (anticyclones). As the air sinks to lower altitudes, it is compressed and consequently its temperature rises. Since turbulence is almost always present near the ground, this lowermost portion of the atmosphere is generally prevented from participating in the general subsidence. Thus, an inversion develops aloft between the lower turbulent zone and the subsiding warmer layers above [Figure 4-19(b)]. Since clear skies are often found in high pressure regions, it is not at all unusual for a surface inversion to form during the night and early morning hours. Such a double inversion is shown in Figure 4-19(c).

sources and types of air pollution

Table 4-4 lists the major sources of air pollution and the types of contaminants that characterize each. The significance of the transportation category is obvious: It accounts for more than one-half of our air pollution (by weight). In addition to motor vehicles, this category includes trains, ships, and airplanes. However, the more than 110 million cars and trucks on U.S. roads are, without a doubt, the greatest sources of air pollution.

Pollutants may be grouped into two categories: primary and secondary.

88

(a)

(b)

(c)

Figure 4-18

(*a*) Los Angeles City Hall on a clear day. (*b*) Smog engulfs City Hall as a temperature inversion at 450 meters traps pollutants below. (*c*) A temperature inversion at 100 meters is visible as smog shrouds the lower portion of City Hall while the upper portion of the building is visible in the clear air above the base of the inversion. (Courtesy of the Los Angeles County Air Pollution Control District.)

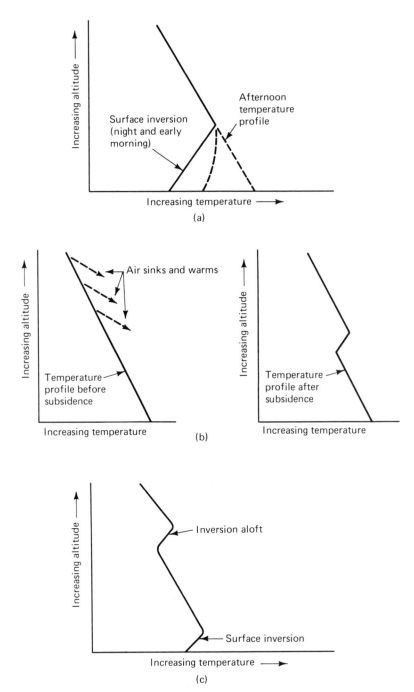

Figure 4-19

Types of temperature inversions. (*a*) The solid line shows the vertical temperature profile as it might appear early in the morning. The dashed lines show the changes that will occur as the surface is heated during the day. (*b*) Subsidence often creates an inversion aloft. (*c*) A double inversion.

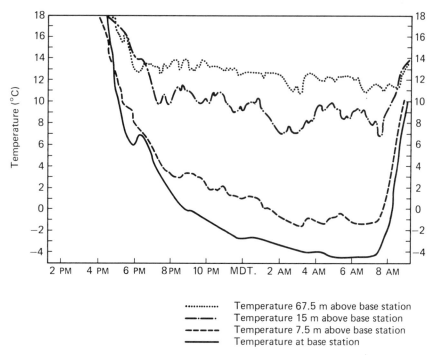

.... Temperature 67.5 m above base station
—·—· Temperature 15 m above base station
- - - - - Temperature 7.5 m above base station
———— Temperature at base station

Figure 4-20

Cold air drainage often results in rather thick surface inversions in valleys.

Primary pollutants are emitted directly from identifiable sources. These are shown in Figure 4-17. *Secondary pollutants* are produced in the atmosphere when certain chemical reactions take place among the primary pollutants. For example, when sunlight acts on nitrogen oxides and certain organic compounds, the toxic gas ozone (O_3) is produced. Reactions like these, which are triggered by sunlight, are called *photochemical*. Sulfuric acid (H_2SO_4) is another common secondary pollutant. It is produced when sulfur dioxide combines with oxygen, yielding sulfur trioxide, which then combines with water to form this irritating and corrosive acid.

The sections that follow briefly describe the major pollutants.

sulfur oxides. Sulfur dioxide, the most abundant of the sulfur oxides, is a colorless gas that corrodes many metals. Sulfur dioxide also creates health problems. It irritates the mucous membranes of the respiratory tract and the eyes and in high concentrations it can produce serious respiratory problems. Almost three-fourths of the sulfur oxides are emitted from the burning of sulfur-bearing fuels, particularly coal and low-grade fuel oil. The primary sources are power plants and industry.

carbon monoxide. Carbon monoxide, an odorless and colorless gas, is the largest single air pollutant and is produced by the incomplete combustion of fuels like gasoline. It is not surprising then that motor vehicles account

Table 4-4

Nationwide Emission Estimates, 1975 (millions of tons)

Source Category	Particulates	Sulfur Oxides	Nitrogen Oxides	Hydro-carbons	Carbon Monoxide
Transportation	1.3	0.8	10.7	11.7	77.4
Highway	0.9	0.4	8.2	10.0	67.8
Nonhighway	0.4	0.4	2.5	1.7	9.6
Stationary fuel combustion	6.6	26.3	12.4	1.4	1.2
Electric utilities	3.5	21.0	6.8	0.1	0.3
Other	3.1	5.3	5.6	1.3	0.9
Industrial processes	8.7	5.7	0.7	3.5	9.4
Chemicals	0.2	1.0	0.3	1.6	3.3
Petroleum refining	0.1	0.9	0.3	0.9	2.2
Metals	1.3	3.2	0	0.2	2.8
Mineral products	4.5	0.6	0.1	0	0
Other	2.6	<0.1	<0.1	0.8	1.1
Solid waste	0.6	<0.1	0.2	0.9	3.3
Miscellaneous	0.8	0.1	0.2	13.4	4.9
Forest wildfires	0.4	0	0.1	0.6	3.3
Forest managed burning	0.1	0	<0.1	0.2	0.5
Agricultural burning	0.1	0	<0.1	0.1	0.6
Coal refuse burning	0.1	0.1	0.1	0.1	0.3
Structural fires	0.1	0	<0.1	<0.1	0.1
Organic solvents	0	0	0	8.3	0
Oil and gas production and marketing	0	0	0	4.2	0
Total	18.0	32.9	24.2	30.9	96.2

SOURCE: *National Air Quality and Emissions Trends Report, 1975,* U.S. Environmental Protection Agency Publication No. EPA-450/1-76-002, November, 1976.

for a very high percentage of the total emissions of this gas. Carbon monoxide reacts with the blood's hemoglobin and diminishes its capacity to transport oxygen to the body's tissues. In large amounts carbon monoxide is lethal; in small quantities it may cause headache, dizziness, and nausea.

hydrocarbons.　Hydrocarbons form a complex group of pollutants that originate primarily from the inefficient combustion of gasoline, coal, and oil. Again, motor vehicles are the single greatest source. Hydrocarbons react with other contaminants to form secondary pollutants, like ozone. When other hydrocarbons react with nitrogen oxides in the presence of sunlight, photo-chemical (Los Angeles-type) smog is the end product. Furthermore, some hydrocarbons have been shown to be cancer-producing agents (carcinogens).

nitrogen oxides.　Nitrogen oxides are yellow-to-brown gases that have a pungent odor and result from high-temperature combustion, such as that which takes place in a gasoline-powered automobile engine. The most sig-nificant impact of nitrogen oxides is that they are prime contributors to photochemical smog and therefore may lead to throat irritation, coughing, and respiratory problems.

particulates. This category consists of a variety of solid and liquid particles having a wide range of sizes. The primary sources are fuel (especially coal) combustion, various industrial processes, and forest fires. The influence of particulate matter in our atmosphere is very complex. Some particles may affect various weather processes and climatic conditions, and others may damage vegetation, corrode metals, or be harmful to human health. Dustfall, especially in industrial areas, soils property and leads to higher costs for cleaning and painting.

For the past several years the concern for clean air has increased, as evidenced by the passage of new laws, the establishment of new govermnent agencies, and increased research on pollution control problems. The costs of controlling pollution may seem high, but when they are compared to the costs of uncontrolled air pollution, both in dollars and in human health, the expenditures are easily justified.

REVIEW

1. Describe the movement of water through the hydrologic cycle.

2. Over the oceans the quantity of water lost to evaporation is not equaled by precipitation. Why then does the sea level not drop?

3. Summarize the processes by which water changes from one state to another. Indicate whether heat energy is absorbed or liberated.

4. After examining Table 4-1, write a generalization relating temperature and the capacity of air to hold water vapor.

5. How do absolute and specific humidity differ? What do they have in common? How is relative humidity different from absolute and specific humidity?

6. Refer to Figure 4-6 and then answer the following questions:
   a. During a typical day when is the relative humidity highest? When is it lowest?
   b. At what time of day would dew most likely form?
   Write a generalization relating air temperature and relative humidity.

7. If the temperature remains unchanged and if the specific humidity decreases, how will the relative humidity change?

8. Explain the principle of the psychrometer. The hair hygrometer.

9. What are the disadvantages of the hair hygrometer? Does this instrument have any advantages over the psychrometer?

10. Using the standard tables (Tables 4-2 and 4-3), determine the relative humidity and dew-point temperature if the dry-bulb thermometer read 22°C and the wet-bulb thermometer reads 16°C. How would the relative humidity and dew point change if the wet-bulb thermometer read 19°C?

11. What is the function of condensation nuclei in the formation of clouds? The function of the dew point?

12. As you drink an ice-cold beverage on a warm day, the outside of the glass or bottle becomes wet while the contents warm rapidly. Explain.

13. Why does air cool when it rises through the atmosphere?

14. If unsaturated air at 20°C were to rise, what would its temperature be at a height of 500 meters? If the dew-point temperature at the condensation level were 11°C, at what elevation would clouds begin to form?

15. Why does the adiabatic rate of cooling change when condensation begins? Why is the wet adiabatic rate not a constant figure?

16. The contents of an aerosol can are under very high pressure. When you push the nozzle on such a can, the spray, feels cold. Explain.

17. How does stable air differ from unstable air?

18. Explain the difference between the lapse rate and adiabatic cooling.

19. How is the stability of air determined?

20. List some weather conditions that would lead you to believe that the air is either stable or unstable.

21. How can the stability of air be altered?

22. Distinguish between subsidence and convergence. How might each influence the stability of the air?

23. How do orographic lifting and frontal wedging act to force air to rise?

24. Explain why the Great Basin area of the western United States is so dry. What term is applied to such a situation?

25. How do temperature inversions influence air pollution?

26. Describe the formation of a surface inversion and compare it with an inversion that occurs aloft.

27. What is the difference between primary and secondary pollutants? What is a photochemical reaction?

VOCABULARY
REVIEW

hydrologic cycle	condensation nuclei
transpiration	hygroscopic nuclei
calorie	adiabatic temperature changes
evaporation	dry adiabatic rate
latent heat	wet adiabatic rate
condensation	stable air
latent heat of condensation	unstable air
melting	absolute stability
freezing	absolute instability
latent heat of fusion	conditional instability
sublimation	specific humidity
humidity	relative humidity
saturation	dew point
vapor pressure	psychrometer
absolute humidity	hygrometer

subsidence

convergence

orographic lifting

rain shadow desert

frontal wedging

temperature inversion

primary pollutants

secondary pollutants

photochemical

Chapter Five

Forms
of condensation
and precipitation

Clouds and fog as well as rain, snow, sleet and hail are among the most conspicuous and observable aspects of the atmosphere and its weather. In this chapter we shall endeavor to gain a basic understanding of each of these phenomena.

clouds

Clouds are a form of condensation best described as visible aggregates of minute droplets of water or tiny crystals of ice. In addition to being prominent and sometimes spectacular features in the sky, clouds are of continual interest to meteorologists because they provide a visible indication of what is going on in the atmosphere. Anyone who observes clouds with the hope of recognizing different types often finds that there is a bewildering variety of these familiar white and gray masses streaming across the sky. However, once one knows the basic classification scheme for clouds, most of the confusion vanishes.

Prior to the beginning of the nineteenth century there were no generally accepted names for clouds. In 1803 Luke Howard, an English naturalist, published a cloud classification that met with great success and subsequently served as the basis of our present-day classification.

Clouds are classified on the basis of their appearance and height (Figure 5-1). Three basic cloud forms are recognized: cirrus, cumulus, and stratus. *Cirrus* clouds are high, white, and thin. They are separated or detached and form delicate veil-like patches or extended wispy fibers and often have a feathery appearance. The *cumulus* form consists of globular individual cloud

96

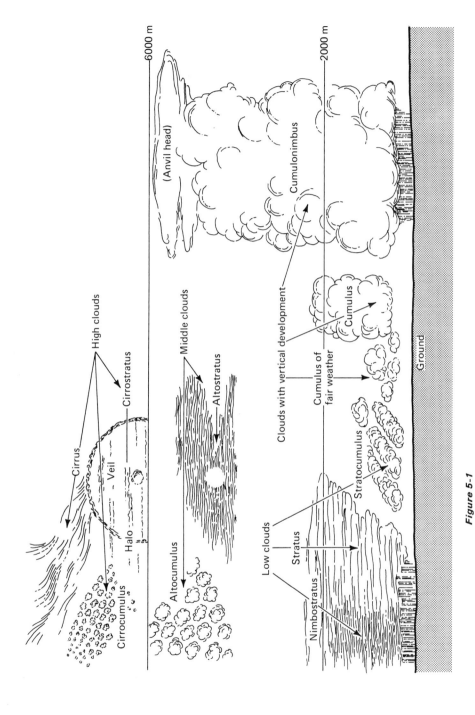

High clouds
Cirrus
Cirrostratus
Veil
Halo
Cirrocumulus

Middle clouds
Altostratus
Altocumulus

6000 m

(Anvil head)

Cumulonimbus

Clouds with vertical development

Cumulus

Cumulus of
fair weather

Cumulus

2000 m

Low clouds
Stratus
Stratocumulus
Nimbostratus

Ground

Figure 5-1

Classification of clouds according to height and form. (After Ward's Natural Science Establishment, Inc., Rochester, N.Y.)

masses. Normally, they exhibit a flat base and have the appearance of rising domes or towers. Such clouds are frequently described as having a cauliflowerlike structure. *Stratus* clouds are best described as sheets or layers that cover much or all of the sky. Although there may be minor breaks, there are no distinct individual cloud units. All other clouds either reflect one of these three basic forms or are combinations or modifications of them.

Looking at the second aspect of cloud classification, height, we see that three levels are recognized: high, middle, and low. *High* clouds normally have bases above 6000 meters; *middle* clouds generally occupy heights from 2000 to 6000 meters; *low* clouds form below 2000 meters. The altitudes listed for each height category are not hard and fast. There is some seasonal as well as latitudinal variation. For example, at high latitudes or during cold winter months in the midlatitudes high clouds are often found at lower altitudes.

Because of the low temperatures and small quantities of water vapor found at high altitudes, all of the high clouds are thin and white and are made up of ice crystals. Since much more water vapor is available at lower altitudes, middle and low clouds are denser and darker.

Layered clouds in any of these height ranges generally indicate that the air is stable. One might not normally expect clouds to grow or persist in stable air. However, cloud growth of this type is common when air is forced to rise as along a front or near the center of a cyclone where converging winds cause air to ascend. Such forced ascent of stable air leads to the formation of a stratified cloud layer that is large horizontally in comparison to its depth.

Some clouds do not fit into any one of these three height categories. Such clouds have their bases in the low height range, but they often extend upward into the middle or high altitudes. Consequently, these clouds are referred to as *clouds of vertical development*. They are all related to one another and are associated with unstable air. Although cumulus clouds are often connected with "fair weather," they may, under the proper circumstances, grow dramatically. Once upward movement is triggered, acceleration is powerful and clouds with great vertical extent are formed. As the cumulus enlarges, its top leaves the low height range and it is called a *cumulus congestus*. Finally, when the cloud becomes even more towering and rain begins to fall, it becomes a *cumulonimbus*.

Since definite weather patterns can usually be associated with certain clouds or certain combinations of cloud types, it is important to become familiar with cloud descriptions and characteristics. Table 5-1 lists the ten basic cloud types, called genera, that are recognized internationally and gives some characteristics of each. The series of photos in Figure 5-2 depicts common forms of each cloud type.

formation of precipitation

Although all clouds contain water, why do some produce precipitation and why do others drift placidly overhead? This seemingly simple question perplexed meteorologists for many years. First, cloud droplets are very small in

Table 5-1

Basic Cloud Types

Cloud Family and Height	Cloud Type	Characteristics
High clouds— above 6000 m	Cirrus	Thin, delicate, fibrous ice-crystal clouds. Sometimes appear as hooked filaments called "mares' tails." [Figure 5-2(a)]
	Cirrocumulus	Thin, white ice-crystal clouds, in the form of ripples, waves, or globular masses all in a row. May produce a "mackerel sky." Least common of the high clouds. [Figure 5-2(b)]
	Cirrostratus	Thin sheet of white ice-crystal clouds that may give the sky a milky look. Sometimes produce halos around the sun or moon. [Figure 5-2(c)]
Middle clouds— 2000–6000 m	Altocumulus	White to gray clouds often made up of separate globules; "sheepback" clouds. [Figure 5-2(d)]
	Altostratus	Stratified veil of clouds that are generally thin and may produce very light precipitation. When thin, the sun or moon may be visible as a "bright spot," but no halos are produced. [Figure 5-2(e)]
Low clouds— below 2000 m	Stratocumulus	Soft, gray clouds in globular patches or rolls. Rolls may join together to make a continuous cloud. [Figure 5-2(f)]
	Stratus	Low uniform layer resembling fog but not resting on the ground. May produce drizzle. [Figure 5-2(g)]
	Nimbostratus	Amorphous layer of dark gray clouds. One of the chief precipitation-producing clouds. [Figure 5-2(h)]
Clouds of vertical development—	Cumulus	Dense, billowy clouds often characterized by flat bases. May occur as isolated clouds or closely packed. [Figure 5-2(i)]
	Cumulonimbus	Towering cloud sometimes spreading out on top to form an "anvil head." Associated with heavy rainfall, thunder, lightning, hail, and tornadoes. [Figure 5-2(j)]

size, averaging less than 10 micrometers in diameter (for comparison, a human hair is about 75 micrometers in diameter).* Because of their small size, the rate at which cloud droplets fall is incredibly slow. An average cloud droplet falling from a cloud base at 1000 meters would require about 48 hours to reach the ground. Of course, it would never complete its journey. Even falling through humid air, a cloud droplet would evaporate before it fell a few meters below the cloud base. Second, clouds are made up of many

---

*One micrometer equals 0.001 millimeter.

(a)

(b)

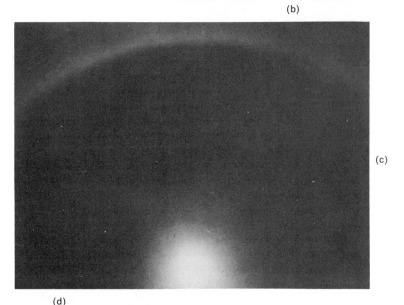

(c)

(d)

Figure 5-2

(*a*) Cirrus. (*b*) Cirrocumulus. (*c*) Cirrostratus with halo. (*d*) Altocumulus. (*e*)
Altostratus. (*f*) Stratocumulus. (*g*) Stratus.
(*h*) Nimbostratus. (*i*) Cumulus. (*j*) Cumulonimbus. (Courtesy of NOAA.)

(e)

(f)

(g)

(h)

(i)

(j)

billions of these droplets, all competing for the available water; thus, their continued growth via condensation is very slow.

A raindrop large enough to reach the ground without evaporating contains roughly a million times the water of a cloud droplet. Therefore, in order for precipitation to form, millions of cloud droplets must somehow coalesce (join together) into drops large enough to sustain themselves during their descent. Two mechanisms have been proposed to explain this phenomenon: the Bergeron process and the collision–coalescence process.

bergeron process

The *Bergeron process*, so named after its discoverer, relies on two interesting properties of water. First, cloud droplets do not freeze at 0°C as expected. In fact, pure water suspended in air does not freeze until it reaches a temperature of nearly −40°C. Water in the liquid state below 0°C is generally referred to as *supercooled*. Supercooled water will readily freeze if it is sufficiently agitated. This explains why airplanes collect ice when they pass through a liquid cloud made up of supercooled droplets. In addition, supercooled droplets will freeze upon contact with solid particles that have a crystal form closely resembling that of ice. These materials have been termed *freezing nuclei*. The need for freezing nuclei to initiate the freezing process is similar to the requirement for condensation nuclei in the process of condensation. However, in contrast to condensation nuclei, freezing nuclei are very sparse in the atmosphere, and they do not generally become active until the temperature reaches −10°C or below. Only at temperatures well below freezing will ice crystals begin to form in clouds, and even at that they will be few and far between. Once ice crystals form, they are in direct competition with the supercooled droplets for the available water vapor.

This brings us to the second interesting property of water. When air is saturated (100 percent relative humidity) with respect to water, it is supersaturated (relative humidity greater than 100 percent) with respect to ice. Table 5-2 shows that at −10°C, when the relative humidity is 100 percent

Table 5-2

Relative Humidity with Respect to Ice When
Relative Humidity with Respect to Water is 100%

Temperature (°C)	Relative Humidity with Respect to:	
	Water (%)	Ice (%)
0	100	100
−5	100	105
−10	100	110
−15	100	115
−20	100	121

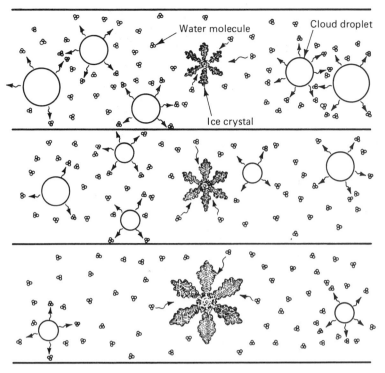

Figure 5-3

The Bergeron process. Ice crystals grow at the expense of cloud droplets until they are large enough to fall. The size of these particles has been greatly exaggerated.

with respect to water, the relative humidity with respect to ice is nearly 110 percent. Thus, ice crystals cannot peacefully coexist with water droplets because the air always "appears" supersaturated to the ice crystals. So the ice crystals begin to consume the "excess" water vapor, which lowers the relative humidity near the surrounding droplets. In turn, the water droplets evaporate in order to replenish the diminishing water vapor, thereby providing a continual source of vapor for the growth of the ice crystals (Figure 5-3).

Because the level of supersaturation with respect to ice can be very great, the growth of ice crystals is generally rapid enough to generate crystals large enough to fall. During their descent these ice crystals enlarge as they intercept cloud drops that freeze upon them. Air movement will sometimes break up these delicate crystals and the fragments will serve as freezing nuclei for other liquid droplets. A chain reaction develops and produces many ice crystals, which, by accretion, will form into larger crystals called snowflakes (Figure 5-4). When the surface temperature is above 4°C, snowflakes usually melt before they reach the ground and continue their descent as rain. Even a summer rain may have begun as a snowstorm in the clouds overhead.

Cloud seeding utilizes the Bergeron process. By adding freezing nuclei

Figure 5-4
Snow crystals are always six-sided, but they come in an infinite variety of forms. (Courtesy of NOAA.)

(commonly silver iodide) to supercooled clouds, the growth of these clouds can be markedly changed. This process is discussed in greater detail later in this chapter.

collision–coalescence

Thirty years ago meteorologists believed that the Bergeron process was responsible for the formation of most precipitation, with the exception of light drizzle. However, it was discovered that copious rainfall is often associated with clouds located well below the freezing level (warm clouds), especially in the tropics. This led to the proposal of a second mechanism thought to produce precipitation, the *collision–coalescence process.*

Clouds made up entirely of liquid droplets must contain droplets larger than 20 micrometers if precipitation is to form. These large droplets form when "giant" condensation nuclei are present and when hygroscopic particles such as sea salt exist. Recall that hygroscopic particles begin to remove water vapor from the air at relative humidities under 100 percent and can grow very large. Since the rate at which drops fall is size-dependent, these "giant" droplets fall most rapidly. As such, they collide with the smaller, slower droplets and coalesce. Becoming larger in the process, they fall even more rapidly (or in an updraft they rise more slowly) and increase their chances of collision and rate of growth (Figure 5-5). After a million such collisions they are large enough to fall to the surface without evaporating.

Because of the number of collisions required for growth to raindrop size, droplets in clouds with great vertical thickness and abundant moisture have a better chance of reaching the required size. Updrafts also aid in this process because they allow the droplets to traverse the cloud repeatedly. Raindrops can grow to a maximum size of 5 millimeters when they fall at the rate of 30 kilometers per hour. At this size and speed the water's surface tension, which holds the drop together, is surpassed by the drag imposed by the air, which, in turn, succeeds in pulling the drops apart. The resulting breakup of a large raindrop produces numerous smaller drops that begin anew the task of sweeping up cloud droplets. Drops that are less than 0.5 millimeter upon reaching the ground are termed drizzle and require about 10 minutes to fall from a cloud 1000 meters overhead.

From the preceding discussion it should be apparent that large condensation nuclei are required if warm clouds are to generate any appreciable precipitation. This fact has led to attempts to "seed" these clouds by adding "giant" nuclei. One material used is common table salt. Water is also sprayed into clouds and acts as giant particles to initiate the collision–coalescence process.

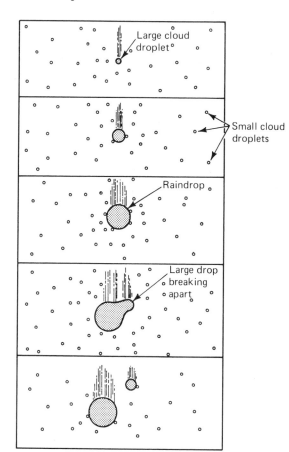

Figure 5-5

The collision–coalescence process. Because large cloud droplets fall more rapidly than smaller droplets, they are able to sweep up the smaller ones in their path and grow. Most cloud droplets are so small that the motion of the air keeps them suspended. Even if these small cloud droplets were to fall, they would evaporate before reaching the surface.

The collision–coalescence process is not that simple, however. First, as the larger droplets descend, they produce an air stream around them similar to that produced by an automobile when it is driven rapidly down the highway. If an automobile is driven at night and we use the bugs that are often out to be analogous to the cloud droplets, it is easy to visualize how most cloud droplets are swept aside. The larger the cloud droplet (or bug), the better chance it will have of colliding with the giant droplet (or car).

Second, collision does not guarantee coalescence. Experimentation has indicated that the presence of atmospheric electricity may be the key to what holds these droplets together once they collide. If a droplet with a negative charge should collide with a positively charged droplet, their electrical attraction may bind them together.

sleet, glaze, and hail

While rain and snow are the most common and familiar forms of precipitation, other forms do eixst and are worthy of mention. Sleet, glaze, and hail fall into this category. Although limited in occurrence and sporadic in both time and space, these forms, especially glaze and hail, may on occasion cause considerable damage.

Sleet is a wintertime phenomenon and refers to the fall of small, clear to translucent particles of ice. In order for sleet to be produced, a layer of air with temperatures above freezing must overlie a subfreezing layer near the ground. When the raindrops, which are often melted snow, leave the warmer air and encounter the colder air below, they solidify and reach the ground as small pellets of ice no larger than the raindrops from which they formed.

On some occasions, when the vertical distribution of temperatures is similar to that associated with the formation of sleet, freezing rain or *glaze* results instead. In such situations, the subfreezing air near the ground is not thick enough to allow the raindrops to freeze. The raindrops, however, do become supercooled as they fall through the cold air and turn to ice upon colliding with solid objects. The result can be a thick coating of ice having sufficient weight to break tree limbs and down power lines as well as make walking or motoring extremely hazardous (Figure 5-6).

Hail is precipitation in the form of hard rounded pellets or irregular lumps of ice. Further, large hailstones often consist of a series of nearly concentric shells of differing densities and degrees of opaqueness (Figure 5-7). Most frequently, hailstones have a diameter of about 1 centimeter, but they may vary in size from 5 millimeters to more than 10 centimeters in diameter. The largest hailstone on record fell on Coffeyville, Kansas, September 3, 1970. With a 14-centimeter diameter and a circumference of 44 centimeters, this "giant" weighed 766 grams! The destructive effects of heavy hail are well known, especially to farmers whose crops can be devastated in a few short minutes and to people whose windows are shattered.

Hail is produced only in cumulonimbus clouds where updrafts are strong and where there is an abundant supply of supercooled water. First, rain is lifted above the freezing level by the rapidly ascending air. Once frozen,

Figure 5-6
Power lines sagging under the weight of ice (glaze). (Courtesy of NOAA.)

Figure 5-7
Section through a large hailstone. Note the concentric layers of ice. (Courtesy of NOAA.)

these small ice granules grow by collecting supercooled cloud droplets as they fall through the cloud. If they encounter another strong updraft, they may be carried upward again and begin the downward journey anew. Each trip above the freezing level may be represented by an additional layer of ice.

Hailstones, however, may also form from a single descent through an updraft. The layered structure is attributed to variations in the rate at which supercooled droplets accumulate and freeze, which, in turn, is related to differences in the amount of supercooled water in different parts of the cumulonimbus tower.

In either case, hailstones grow by the addition of supercooled water upon growing ice pellets. The ultimate size of the hailstone depends primarily upon three factors: (*1*) the strength of the updrafts, (*2*) the concentration of supercooled water, and (*3*) the length of the path through the cloud.

precipitation measurement

The most common form of precipitation, rain, is probably the easiest to measure. Any open container that has a consistent cross section throughout can be a rain gauge. In general practice, however, more sophisticated devices are used in order to measure small amounts of rainfall more accurately, as well as reduce losses resulting from evaporation. The *standard rain gauge* (Figure 5-8) has a diameter of about 20 centimeters at the top. Once the water is caught, a funnel conducts the rain through a narrow opening into a cylindrical measuring tube that has a cross-sectional area only one-tenth as

Figure 5-8
Standard rain gauge. (Courtesy of Weather-Measure Corporation)

Figure 5-9

An interval view of a tipping-bucket precipitation gauge. (Courtesy of Weather-Measure Corporation.)

large as the receiver. Consequently, rainfall depth is magnified ten times, which allows for accurate measurements to the nearest .025 centimeter, while the narrow opening minimizes evaporation. When the amount of rain is less than .025 centimeter, it is generally reported as being a *trace* of precipitation.

In addition to the standard rain gauge, several types of recording gauges are routinely used. These instruments not only record the amount of rain but also its time of occurrence and intensity (amount per unit of time). Two of the most common gauges are the tipping-bucket gauge and the weighing gauge.

As can be seen in Figure 5-9, the *tipping-bucket gauge* consists of two compartments, each one capable of holding .025 centimeter of rain, situated at the base of a 25-centimeter funnel. When one "bucket" fills, it tips and empties its water. Meanwhile the other "bucket" takes its place at the mouth of the funnel. Each time a compartment tips, an electrical circuit is closed and the amount of precipitation is automatically recorded on a graph.

The *weighing gauge*, as the name would indicate, works on a different principle. The precipitation is caught in a cylinder that rests upon a spring balance. As the cylinder fills, the movement is transmitted to a pen that records the data.

No matter which rain gauge is used, proper exposure is critical. Errors arise when the gauge is shielded from obliquely falling rain by buildings, trees, or other high objects. Hence, the instrument should be at least as far away from such obstructions as the objects are high. Another cause for error is the wind. It has been shown that with increasing wind and turbulence, it becomes more difficult to collect a representative quantity of rain. To offset this effect, a windscreen is often placed about the instrument so that rain falls into the gauge and is not carried across it.

When snow records are kept, two measurements are generally taken: depth and water equivalent. Usually, the depth of snow is measured with a calibrated stick. The actual measurement is not difficult, but choosing a representative spot often poses a dilemma. Even when winds are light or moderate, snow drifts freely. Usually, it is best to take several measurements in an open place away from trees and obstructions and then average them. To obtain the water equivalent, samples may be melted and then weighed or measured as rain.

Sometimes large cylinders are used to collect snow. A major problem that hinders accurate measurement by snow gauges is the wind. Snow will blow around the top of the cylinder instead of falling into it. Therefore, the amount caught by the gauge is generally less than the actual fall. As is often the practice with rain gauges, shields designed to break up wind eddies are placed around the snow gauge to insure a more accurate catch.

The quantity of water in a given volume of snow is not constant. A general ratio of 10 units of snow to 1 unit of water is often used when exact information is not available, but the actual water content of snow may deviate widely from this figure. It may take as much as 30 centimeters of light and fluffy dry snow or as little as 4 centimeters of wet snow to produce 1 centimeter of water.

fog

Fog is generally considered to be an atomospheric hazard. When it is light, visibility is reduced to 2 or 3 kilometers. When it is dense, visibility may be cut to a few tens of meters or less, making travel by any mode not only difficult but also often dangerous. Officially, visibility must be reduced to 1 kilometer or less before it is reported. Although this figure is arbitrary, it does permit a more objective criterion for comparing fog frequencies at different locations.

Fog is defined as a cloud with its base at or very near the ground. Physically, there is basically no difference between a fog and a cloud; the appearance and structure of both are the same. The essential difference is the method and place of formation. While clouds result when air rises and cools adiabatically, fogs (with the exception of upslope fogs) are the consequence of radiation cooling or the movement of air over a cold surface. In other circumstances, fogs are formed when enough water vapor is added to the air to bring about saturation (evaporation fogs).

fogs formed by cooling

Radiation fog, as the name implies, results from radiation cooling of the ground and adjacent air. Most common in the fall and winter, it is a night-time phenomenon that requires clear skies and a fairly high relative humidity. Under these circumstances, the ground and the air immediately above will cool rapidly. Since the relative humidity is high, just a small amount of cooling will lower the temperature to the dew point. If the air is calm, the fog may be patchy and less than a meter deep. For radiation fog to be more extensive vertically, a light breeze of 3 to 4 kilometers per hour is necessary. Under these conditions, the light wind creates enough turbulence to carry the fog upward 10 to 30 meters without dispersing it.

Since the air containing the fog is relatively cold and dense, it drains downslope. As a result, radiation fog is thickest in valleys while the sur-rounding hills are clear (Figure 5-10). Normally, these fogs dissipate within 1 to 3 hours after sunrise. Often the fog is said to "lift." However, it does not actually rise. Instead, the sun warms the earth, which, in turn, heats the surface air first. Consequently, the fog evaporates from the bottom up, giving the impression of lifting and the last vestiges may appear to be a low white cloud layer.

When warm moist air is blown over a cold surface, it becomes chilled by contact and to a certain extent by mixing with the cold air associated with the cold surface below. If the cooling is sufficient, the result will be a blanket of fog called *advection fog*. Since the term advection refers to air moving horizontally, such fogs are a consequence of air giving up heat to the surface below during horizontal movement.

A certain amount of turbulence is needed for proper development, thus winds between 10 and 30 kilometers per hour are usually associated with advection fog. Not only does the turbulence facilitate cooling through a thicker layer of air, but it also carries the fog to greater heights. Unlike radiation fogs, advection fogs are often thick (300 to 600 meters deep) and persistent.

Figure 5-10
Early morning radiation fog in a valley. (Photograph by Jack L. Bradley. Cour-tesy of the *Peoria Journal Star.*)

Examples of such fogs are very common. The foggiest location in the United States, and perhaps in the world, is Cape Disappointment, Washington. The name is indeed appropriate because the station averages 2552 hours of fog each year. The fog experienced at Cape Disappointment, as well as that at other West Coast locations during the summer and early autumn, is produced when warm, moist air from the Pacific Ocean moves over the cold California Current. It is then carried onshore by westerly winds or a local sea breeze. Advection fogs are also common in the northern tier of states in the winter when warm air from the Gulf of Mexico is blown over cold, often snow-covered surfaces.

As its name implies, *upslope fog* is created when relatively humid air moves up a gradually sloping plain or, in some cases, up the steep slopes of a mountain. Because of the upward movement, air expands and cools adiabatically. If the dew point is reached, an extensive layer of fog may form.

In the United States the Great Plains offers an excellent example. When humid easterly or southeasterly winds move westward from the Mississippi River toward the Rocky Mountains, the air gradually rises, resulting in an adiabatic decrease of about 13°C. When the difference between the air temperature and dew point of westward moving air is less than 13°C, an extensive fog often results in the western plains.

evaporation fogs

When cool air moves over warm water, enough moisture may evaporate from the water surface to produce saturation. As the rising water vapor meets the cold air, it immediately recondenses and rises with the air that is being warmed from below. Since the water has a steaming appearance, the phenomenon is called *steam fog* (Figure 5-11). It is a fairly common occurrence over lakes and rivers in the fall and early winter when the water may

Figure 5-11

Steam fog rising from a small lake. (Courtesy of Ward's Natural Science Establishment, Inc., Rochester, N.Y.)

Cloud Chart

Cumulus—*Good weather*

Stratocumulus—*Good weather*

Cirrus uncinus—*Good weather*

Stratocumulus—*Rain warning*

Cumulus congestus—*Rain warning*

Cirrostratus—*Rain warning*

Altostratus—*Sun or rain*

Stratus—*Rain and squall*

Cumulonimbus—*Rain and squall*

Courtesy of C.C. Marketing, Glen Allen, VA.

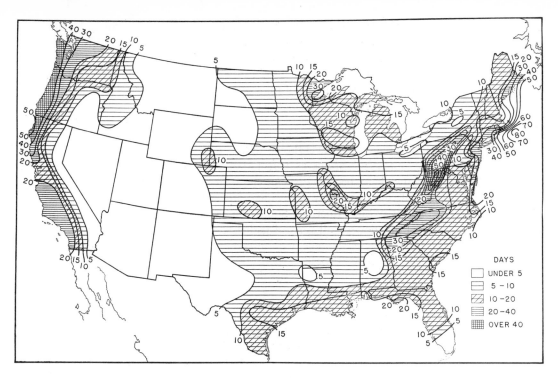

DAYS

☐	UNDER 5
☰	5 – 10
▨	10 –20
▤	20 –40
▦	OVER 40

Figure 5-12

Average number of days per year with dense fog in the United States. (Courtesy of NOAA.)

still be relatively warm and when the air is rather crisp. Steam fog is often very shallow, for as the steam rises, it reevaporates in the unsaturated air above.

Steam fogs, however, can be dense. During the winter when cold air pours off the continents and ice shelves of the north into the open ocean, the air may be from 20°C to 30°C colder than the water. Steaming is intense and it saturates a large volume of air. Because of its source and appearance, this type of steam fog is given the name *arctic sea smoke*.

When frontal wedging occurs, warm air is lifted over colder air. If the resulting clouds yield rain, and the cold air below is near the dew point, enough rain will evaporate to produce fog. A fog formed in this manner is called *frontal* or *precipitation fog*. The result is a more or less continuous zone of condensed water droplets reaching from the ground up through the clouds.

In summary, both steam fog and frontal fog result from the addition of moisture to a layer of air. As you saw, the air is usually cool or cold and already near saturation. Since air's capacity to hold water vapor at low temperatures is small, only a relatively modest amount of evaporation is necessary to produce saturated conditions and fog.

The frequency of occurrence of dense fog varies considerably from place to place. Figure 5-12 is a generalized map showing the average number of days per year with dense fog in the 48 contiguous states. As might be expected, fog incidence is highest in coastal areas, especially where cold currents prevail, as along the Pacific and New England coasts. Relatively high frequencies are also found in the Great Lakes region and in the humid Appalachian Mountains of the East. By contrast, fogs are rare in the interior of the continent, especially in the arid and semiarid areas of the West.

intentional weather modification

The old adage, "Everybody talks about the weather but nobody does anything . . . ," seems as true today as it ever was. Attempts to alter the weather have been made, but the fruits of these labors have been meager. People have probably always had a desire to change the weather. The Hopi Indians of the American Southwest performed rain dances, and European inventors devised all sorts of gadgetry, including cannons that were fired to suppress the development of hailstones. The success of these early attempts is questionable, at best.

cloud seeding

The first scientific breakthrough in weather modification came in 1946 when V. J. Schaeffer discovered that Dry Ice dropped into a supercooled cloud spurred the growth of ice crystals. This discovery sparked the practice of *cloud seeding* for the purpose of rainmaking. Recall that once ice crystals form, they grow larger at the expense of the remaining liquid cloud droplets and, upon reaching a sufficient size, they fall as precipitation. Shortly after Schaeffer's discovery it was learned that silver iodide could also be used for cloud seeding. The similarity in the crystalline structures of silver iodide and ice accounts for silver iodide's ability to initiate the growth of ice crystals. Thus, unlike Dry Ice, silver iodide crystals act as freezing nuclei rather than as a cooling agent. This substance has an advantage over Dry Ice in that it can be supplied to the clouds from burners located on the ground. However, for either substance to be successful, certain atmospheric conditions must exist. Clouds must be present, for seeding cannot generate them. In addition, at least the top portion of the cloud must be supercooled, that is, it must be made of liquid droplets having a temperature below freezing. The object, then, is to produce just the correct number of ice crystals. Overseeding will simply produce billions of minute ice crystals, none having sufficient size to fall.

Attempts have also been made to trigger rainfall from warm clouds. The precipitation-producing mechanism in warm clouds is the collision–coalescence process. In seeding these clouds the object is to produce large droplets that will collide with smaller cloud droplets and grow by coalescence. Large hygroscopic particles like salt, and even water itself, have been used for this purpose. These "giant" condensation nuclei are introduced at the base of the cloud where updrafts carry them into the cloud. The results obtained by using this method are open to question at this time.

Numerous studies have been conducted to determine the effectiveness of silver iodide in cloud seeding. In a test conducted in Missouri it was actually determined that unseeded clouds produced more precipitation than seeded ones. In other tests a 10 to 20 percent increase in precipitation could be attributed to cloud seeding. It seems that more knowledge in the field of cloud physics is needed before cloud seeding can become as successful as first thought possible. Nevertheless, it is certain that seeding of supercooled clouds does alter their "normal" development. Because of this, cloud seeding has many other applications, including the dispersal of clouds and fogs, the suppression of hail and lightning, and hurricane modification. Cloud seeding's role in hurricane modification will be discussed in Chapter 10.

fog dispersal. The most successful use of cloud seeding has been in the dispersal of supercooled fog at airports. An application of Dry Ice triggers the formation of ice crystals that fall out, leaving a "hole" in the fog and providing a clear runway for takeoffs and landings (Figure 5-13). Other cold substances, such as liquid propane, have been used with similar success and are somewhat easier to apply. Unfortunately, most fog is of the warm type and is harder to combat. Warm fog can be dispersed by mechanical mixing of the fog with drier, warmer air from above or by heating the air. When the layer of fog is very shallow, helicopters are sometimes used to disperse the fog. By flying just above the fog the helicopter creates a strong downdraft that pulls dry air down and mixes it with the saturated air near the ground. If the air aloft is dry enough, sufficient evaporation will take place and will eliminate the foggy condition. At many airports where fog is a common hazard it has become more common to heat and hence evaporate the fog. At Orly Airport in Paris a sophisticated thermal fog dissipation system, called Turboclair, was installed in 1970. This system consists of eight jet engines located in underground chambers alongside the upwind edge of the

Figure 5-13
Results of Dry Ice seeding at Elmendorf AFB, Alaska, on January 24, 1969. One hour and 20 minutes after seeding visibility was greater than 1 mile. (Courtesy of NOAA.)

Figure 5-14

Effects produced by seeding a cloud deck with Dry Ice. Within 1 hour a hole developed over the seeded area. (Courtesy of General Electric.)

runway. Although this system is expensive to install, it is capable of improving visibility for a distance of about 900 meters along the approach and touchdown zones.

cloud dispersal. Cloud seeding used for cloud dispersal may be of far greater importance in the future than it has been in the past. Because of the energy crisis, the use of solar radiation for such things as home heating is becoming a more viable option. One of the major obstacles to the use of solar heat is the layer of supercooled stratus clouds often present in the winter season. Figure 5-14 shows how this type of supercooled cloud can be dissipated by cloud seeding. The increase in the amount of winter sunshine provided by extensive seeding programs would not only increase the effectiveness of solar heating, but it might also improve our psychological outlook.

hail suppression. Hail damage to crops is a concern of farmers worldwide. In France a cloud-seeding project that used silver-iodide burners to combat hail damage had encouraging results. The object is to prevent the accumulation of large amounts of supercooled water by providing freezing nuclei that will convert the water to ice. Without ample supplies of supercooled water, hailstones cannot grow. Russian scientists, using rockets and artillery shells to carry the freezing nuclei to the clouds, have claimed great success with this method.

116

lightning suppression. Because lightning is a primary cause of forest fires, the U.S. Forest Service has conducted tests to determine whether or not cloud seeding will reduce the number of cloud-to-ground discharges. For unknown reasons, it appears that cloud seeding did reduce the number of lightning strokes, but the actual number of forest fires did not seem to change appreciably. Other attempts to discharge clouds by dropping metallic objects into them have had varying degrees of success. Nevertheless, it has become common practice to drop aluminum-coated streamers prior to the launching of a spacecraft if the weather appears threatening.

frost prevention

Frost, the fruit growers' plight, occurs when the air temperature falls to 0°C or below. It may be accompanied by deposits of ice crystals commonly called *white frost*. This, however, happens only if the air becomes saturated. White frost is not a requirement for crop damage.

Frost hazards exist when a cold air mass moves into a region or when ample radiation cooling occurs on a clear night. The conditions accompanying the invasion of cold air are characterized by low daytime temperatures, long periods of effective frost, strong winds, and widespread damage. Frost induced by radiation loss is a nighttime phenomenon associated with a surface temperature inversion and is much easier to combat.

Several methods of frost prevention have been used with varying degrees of success. Generally, these attempts are directed at reducing the amount of heat lost during the night or at adding heat to the lowermost layer of air. Heat conservation methods include covering plants with material having a low thermal conductivity, such as paper and cloth, and producing particles which, when suspended in air, reduce the radiation loss. Smudge fires have been used for particle production but have often proven unsatisfactory. In addition to the pollution problems created by the dense clouds of black smoke, the carbon particles impede daytime warming by reducing the amount of solar radiation that can reach the surface. This reduction in daytime warming may offset the benefits gained during the night.

Methods of warming include sprinkling, air mixing, and the use of orchard heaters (Figure 5-15). Sprinklers distribute water to the plants and add heat in two ways: first, from the warmth of the water and, second, but more important, from the latent heat of fusion that is released when the water freezes. As long as an ice-water mixture remains on the plant, the latent heat released will keep the temperature from dropping below 0°C. Air mixing is successful when the temperature at 15 meters above the ground is 5°C higher than the surface temperature. By using a wind machine, the warmer air aloft is mixed with the colder surface air. Orchard heaters probably produce the most successful results. Since as many as 30 or 40 heaters per acre are required, the fuel cost can be significant, but usually the effectiveness of this method seems to warrant the cost.

117

Figure 5-15

Two types of freeze controls used in Florida citrus groves. The wind machine gives protection down to $-2°C$ and the sprinkler system (thin pole) gives protection down to $-4°C$. Notice the ice on the plants. (Courtesy of the U.S. Department of Agriculture.)

REVIEW

1. What is the basis for the classification of clouds?

2. Why are high clouds always thin in comparison to low and middle clouds?

3. Which cloud types are associated with the following characteristics? Thunder, halos, precipitation, hail, mackerel sky, lightning, mare's tails.

4. What do layered clouds indicate about the stability of the air? What do clouds of vertical development indicate about the stability of air?

5. What is the difference between condensation and precipitation?

6. Describe the steps in the formation of precipitation according to the Bergeron process. Be sure to include the following: (*a*) the importance of supercooled cloud droplets, (*b*) the roll of freezing nuclei, and (*c*) the difference in saturation between liquid water and ice.

7. How does the collision–coalescence process differ from the Bergeron process?

8. If snow is falling from a cloud, which process must have produced it? Explain.

9. Describe sleet and glaze and the circumstances under which they form. Why does glaze result on some occasions and sleet on others?

10. How does hail form? What factors govern the ultimate size of hailstones?

11. Although an open container can serve as a rain gauge, what advantages does a standard rain gauge provide?

12. How do recording rain gauges work? Do they have advantages over a standard rain gauge?

13. Describe some of the factors that could lead to an inaccurate measurement of rain or snow.

14. How does the amount of snowfall relate to the amount of rain?

15. Distinguish between clouds and fog.

16. List five types of fog and discuss the details of their formation.

118

17. What actually happens when a radiation fog "lifts"?

18. Identify the fogs described in the following situations:
    a. You have stayed the night in a motel and decide to take an early morning swim. As you approach the heated swimming pool, you notice a fog over the water.
    b. You are located in the western Great Plains, and the winds are from the east, and fog is extensive.
    c. You are driving through hilly terrain during the early morning hours and experience fog in the valleys and clearing on the hills.

19. Why is there a relatively high frequency of dense fog along the Pacific Coast (Figure 5-12)?

20. Why are silver iodide crystals often used to seed supercooled clouds.

21. If cloud seeding is to be successful (or have a chance of being successful) certain atmospheric conditions must exist. What are these conditions?

22. How do frost and white frost differ?

23. Describe how smudge fires, sprinkling, and air mixing are used in frost prevention.

VOCABULARY
REVIEW

clouds	hail
cirrus	standard rain gauge
cumulus	trace (of precipitation)
stratus	tipping-bucket gauge
high (clouds)	weighing gauge
middle (clouds)	fog
low (clouds)	radiation fog
clouds of vertical development	advection fog
cumulus congestus	upslope fog
cumulonimbus	steam fog
Bergeron process	arctic sea smoke
supercooled	frontal or precipitation fog
freezing nuclei	cloud seeding
collision–coalescence process	frost
sleet	white frost
glaze	

Chapter Six

air Pressure
anD winDs

Of the various elements of weather and climate, changes in air pressure are the least noticeable. When we listen to a weather report, we are generally interested in moisture conditions (humidity and precipitation), temperature, and perhaps wind. It is the rare individual, however, who wonders about air pressure. Although the hour-to-hour and day-to-day variations in air pressure are not perceptible to human beings, they are very important in producing changes in our weather. Variations in air pressure from place to place are responsible for the movement of air (wind), as well as being among the most significant factors in weather forecasting. As we shall see, air pressure is tied very closely to the other elements of weather in a cause-and-effect relationship.

behavior of gases

The concept of air pressure and the fact that it decreases with altitude can be more fully realized if we examine the behavior of gases and the principles that govern this behavior. Gas molecules, unlike those of the liquid and solid phases, are not "bound" to one another, but are freely moving about, filling all space available to them. When two gas molecules collide, which frequently happens under the conditions normally experienced near the earth's surface, they bounce off each other as if they were very elastic balls. If a gas is confined to a container, this motion is restricted by the sides of the container, much the way that the walls of a handball court redirect the motion of a handball. The continuous bombardment of gas molecules against the wall of the container exerts an outward push we call air pressure. Although the atmosphere

definition of air pressure

is without walls, it is confined from below by the earth's land–sea surface and from above by the force of gravity that prevents its outward escape. Here we define *air pressure* as the force exerted against a surface by the continuous collision of gas molecules.

Two factors, namely, temperature and density, largely determine the amount of pressure a particular gas will exert. We will first examine the effect of temperature on air pressure when density is kept constant by observing the behavior of a gas in a closed container (constant volume). As we shall see later, a change in temperature is nearly always accompanied by a change in density, which complicates the matter considerably. Nevertheless, when the density is kept constant and the temperature of the air is raised, the speed of the gas molecules, and therefore their force, is increased. From this observation we conclude that an increase in temperature results in an increase in pressure and conversely a decrease in temperature causes a decrease in pressure. The fact that air pressure is proportional to temperature is precisely why aerosol spray cans have a warning that cautions to keep them away from heat. Overheating of these containers can result in a dangerous explosion if the internal gas pressure were to exceed the strength of the container.

It may appear from the preceding discussion that on warm days the air pressure will be high and that on cold days it will be low. This however, is not necessarily the case. Over the continents in the midlatitudes, for example, the highest pressures are recorded in the winter when the temperatures are lowest. This results because air pressure is also proportional to density (the number of gas molecules involved) as well as temperature. On cold days the air molecules are more closely packed (greater density) than on warm days. Often the decrease in molecular motion associated with low temperatures is more than offset by the increased number of molecules exerting pressure. Thus, low temperatures mean greater densities and often greater surface pressures. Conversely, when air is heated in the atmosphere, it expands (increases its volume) because of the increased speed of the gas molecules. Consequently, an increase in temperature is generally accompanied by a decrease in density and therefore a decrease in pressure.

The relationships between pressure, temperature, and density described in this section can be expressed by the following equation called the *gas law*:

$$\text{pressure} = \text{density} \times \text{temperature} \times \text{constant}$$

In a verbal form the gas law states that the pressure exerted by a gas is proportional to its density and absolute temperature. Thus, an increase in either temperature or density will cause an increase in pressure as long as the other variable (density or temperature) remains constant. Further, the gas law predicts the relationship between temperature and density we observed earlier. When the pressure remains constant, a decrease in temperature results in increased density and vice versa.*

use in lecture

---

*A mathematical treatment of the ideal gas law is provided in Appendix E.

121

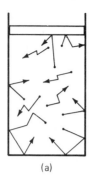 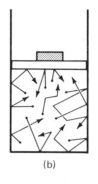 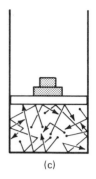

(a)   (b)   (c)

Figure 6-1

Schematic drawing showing the relationship between air pressure and density. As more weight is added (*B* and *C*) to the cylinder, the increased number of molecules per unit volume (density) causes an increase in the pressure exerted on the walls of the cylinder.

Let us now consider the decrease in pressure with altitude that was mentioned in Chapter 1. The relationship between air pressure and density largely explains the observed decrease. To illustrate, imagine a cylinder fitted with a movable piston as shown in Figure 6-1. If the temperature is kept constant and a weight is placed upon the piston, the downward force exerted by gravity will begin to squeeze (compress) the gas molecules together. The result is to increase the density of the gas and hence the number of gas molecules (per unit area) bombarding the cylinder walls as well as the bottom of the piston. In accordance with the gas law, the increase in density causes an increase in pressure. The piston will continue to squeeze the air molecules until the downward force is balanced by the ever increasing gas pressure. When this balance is reached, the gas pressure within the cylinder will equal the weight of the piston and no further compression will occur. If more weight is added to the piston, the air will compress further until the pressure of the gas once again equals the new weight of the piston. In a similar manner, the pressure at any given altitude in the atmosphere is equal to the weight of the air directly above that point. At sea level a standard column of air weighs slightly more than 1 kilogram per square centimeter and as such exerts an equivalent amount of pressure. As we ascend through the atmosphere we find that the air becomes less dense, a consequence of the lesser amount (weight) of air above and, as would be expected, this results in a corresponding decrease in pressure.

Recall from Chapter 1 that the rate at which pressure decreases with altitude is not a constant, that is, the rate of decrease is much greater near the earth's surface where pressure is high than aloft where air pressure is low. The "normal" decrease in pressure experienced with increased altitude is provided by the U.S. Standard Atmosphere in Table 6-1. We see from Table 6-1 that pressure is reduced by approximately one-half for each 5-kilometer increase in altitude. Hence, at 5 kilometers the pressure is one-half its sea-level value, at 10 kilometers it is one-fourth, at 15 kilometers it is one-eighth, and so

forth. Thus, at the altitude at which commercial jets fly (10 kilometers), the
air exerts a pressure equal to only one-fourth that at sea level.

Table 6-1

U.S. Standard Atmosphere

Height (km)	Temperature (°K)	Pressure (mb)
0	288	1013.2
.5	285	954.6
1.0	282	898.8
1.5	278	845.6
2.0	275	795.0
2.5	272	746.9
3.0	269	701.2
3.5	265	657.8
4.0	262	616.6
5.0	256	540.4
6.0	249	472.2
7.0	243	411.1
8.0	236	356.5
9.0	230	308.0
10.0	223	265.0
12.0	216	194.0
14.0	216	141.7
16.0	216	103.5
18.0	216	75.65
20.0	216	55.29
25.0	221	25.49
30.0	226	11.97
35.0	236	5.75
40.0	250	2.87
50.0	270	0.78
60.0	256	0.23
70.0	220	0.06

In summary, the gas law explains how air pressure is affected by density
and temperature. However, it must be applied to the atmosphere with caution since pressure is dependent upon both variables. Recall that an increase
in temperature is often accompanied by an increase in volume and hence a
decrease in density, and vice versa; therefore, it is not always possible to
conclude how a change in temperature will influence pressure.

**measuring air
pressure**

When meteorologists measure atmospheric pressure, they use the unit of
force used in physics called the *newton*. One newton is the force that would
accelerate 1 kilogram of mass 1 meter per second. At sea level the standard
atmosphere exerts a force of 101,325 newtons per square meter. To simplify
this large number, the United States National Weather Service has adopted

123

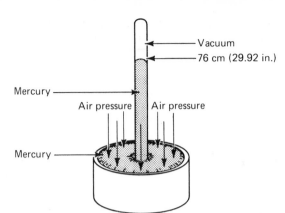

Vacuum

76 cm (29.92 in.)

Mercury

Air pressure   Air pressure

Mercury

Figure 6-2

Simple mercurial barometer. The weight of the column of mercury is balanced by the pressure exerted on the dish of mercury by the air above. If the pressure decreases, the column of mercury falls; if the pressure increases, the column rises.

the *millibar* (mb) which equals 100 newtons per square meter. Thus, standard sea-level pressure is given as 1013.25 millibars.* The millibar has been the unit of measure on all United States weather maps since January, 1940.

Although millibars are used almost exclusively in meteorology, you might be better acquainted with the expression "inches of mercury" that is used by the media to describe atmospheric pressure. In the United States the National Weather Service converts millibar values to inches of mercury for public and aviation use. This expression dates from 1643 when Torricelli, a student of the famous Italian scientist Galileo, invented the *mercurial barometer*. Torricelli correctly described the atmosphere as a vast ocean of air that exerts pressure on us and all things about us. To measure this force, he filled a glass tube that was closed at one end with mercury. The tube was then inverted into a dish of mercury (Figure 6-2). Upon doing this, Torricelli found that the mercury flowed out of the tube until the weight of the column was balanced by the pressure exerted on the surface of the mercury by the air above. In other words, the weight of the mercury in the column equaled the weight of a similar diameter column of air that extended from the ground to the top of the atmosphere. Torricelli noted that when air pressure increased, the mercury in the tube rose; conversely, when air pressure decreased, so did the height of the column of mercury. The length of the column of mercury, therefore, became the measure of the air pressure. With some refinements the mercurial barometer invented by Torricelli is still the standard pressure-measuring instrument used today. Standard atmospheric pressure at sea level equals 29.92 inches of mercury.

The need for a smaller and more portable instrument for measuring air pressure led to the development of the *aneroid* (without liquid) *barometer*. Using a different principle from the mercurial barometer, this instrument consists of partially evacuated metal chambers that have a spring inside,

---

*The standard unit of pressure in the SI system is the pascal, which is the name given to a newton per square meter (N/m^2). In this notation a standard atmosphere has a value of 101,325 pascals or 101.325 kilopascals. When the National Weather Service officially converts to the metric system, it will probably adopt this unit.

keeping them from collapsing. The metal chambers, being very sensitive to variations in air pressure, change shape, compressing as the pressure increases and expanding as the pressure decreases. Aneroids are often used in making *barographs*, instruments that continuously record pressure changes (Figure 6-3). Another important adaptation of the aneroid has been its use as an *altimeter* in aircraft. Recall that air pressure decreases with altitude and that the pressure distribution with height is well established. By marking an aneroid in meters instead of millibars we have an altimeter. For example, we see in Table 6-1 that a pressure of 265 millibars "normally" occurs at a height of 10,000 meters and as such would indicate that altitude. Because of temperature variations the standard pressures are generally slightly different from the actual conditions; therefore, altimeter corrections are made when accurate altitudes are required. Above 10 kilometers, where commercial jets frequent and pressure changes are more gradual, these corrections cannot be made accurately enough. Consequently, these aircraft have their altimeters set by the standard atmosphere and fly paths of constant pressure instead of constant altitude. Stated another way, when an aircraft flies a constant altimeter setting, a pressure variation will result in a change in height. When pressure increases along a flight path, the plane will climb, and it will descend with decreasing pressure. This causes little risk for midair collisions because all aircraft will be adjusting their altitude a comparable amount. Large commercial aircraft also use radio altimeters to measure heights above the terrain. The time required for a radio signal to reach the surface and return accurately determines height. Nevertheless, this system is not without its drawbacks.

Figure 6-3
Aneroid barograph. (Courtesy of Weather-Measure Corporation.)

A knowledge of the elevation of the land surface is required and this can pose a major problem in very rugged terrain.

As we shall see later, meteorologists are most concerned with pressure differences that occur horizontally across a given region. To enable a comparison to be made between readings obtained at various weather stations, compensation must be made for the elevation of the recording station. This is accomplished by reducing all pressure measurements to sea-level equivalents. It requires that meteorologists determine the pressure that would be exerted by an imaginary column of air equal in height to the elevation of the recording station and adding this to the pressure reading. Since temperature greatly affects the density, and therefore the weight of this imaginary column, it must be considered in the calculations. Thus, the corrected reading would give the pressure, at that time, as if it were taken at sea level under the same conditions.* A comparison of pressure records taken over the globe reveals that horizontal variations in pressure are rather small. Extreme pressure readings are rarely greater than 30 millibars (one inch of mercury) over the standard pressure or less than 60 millibars (2 inches) below the standard pressure. On occasion severe storms such as hurricanes are associated with even lower pressures.

factors affecting wind

We have discussed the upward movement of air and its importance in cloud formation. As important as vertical motion is, far more air is involved in horizontal movement, the phenomenon we call *wind*. Although we know that air will move vertically if it is warmer, and consequently more buoyant, than the surrounding air, what causes air to move horizontally? Simply stated, wind is the result of horizontal differences in air pressure. Air flows from areas of high pressure to areas of low pressure. You may have experienced this when opening a vacuum-packed can of coffee. The noise you hear is caused by air rushing from the higher pressure outside the can to the lower pressure inside. Wind is nature's attempt to balance similar inequalities in air pressure. Because unequal heating of the earth's surface generates these pressure differences, solar radiation is the ultimate driving force of wind.

If the earth did not rotate, and if there were no friction, air would flow directly from areas of high pressure to areas of low pressure. But because both of these factors exist, wind is controlled by a combination of forces. These are: (1) the pressure gradient force; (2) gravity; (3) the Coriolis force; (4) friction; and (5) the tendency of a moving object to continue moving in a straight line. The last factor is often referred to as *centrifugal force*. The magnitude of centrifugal force is small when compared with the other forces and thus of minor importance, except in rapidly rotating storms such as tornadoes and hurricanes. Discussions of these factors follow.

---

*Appendix C explains how barometer corrections may be computed.

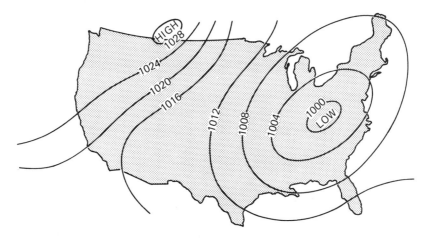

Figure 6-4

Isobars. Isobars, which are lines connecting places of equal barometric pressure, are used to show the distribution of pressure on daily weather maps. The lines usually curve and often join where cells of high and low pressure exist.

pressure gradient force

To get anything to accelerate (change its velocity) requires a net unbalanced force in one direction.* The force that drives the winds results from horizontal pressure differences. If air is subjected to greater pressure on one side than on another, this imbalance will produce a net force from the region of higher pressure toward the area of lower pressure. Thus, pressure differences cause the wind to blow, and the greater these differences, the greater the wind speed.

Over the earth's surface variations in air pressure are determined from barometric readings taken at hundreds of weather stations. These pressure data are shown on a weather map by means of *isobars*, which are lines connecting places of equal air pressure (Figure 6-4). The spacing of the isobars indicates the amount of pressure change occurring over a given distance and is expressed as the *pressure gradient*. You might find it easier to visualize the concept of a pressure gradient if you think of it as being analogous to the slope of a hill. A steep pressure gradient, like a steep hill, causes greater acceleration of a parcel than does a weak pressure gradient. Thus, the relationship between wind speed and the pressure gradient is rather simple. Closely spaced isobars indicate a steep pressure gradient and high winds; widely spaced isobars indicate a weak pressure gradient and light winds. Figure 6-5 illustrates the relationship between the spacing of isobars and wind speed.

It should be noted that the greatest pressure differences (steepest pressure gradients) are found vertically in the atmosphere, not horizontally. (Recall

*A definition of force as stated by Newton's second law is given in Appendix D.

127

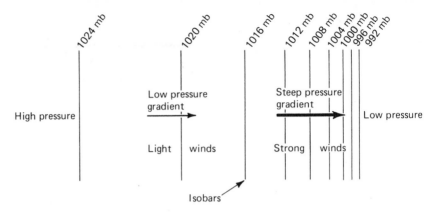

Figure 6-5

Pressure gradient force. Closely spaced isobars indicate a steep pressure gradient and high wind speeds; widely spaced isobars indicate low wind speeds.

that pressure decreases rapidly with an increase in altitude.) We might expect that very strong vertical currents are generated in the atmosphere because of these pressure differences. However, this does not occur because the vertical pressure gradient is almost always balanced by the force of gravity that acts in the opposite direction. Hence, although the vertical pressure force is very strong, it produces little motion while the much weaker horizontal pressure gradient that is not opposed by gravity can generate winds of gale intensity.

The factors that contribute to the pressure differences we observe on the daily weather map are generally very complex, but the underlying cause of these differences is simply unequal heating of the earth's land–sea surface. To illustrate how temperature differences can generate a pressure gradient and thereby create winds, we will examine a common example, the *sea breeze*. Figure 6-6(*a*) shows a vertical cross section of a coastal location just before sunrise. At this time we are assuming that the temperatures and pressures do not vary horizontally at any level. This assumption is shown in Figure 6-6(*a*) by the horizontal orientation of the lines (isobars) that indicate the pressure at various heights. Because there is no horizontal variation in pressure (zero pressure gradient) there would be no wind. However, after sunrise the unequal rates at which land and water heat will initiate air flow. Recall from Chapter 3 that surface temperatures over the ocean change only slightly on a daily basis. On the other hand, land surfaces and the air above can be substantially warmed during a single daylight period. As the air over the land warms, it expands, causing the isobars to bend upward as shown in Figure 6-6(*b*). Although this warming does not by itself produce a surface pressure change, the pressure aloft does become higher over the land than at comparable altitudes over the ocean. The resultant pressure gradient aloft causes the air there to move from over the land toward the ocean. The mass transfer of air seaward creates a surface high over the ocean where the air is collecting and a surface low over the land. The surface circulation that develops from this redistribution of mass aloft is from the sea toward the land (sea breeze)

Use in lecture

Important

Sea breeze

Press. grad.

128

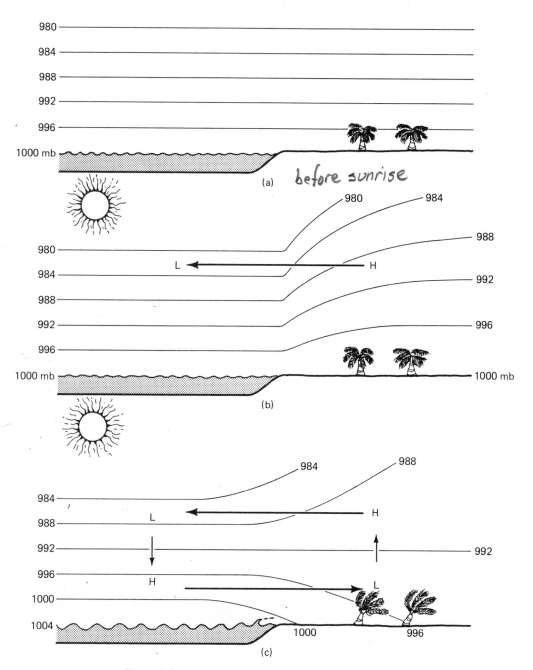

Figure 6-6
Cross-sectional view illustrating the formation of a sea breeze.

as shown in Figure 6-6(*c*). Thus, a simple thermal circulation has been developed with flow seaward aloft and landward at the surface. Note that vertical movement is required to make the circulation complete.

A very important relationship exists between pressure and temperature, as we saw in the preceding discussion. Temperature variations create pressure differences and therefore wind, and the greater these temperature differences, the stronger the pressure gradient and resultant wind. Daily temperature differences and the pressure gradients so generated are confined to a rather shallow layer of the atmosphere. On a global scale, however, the latitudinal variations in solar incidence generates, in a similar manner, the much larger general atmospheric circulation. This is the topic of the next chapter.

In summary, the horizontal pressure gradient is the driving force of wind. It has both magnitude and direction. Its magnitude is determined from the spacing of isobars and the direction of force is always from areas of high pressure to areas of low pressure and at right angles to the isobars.* Once the air starts to move, the Coriolis force and friction come into play, but then only to modify the movement, not to produce it.

coriolis effect

Figure 6-14 shows the typical air movements associated with high- and low-pressure systems. As expected, the air moves out of the regions of high pressure and into the regions of low pressure. However, the wind does not cross the isobars at right angles as the pressure gradient force directs. This deviation is the result of the earth's rotation and has been named the *Coriolis effect* after the French scientist who first expressed its magnitude quantitatively. All freemoving objects, including wind, are deflected to the right of their path of motion in the Northern Hemisphere and to the left in the Southern Hemisphere. The reason for this deflection can be illustrated by imagining the path of a rocket launched from the North Pole toward a target located on the equator (Figure 6-7). If the rocket took an hour to reach its target, during its flight the earth would have rotated 15 degrees to the east. To someone standing on the earth, it would look as if the rocket veered off its path and hit the earth 15 degrees west of its target. The true path of the rocket was straight and would appear so to someone out in space looking down at the earth. It was the earth turning under the rocket that gave it its apparent deflection. Note that the rocket was deflected to the right of its path of motion because of the counterclockwise rotation of the Northern Hemisphere. In the Southern Hemisphere clockwise rotation produces a similar deflection, but to the left of the path of motion.

Although it is usually easy for students to visualize the Coriolis effect when the motion is from north to south, as in our rocket example, it is not so easy to see how a west to east flow would be deflected. Figure 6-8 illustrates this situation by examining a wind that is blowing eastward along the fortieth parallel. Several hours later, that which started as a west wind will be coming from a more northwesterly direction. Note that it was really the observer's orientation that has changed, not the wind direction.

It is of interest to point out that any "free-moving" object will experience

---

*The mathematical expression of the pressure gradient force is provided in Appendix D.

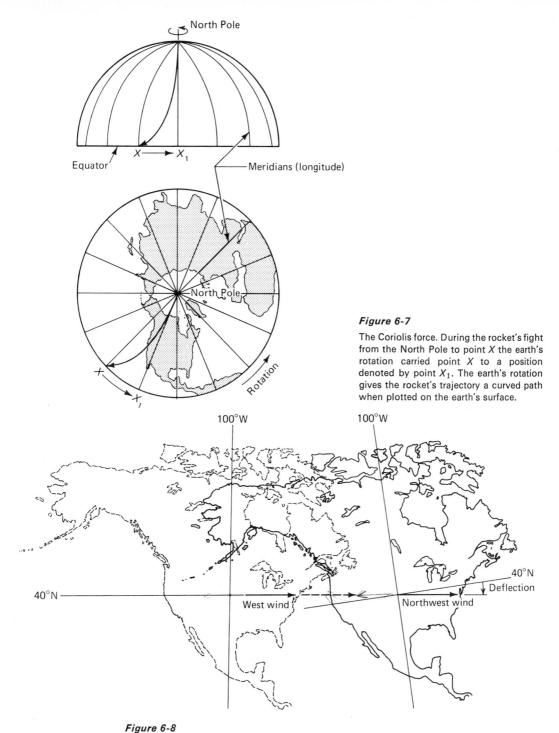

North Pole

Equator

X ———→ X₁

Meridians (longitude)

North Pole

North Pole

Rotation

X

X₁

Figure 6-7

The Coriolis force. During the rocket's fight from the North Pole to point X the earth's rotation carried point X to a position denoted by point X_1. The earth's rotation gives the rocket's trajectory a curved path when plotted on the earth's surface.

100°W

100°W

40°N

Deflection

40°N

West wind

Northwest wind

Figure 6-8

Coriolis deflection of a west wind. After a few hours the earth's rotation changes the position of the surface over which the wind blows, thus causing the apparent deflection.

131

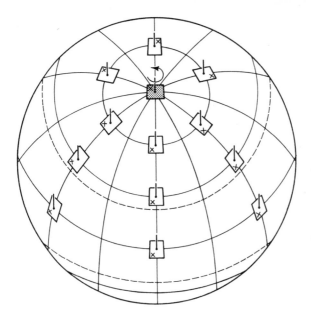

Figure 6-9

Illustration of the amount of rotation of a horizontal surface about a vertical axis at various latitudes. The magnitude of the Coriolis force is proportional to the amount of rotation experienced.

a similar deflection. This was dramatically discovered by our Navy during World War II. During target practice long-range rifles on battleships continually missed their targets by as much as several hundred yards until ballistic corrections were made for the changing position of a seemingly stationary target. Over a short distance the Coriolis effect can be ignored. For example, the Roman Legions did not have to account for it when they used their catapults.

For convenience, we attribute this apparent shift in wind direction to the Coriolis force. It is hardly a "real" force, but rather the effect of the earth's rotation on a moving body. This deflecting force (1) is always directed at right angles to the direction of airflow; (2) affects only wind direction, not wind speed; (3) is affected by wind speed (the stronger the wind, the greater the deflecting force); and (4) is strongest at the poles and weakens equatorward where it eventually becomes nonexistent.*

The fourth condition is the most difficult to understand. It results because the amount of reorientation of the surface over which the wind blows is dependent upon latitude. For example, at the poles where the land surface is perpendicular to the earth's axis, the daily rotation causes the horizon to make one complete turn about a vertical axis each day (Figure 6-9). Restated more simply, the surface over which the wind blows makes one complete rotation each day. At the equator the earth's surface is parallel to the earth's axis of rotation and consequently it has no rotation about an axis oriented vertically to this surface. Therefore, at the equator the surface over which the air blows does not rotate in a horizontal sense. This difference might be

*A mathematical treatment of the Coriolis effect is provided in Appendix D.

easier to visualize if you were to imagine a vertical post situated at the North Pole and one situated at the equator. During the course of a single day the post located at the North Pole would make one complete rotation about its vertical axis, but the post situated at the equator would not spin at all. Here the post would tumble end over end as the earth rotates. Posts situated between these extremes would experience intermediate rates of rotation about their vertical axis each day. Consequently, because the horizontal orientation (rotation about a vertical axis) of the earth's surface would change more rapidly at high latitudes than at low latitudes the Coriolis force would be greater there as well. To summarize, the magnitude of the Coriolis force is dependent upon latitude; it is strongest at the poles, and it weakens equatorward where it eventually becomes nonexistent.

Although the magnitude of the Coriolis force varies considerably with both wind speed and latitude, calculations of its strength indicate that at reasonable wind speeds it can approximate the strength of the pressure gradient force. This fact is the basis of the following section.

the geostrophic wind

Previously we stated that the pressure gradient force is the primary driving force of the wind. As an unbalanced force it causes air to accelerate from regions of high pressure to regions of low pressure. Thus, the wind speeds should continually increase (accelerate) for as long as this imbalance exists. But we know from our own personal experiences that winds do not get faster indefinitely. Hence, some other force, or forces, must oppose the pressure gradient force in order to moderate airflow. From our everyday experiences we know that friction acts to slow a moving object. Although friction significantly influences airflow near the earth's surface, its effect is negligible above a height of a few kilometers. For this reason, we will divide our discussion of surface winds where friction is significant from that of flow aloft where its effect is small. This section will deal only with airflow above a few kilometers where the effects of friction are significantly small enough to be neglected.

Aloft the Coriolis force is responsible for balancing the pressure gradient force and thereby directing airflow.* Figure 6-10 helps to show how a balance is reached between these opposing forces. For purposes of illustration only, we assume a nonmoving parcel of air at the starting point in Figure 6-10. It should be remembered that air is rarely stationary in the "real" atmosphere. Since our parcel of air has no motion, the Coriolis force exerts no influence; only the pressure gradient force can act. Under the influence of the latter force, which is always directed perpendicular to the isobars, the parcel begins to accelerate directly toward the area of low pressure. As soon as the flow begins, the Coriolis force causes a deflection to the right for winds in the Northern Hemisphere. As the parcel continues to accelerate,

---

*The magnitude of the pressure gradient force and the magnitude of the Coriolis effect are compared in Appendix D.

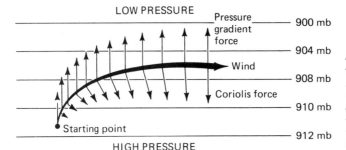

LOW PRESSURE

Pressure gradient force

Wind

Coriolis force

Starting point

HIGH PRESSURE

900 mb
904 mb
908 mb
910 mb
912 mb

Figure 6-10

The geostrophic wind. Upper-level winds are deflected by the Coriolis force until the Coriolis force just balances the pressure gradient force. Above 600 meters, where friction is negligible, these winds will flow nearly parallel to the isobars and are called geostrophic winds.

the Coriolis force intensifies (recall that the magnitude of the Coriolis force is proportional to wind speed). Thus, the increased speed results in further deflection. Eventually, the wind will be turned so that it is flowing parallel to the isobars, with the pressure gradient force directed toward the area of low pressure and opposed by the Coriolis force which is directed toward the region of high pressure (Figure 6-10). When these two opposing forces are equal, the wind will continue to flow parallel to the isobars at a constant speed. This results because when the airflow is parallel to the isobars, the pressure gradient is oriented perpendicular to the flow where it no longer causes further acceleration, and since the Coriolis force is proportional to wind speed, it too will remain constant. Consequently, the balance that exists between these forces will be maintained and airflow will continue along the isobars. Under these idealized conditions, the wind can be considered to be coasting (not accelerating or decelerating) along a pathway defined by the isobars.

So far we have assumed that these two opposing forces just balance each other. In order for this to occur, the pressure gradient force must produce a wind with the exact velocity needed to generate a Coriolis force equal to itself. But there is no reason why this should occur, at least not on the first attempt. Theory predicts that the speed of the wind generated in the manner described above would result in a Coriolis force that exceeds the pressure gradient force. This would cause the wind to be deflected beyond the middle ground and to cross the isobars toward the region of high pressure (see Figure 6-10). Or stated another way, the winds would move against the pressure gradient force. However, this would result in an immediate weakening of the wind and a reduced Coriolis force. Hence, the wind would be deflected back toward the region of low pressure by the pressure gradient force and again accelerated. Thus, instead of flowing directly along the isobars, the wind is constantly adjusting its speed and direction in an attempt to reach a balance between these opposing forces.

The preceding notwithstanding, the pressure gradient and Coriolis forces can be considered to reach a state of equilibrium, called the geostrophic balance. The winds generated by this balance are called *geostrophic winds* and flow nearly parallel to the isobars with velocities proportional to the

134

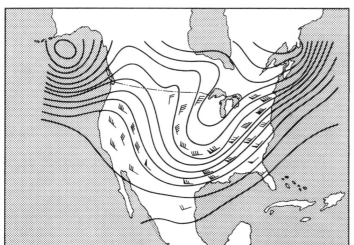

ff	Miles (statute) per hour	Knots
◎	Calm	Calm
——	1–2	1–2
⌐	3–8	3–7
⌐	9–14	8–12
⌐	15–20	13–17
⌐	21–25	18–22
⌐	26–31	23–27
⌐	32–37	28–32
⌐	38–43	33–37
⌐	44–49	38–42
⌐	50–54	43–47
⌐	55–60	48–52
⌐	61–66	53–57
⌐	67–71	58–⁻2
⌐	72–77	63–67
⌐	78–83	68–72
⌐	84–89	73–77
⌐	119–123	103–107

Figure 6-11

Upper-air winds. This map shows the direction and speed of the upper-air wind of February 24, 1974. Note that the airflow is almost parallel to the contours. These isolines are height contours for the 500-millibar level.

pressure gradient force.* A steep pressure gradient creates strong winds that generate an equally strong Coriolis force.

Referring back to Figure 6-10, we see that for geostrophic flow the pressure gradient force is directed at right angles to the wind toward the area of low pressure and is opposed by the Coriolis force that is directed toward the area of high pressure. This rather simple relationship between wind direction and pressure distribution was first formulated by a Dutch meteorologist, Buys Ballott, in 1857. Essentially, *Buys Ballott's law* states: In the Northern Hemisphere if you stand with your back to the wind, low pressure will be to your left and high pressure will be to your right. In the Southern Hemisphere the situation is reversed since the Coriolis deflection is to the left. Although Buys Ballott's law holds for airflow aloft, it must be used with some caution when considering surface winds. Numerous geographic effects can generate local disturbances that interfere with the larger circulation.

In the real atmosphere the winds are never purely geostrophic. Nonetheless, the importance of the idealized geostrophic flow lies in the fact that it gives a good approximation of the actual winds aloft. Thus, by measuring the pressure field (orientation and spacing of isobars) aloft, meteorologists can determine both the wind direction and speed. The idealized geostrophic flow predicts winds parallel to the isobars with speeds dependent upon isobaric spacing (Figure 6-11). Of even greater use to meteorologists is that

---

*The speed of geostrophic flow is also inversely proportional to the magnitude of the Coriolis force. When influenced by equal pressure gradients, geostrophic flow is faster near the equator where the Coriolis force is weak than at higher latitudes where it is stronger.

the same method can be used in reverse to determine the distribution of pressure from airflow measurements. This interrelationship between pressure and winds greatly enhances the reliability of upper-air weather charts by providing checks and balances. In addition, it minimizes the number of direct observations required to adequately describe the conditions aloft where accurate data are most expensive and difficult to obtain.

In summary, above a few kilometers the winds can be considered to be almost geostrophic, that is, they flow almost parallel to the isobars at speeds that can be calculated from the pressure gradient and Coriolis forces. The major discrepancy involves the flow along highly curved isobars, a topic to be considered next.

curved flow

Even a casual glance at a weather map reveals that the isobars are not generally straight; instead, they make broad sweeping curves. Occasionally, the isobars connect to form roughly circular cells of either high or low pressure. The curved nature of isobars is one of the factors that tends to modify true geostrophic flow mainly by affecting wind speed, but not direction which remains roughly parallel to the isobars. Since the geostrophic wind is a reasonable approximation of the wind direction, we will first examine the flow around circular pressure centers as if they were geostrophic, that is, we will ignore the effect of curvature.

Figure 6-12 shows that for geostrophic flow around a center of low pressure, the pressure gradient force is directed inward and is balanced by the outward-directed Coriolis force. In the Northern Hemisphere where the Coriolis force deflects the flow to the right, the resultant wind blows counterclockwise about a low. Conversely, around a high-pressure cell the outward-directed pressure gradient force that is balanced by the inward-directed Coriolis force results in clockwise flow. Because the Coriolis force deflects the winds to the left in the Southern Hemisphere, the flow is reversed there—clockwise around low-pressure centers and counterclockwise around high-pressure centers.

It is common practice to call all centers of low pressure *cyclones* and the flow around them cyclonic. *Cyclonic flow* has the same direction of rotation as the earth: counterclockwise in the Northern Hemisphere and clockwise in the Southern Hemisphere. Centers of high pressure are called *anticyclones* and have *anticyclonic flow* (opposite that of the earth's rotation). Whenever the isobars curve to form elongated regions of low and high pressure, these areas are called *troughs* and *ridges*, respectively. The flow about a trough is cyclonic; the flow around a ridge is anticyclonic.

Now let us consider the affects of curvature on the speed of cyclonic and anticyclonic flow. Whenever the flow is curved, a force is required to turn the air (change its direction) even when no change in speed results. This is a consequence of a portion of Newton's first law of motion which states that a moving object will continue to move in a straight line unless acted upon by an unbalanced force. You have undoubtedly experienced the effect of this

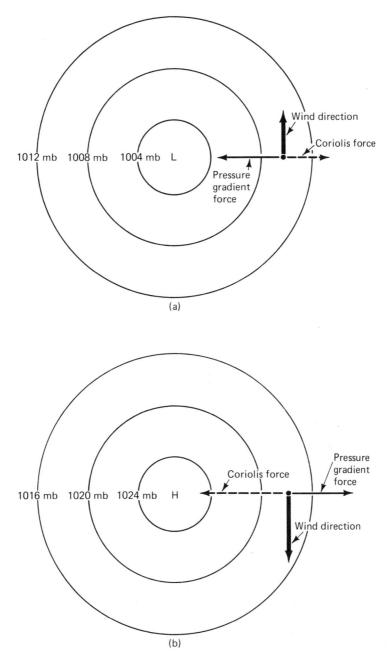

Figure 6-12
Curved flow about (*a*) low- and (*b*) high-pressure centers.

law when the auto you were riding in made a sharp turn and your body continued straight ahead. Referring back to Figure 6-12, we see that in a high-pressure center the outward-directed pressure gradient force is balanced by

137

the inward-directed Coriolis force. But in order to keep the path curved (parallel to the isobars), the inward pull of the Coriolis force must be strong enough to balance the pressure gradient force as well as turn (accelerate) the air inward. (The inward turning of the air is termed *centripetal accelera-tion.*) Stated another way, the Coriolis force must exceed the pressure gradient force in order to overcome the air's tendency to continue in straight line.* The opposite situation exists in cyclonic flow where the inward-directed pressure gradient force must balance the Coriolis force as well as provide the inward acceleration needed to turn the air. A consequence of these differences is that with equal pressure gradients anticyclonic winds exceed cyclonic winds. This results because the magnitude of the Coriolis force is proportional to wind speed. Thus, greater wind speeds are a requirement for anticyclonic flow where the Coriolis force must provide the inward acceleration. Notice in Figure 6-12 that the pressure gradient and Coriolis forces are not balanced as they are in geostrophic flow. This imbalance is necessary in order to provide the change in direction (centripetal acceleration) needed for curved flow.

Although we concluded that for similar pressure gradients airflow around a high-pressure center exceeds that around a low, we commonly experience our strongest winds while under the influence of low-pressure centers. This apparent conflict is easily resolved when we note that the pressure gradients of cyclones often exceed those associated with anticyclones. It is also important to note that the magnitude of centripetal acceleration is small compared to the other forces involved. Consequently, except for very tight storms such as hurricanes, the geostrophic wind provides a good approximation of the true wind speed as well as the wind direction.

friction layer winds

Friction as a factor affecting wind is only important within the first few kilometers of the earth's surface. We know that it acts to slow the movement of air, but as a consequence of this, it also alters wind direction. Recall that the Coriolis force is proportional to wind speed. By lowering the wind speed, friction reduces the Coriolis force. Since the pressure gradient force is not affected by wind speed, it wins the tug of war against the Coriolis force, as shown in Figure 6-13. The result is the movement of air at an angle across the isobars toward the area of low pressure. The roughness of the terrain determines the angle at which the air will flow across the isobars as well as the speed at which it will move. Over the relatively smooth ocean surface, where friction is low, the air moves at an angle of 10 to 20 degrees to the isobars and at speeds roughly two-thirds that of geostrophic flow. Over rugged terrain, where friction is high, the angle could be as great as 45 degrees from the isobars, with wind speeds reduced by as much as 50 percent.

By changing the direction of airflow, friction plays a major role in redistributing the air within the atmosphere. This is especially noticeable when

---

*The tendency of a particle to move in a straight line when rotated creates an imaginary outward force called the centrifugal force.

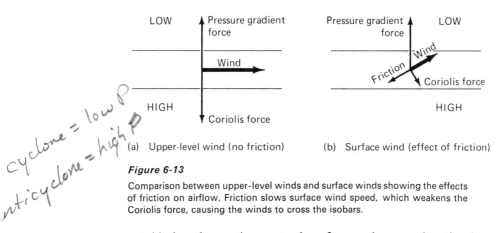

(a) Upper-level wind (no friction)   (b) Surface wind (effect of friction)

Figure 6-13

Comparison between upper-level winds and surface winds showing the effects of friction on airflow. Friction slows surface wind speed, which weakens the Coriolis force, causing the winds to cross the isobars.

[handwritten margin note: Cyclone = low P; anticyclone = high P]

considering the motion around surface cyclones and anticyclones, two of the most common features on surface weather maps. In the preceding section we learned that above the friction layer in the Northern Hemisphere winds blow counterclockwise around a cyclone and clockwise around an anticyclone with winds nearly parallel to the isobars. When we add the effects of friction, we notice that the airflow will cross the isobars at varying angles depending upon the roughness of the terrain, but always from high to low pressure. In a cyclone in which pressure decreases inward, friction causes a net flow toward its center (Figure 6-14), but in an anticyclone just the opposite is the case; the pressure decreases outward, and the resultant flow is also outward. We thus conclude that in the Northern Hemisphere winds blow into and counterclockwise about a surface cyclone and outward and clockwise about a surface anticyclone. Of course, in the Southern Hemisphere the Coriolis force deflects the winds to the left and reverses the direction of flow. However, in whatever hemisphere, friction causes a net inflow (*convergence*) around a cyclone and a net outflow (*divergence*) around an anticyclone. This very important relationship between cyclonic flow and convergence and anticyclonic flow and divergence will be considered again.

how winds generate vertical motion

Up to now we have discussed the factors that affect wind without regard to how the movement of air might alter the state of the atmosphere. In other words, how might wind in one zone of the atmosphere affect airflow elsewhere? Of particular importance is the following question: How do winds relate to vertical flow? It should be remembered that although vertical transport is small in comparison to horizontal motion, it is very important as a weather maker. Rising air is associated with cloudy conditions and precipitation while subsidence produces adiabatic heating and clearing conditions. In this section we will discern how the movement of air (dynamic affect) can itself create pressure change and hence generate winds. Upon doing so, we will examine the interrelationship between horizontal and vertical flow and its effects on the weather.

Let us first consider the situation around a surface low-pressure system

139

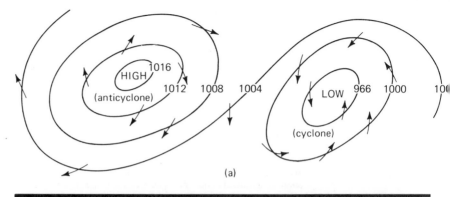

(a)

(b)

Figure 6-14

(*a*) Cyclonic and anticyclonic winds in the Northern Hemisphere. Arrows show the winds blowing in and counterclockwise about a cyclone and out and clockwise about an anticyclone. (*b*) Apollo IX photograph of a cyclonic storm located about 2000 kilometers due north of Hawaii (courtesy of NASA).

in which the air is spiraling inward. Here, the net inward transport of air causes a shrinking of the area occupied by the air mass, a process called horizontal convergence. Whenever air converges horizontally, it must pile up, that is, it must increase in height to allow for the decreased area it now occupies. This generates a "taller" and therefore heavier air column. Yet, a surface low can exist only as long as the column of air above remains light. We seem to have encountered a paradox—low-pressure centers cause a net accumulation of air, which increases their pressure. Consequently, a surface cyclone should quickly eradicate itself in a manner not unlike what happens to the vacuum in a coffee can upon being opened.

140

It should be apparent in light of the preceding discussion that in order for a surface low to exist for any reasonable time, compensation must occur at some layer aloft. For example, surface convergence could be maintained if divergence (spreading out) aloft occurred at a rate equal to the inflow below. Figure 6-15(a) diagrammatically shows the relationship between surface convergence (inflow) and divergence (outflow) aloft that is needed to maintain a low-pressure center. Note that surface convergence about a cyclone causes a net upward movement. The rate of this vertical movement is very slow, generally less than 1 kilometer per day. Nevertheless, since rising air often results in cloud formation and precipitation, the passage of a low is generally related to unstable conditions and "stormy" weather. Divergence aloft may occasionally even exceed surface convergence. This results in intensified surface inflow and increased vertical motion. Thus, divergence aloft can intensify these storm centers as well as maintain them. On the other hand, inadequate divergence aloft permits surface flow to "fill" and weaken the accompanying cyclone. As often as not, it is divergence aloft that first creates the surface low. Spreading out (divergence) aloft initiates upflow in the layer directly below, and this eventually works its way to the surface where inflow is encouraged.

Like their counterparts, anticyclones, which are associated with surface divergence, must also be maintained from above. The mass outflow near the surface is accompanied by convergence aloft and general subsidence of the air column [Figure 6-15(b)]. Since descending air is compressed and warmed, cloud formation and precipitation are unlikely in an anticyclone, and "fair" weather can usually be expected with the approach of a high.

For reasons that should now be obvious, it has been common practice to write on barometers intended for household use the words "stormy" at the low end and "fair" at the high end. By noting whether the pressure is rising, falling, or steady, we have a good indication of what the forthcoming weather will be. Such a determination, called the *pressure* or *barometric tendency*, is a very useful aid in short-range weather prediction. The generalizations relating cyclones and anticyclones to the weather conditions just con-

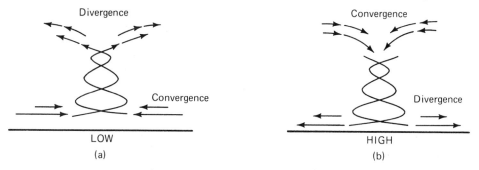

Figure 6-15

Schematic of the airflow associated with cyclones and anticyclones. (*a*) Converging winds and rising air are associated with a low, or cyclone. (*b*) Highs, or anticyclones, are associated with descending air and diverging winds.

141

sidered are stated rather poetically in the rhyme that follows. Note that glass refers to the barometer.

> *When the glass falls low,*
> *Prepare for a blow;*
> *When it rises high,*
> *Let all your kites fly.*

Because of the close tie between winds and weather systems, we will consider some of the factors that contribute to horizontal convergence and divergence. Earlier we mentioned the effect of friction on curved flow causing the winds to cross the isobars toward the area of low pressure. To recapitulate, cyclonic flow encourages convergence and uplifting and anticyclonic flow encourages divergence and subsidence. Friction can also cause mass convergence when the flow is straight. For example, when air moves from the relatively smooth ocean surface onto land, the increased friction causes an abrupt drop in wind speed. This reduction of wind speed downstream results in a pile up of air upstream. Thus, converging winds and ascending air accompany flow off the ocean. This effect contributes to the cloudy conditions over the land often associated with a sea breeze in a humid region such as Florida. As expected, general divergence and subsidence accompany the flow of air seaward because of increasing wind speeds.

Mountains, which also hinder the flow of air, cause divergence and convergence in yet another way. As air passes over a mountain range, it must shrink vertically, which produces horizontal spreading (divergence) aloft. Upon reaching the lee side of the mountain, the air experiences vertical stretching which causes horizontal convergence. This effect greatly influences the weather in the United States east of the Rocky Mountains, as we shall examine later. When air flows equatorward where the Coriolis force is weakened, divergence and subsidence prevail; during poleward migration convergence and slow uplift are favored.

In conclusion, you should now be better able to understand why local television weather broadcasters emphasize the positions and projected paths of cyclones and anticyclones. The "villain" on these weather programs is always the cyclone that produces "bad" weather in any season. Lows move in roughly a west-to-east direction across the United States and require as little as a few days to more than a week for the journey. Since their paths can be somewhat erratic, accurate prediction of their migration is difficult, yet essential for short-range forecasting. Meteorologists must also determine if the flow aloft will intensify an embryo storm or act to suppress its development. Because of the close tie between surface conditions and those aloft, a great deal of emphasis has been placed on the importance and understanding of the total atmospheric circulation, especially in the midlatitudes. Once we have examined the workings of the general circulation, we will again consider the development and structure of the cyclone in light of these findings.

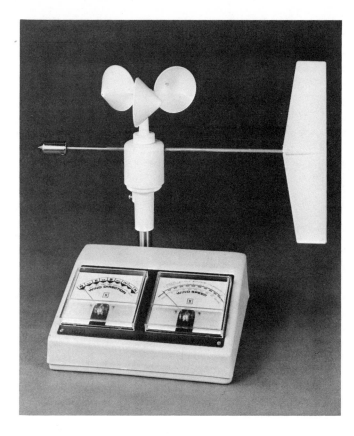

Figure 6-16
Wind vane and cup anemometer. (Courtesy of WeatherMeasure Corporation.)

wind measurement

Two basic wind measurements—direction and speed—are particularly significant to the weather observer. Winds are always labeled by the direction from which they blow. A north wind blows from the north toward the south; an east wind blows from the east toward the west. The instrument most commonly used to determine wind direction is the *wind vane* (Figure 6-16). This instrument, which is a common sight on many buildings, always points into the wind. Often the wind direction is shown on a dial that is connected to the wind vane. The dial indicates the direction of the wind either by points of the compass, that is, N, NE, E, SE, etc., or by a 0-to 360-degree scale. On the latter scale 0 degrees or 360 degrees is north, 90 degrees is east, 180 degrees is south, and 270 degrees is west. When the wind consistently blows more often from one direction than from any other, it is termed a *prevailing wind.*

Wind speed is commonly measured by using a *cup anemometer* (Figure 6-16). The wind speed is read from a dial much like the speedometer of an automobile. When instruments are not available, wind speed can be estimated by using the *Beaufort scale* (Table 6-2).

Table 6-2

Beaufort Scale

Beaufort Number	m/s	mph	International Description	Specifications
0	< 1	< 1	Calm	Calm; smoke rises vertically
1	1	1–3	Light air	Direction of wind shown by smoke drift but not by wind vanes
2	2	4–7	Light breeze	Wind felt on face; leaves rustle, vanes moved by wind
3	4	8–12	Gentle breeze	Leaves and small twigs in constant motion; wind extends light flag
4	7	13–18	Moderate	Raises dust, loose paper; small branches moved
5	10	19–24	Fresh	Small trees in leaf begin to sway; crested wavelets form on inland waters
6	12	25–31	Strong	Large branches in motion; whistling heard in telegraph wires; umbrellas used with difficulty
7	15	32–38	Near gale	Whole trees in motion; inconvenience felt walking against wind
8	18	39–46	Gale	Breaks twigs off trees; impedes progress
9	20	47–54	Strong gale	Slight structural damage occurs
10	26	55–63	Storm	Trees uprooted; considerable damage occurs
11	30	64–72	Violent storm	Widespread damage
12	≥ 33	> 73	Hurricane	

If you know the placement of cyclones and anticyclones in relation to where you are, you can predict the changes in wind direction that will be experienced as the pressure center moves past. Since changes in wind direction often bring changes in temperature and moisture conditions, the ability to predict the winds can be very useful. In the Midwest, for example, a north wind may bring cool, dry air from Canada, while a south wind may bring warm, humid air from the Gulf of Mexico. Sir Francis Bacon summed it up nicely when he wrote, "Every wind has its weather."

REVIEW

1. When density remains constant and the temperature is raised, how will the pressure of a gas change?

2. When the gases in the atmosphere are heated, air pressure normally decreases. In light of your answer to the first question, explain this apparent paradox.

3. What is standard sea-level pressure in millibars? In inches of mercury?

4. Describe the principle of the mercurial barometer and the aneroid barometer.

5. What force is responsible for generating wind?

6. Write a generalization relating the spacing of isobars to the speed of wind.

7. Although vertical pressure differences may be great, such variations do not generate strong vertical currents. Explain.

144

8. Temperature variations create pressure differences which, in turn, produce winds. On a small scale the sea breeze illustrates this principle very nicely. Describe how a sea breeze forms.

9. Briefly describe how the Coriolis effect modifies the movement of air.

10. What two factors influence the magnitude of the Coriolis effect?

11. Explain the formation of a geostrophic wind.

12. Answer the following question by applying Buys Ballott's law: If you are facing north and the wind is at your back, where is the low pressure?

13. Unlike winds aloft, which blow nearly parallel to the isobars, surface winds generally cross the isobars. Explain why this difference exists.

14. Sketch a diagram (isobars and wind arrows) showing the winds associated with cyclones and anticylones in both the Northern and Southern Hemispheres.

15. In order for surface low pressure to exist for an extended period of time, what condition must exist aloft?

16. Describe the general weather conditions one should expect when the pressure tendency is rising and when the pressure tendency is falling.

17. Converging winds and ascending air are often associated with the flow of air from the oceans onto land. Conversely, divergence and subsidence often accompany the flow of air from land to sea. What causes this convergence over land and divergence over the ocean?

18. A Southwest wind is blowing from the _____ (direction) toward the _____ (direction).

19. The wind direction is 315 degrees. From what compass direction is the wind blowing?

VOCABULARY
REVIEW

air pressure	Buys Ballott's law
gas law	cyclone
newton	cyclonic flow
millibar	anticyclone
mercurial barometer	anticyclonic flow
aneroid barometer	trough
barograph	ridge
altimeter	centripetal acceleration
wind	convergence
centrifugal force	divergence
isobar	pressure or barometric tendency
pressure gradient	wind vane
sea breeze	prevailing wind
Coriolis effect	cup anemometer
geostrophic wind	Beaufort scale

Chapter Seven

GLOBaL CIRCULATION

Those of us who live in the United States are familiar with the term "westerlies" to describe our winds. But all of us have experienced winds from the south and north, and even directly from the east. You may even recall being in a wind storm when shifts in wind direction and speed came in such rapid succession that it was impossible to determine the wind's direction. With all of this variety, how can we describe our winds as westerly? The answer to this question lies in our attempt to simplify descriptions of the atmospheric circulation by sorting out events according to size. For example, on the scale of a weather map, where observing stations are spaced about 150 kilometers apart, small whirlwinds that carry dust skyward are far too small to show up. Instead, these maps reveal larger scale wind patterns such as those associated with traveling cyclones and anticyclones. Not only do we separate winds according to the size of the system, but equal consideration is given to the time frame in which they occur. A small eddy, like a dust devil, may only last a few moments while a larger vortex like a hurricane may last for days.

The time and space scales we will use for atmospheric motions are provided in Table 7-1. The largest scale wind pattern is called *macroscale circulation* and is exemplified by the global circulation, that is, the mean airflow over the entire earth. This mean airflow is a statistical average of worldwide wind observations made over many years. Because of the marked seasonal variations in atmospheric motion, separate global circulation patterns have been established for both summer and winter. A somewhat smaller type of macroscale circulation is commonly called the *synoptic scale* and consists mainly of individual traveling cyclones and anticyclones. These vortices are a very important part of the circulation in the middle latitudes. The airflow in these less than global-sized macroscale circulations consists primarily of horizontal flow, with only modest amounts of vertical motion. By contrast, *mesoscale* and *microscale* winds influence smaller areas and exhibit extensive vertical flow, which can be rapid, as in a developing thunderstorm. Although the primary emphasis of this chapter will be upon developing an understanding of the largest type of macroscale circulation, global wind patterns, it will also deal with smaller local wind systems, which are classed as mesoscale phenomena.

Table 7-1

Time and Space Scales for Atmospheric Motions

	Name of Scale	Time Scale	Length Scale	Examples
Macroscale	General Circulation	Weeks to Years	1000 to 40,000 km	Waves in the Westerlies
	Synoptic Scale	Days to Weeks	100 to 5000 km	Cyclones, Anticyclones, and Hurricanes
	Mesoscale	Minutes to Days	1 to 100 km	Land-sea Breeze, Thunderstorms, and Tornadoes
	Microscale	Seconds to Minutes	< 1 km	Turbulence

Although it is common practice to divide atmospheric motions according to scale, it is important to remember that this flow is very complex—much like a turbulent river with smaller eddies within larger eddies within still larger eddies. Also, like the flow of a turbulent river, each scale of motion is related to the other. For example, let us examine the flow associated with the hurricanes that form over the North Atlantic. When we view one of these tropical cyclones on satellite photographs, the storm appears as a large whirling cloud migrating slowly across the ocean (Figure 7-1). From this perspective (synoptic scale) the general counterclockwise rotation can be easily seen, and the storm's path, which persists for a week or so, can be easily followed. However, when we average the winds of a hurricane, we find that they have a net motion from east to west. This indicates that these rather large eddies are embedded in a still larger flow (the general circulation) that is moving westward across the tropical portion of the North Atlantic. When we examine a hurricane more closely by flying an airplane through it, some of the small scale aspects of the storm become noticeable. As the plane approaches the outer edge of the system, it becomes evident that the large rotating cloud that we saw in the satellite photographs is made of numerous towering cumulonimbus clouds. Each of these mesoscale phenomena (thun-

Figure 7-1
Satellite view of Hurricane Ione (left) and Hurricane Kirsten (right).

derstorms) lasts for only a few hours and must be continually replaced by new ones if the hurricane is to persist. When we fly into these storms, we quickly realize that the individual clouds are also made up of even smaller scale turbulences. The small thermals of rising air that comprise these clouds make for a rather rough trip. Thus, a typical hurricane exhibits several different scales of motion, including many mesoscale thunderstorms, which, in turn, consist of numerous microscale turbulences, while the counterclockwise circulation of the hurricane is itself part of the larger general circulation in the tropics.

idealized global circulation

Our knowledge of global winds comes from both the observed distribution of the pressure–wind regimes and theoretical studies of fluid motion. We will begin our discussion by first considering the classical model of global circulation that was developed largely from the mean worldwide pressure distribu-

148

tion. We will then add to the idealized circulation some of the more recently discovered aspects of the atmosphere's very complex motions.

One of the first contributions to the classical model of the general circulation came from George Hadley in 1735. Hadley was well aware of the fact that solar energy drives the winds. He proposed that the large temperature contrast between the poles and the equator would create a thermal circulation very similar to that of the sea breeze that we discussed in Chapter 6. As long as the earth's surface is heated unequally, air will move in an attempt to balance the inequalities. Hadley suggested that on a nonrotating earth the air movement would take the form of one large *convection cell* in each hemisphere, as shown in Figure 7-2. The more intensely heated equatorial air would rise and move poleward. Eventually, this upper-level flow would reach the poles where it would sink and spread out at the surface and return to the equator. As the cold polar air approached the equator, it would be reheated and rise again. Thus, the proposed Hadley circulation for a nonrotating earth has upper-level air flowing poleward and surface air flowing equatorward. When we add the effect of the earth's rotation, the Coriolis effect causes the surface flow to become more or less easterly (toward the west) and the flow aloft to be westerly.

There are several reasons why the simple convectional circulation developed by Hadley cannot exist. Perhaps the easiest reason to understand hinges on the fact that the atmosphere is connected to the earth by the force of gravity and, therefore, it must rotate with the earth. In the Hadley model the surface winds blow toward the west and would, because of friction, oppose the earth's rotation, which is toward the east. Since the atmosphere is attached to the earth, however, it cannot, over the long haul, slow the earth's rotation. Thus, easterly flow at one latitude must be balanced by westerly flow at another. Further, Hadley's simple convective system did not fit the observed pressure distribution that was subsequently established for the earth. Consequently, the popular Hadley model was replaced by a version that better fit the observed situation.

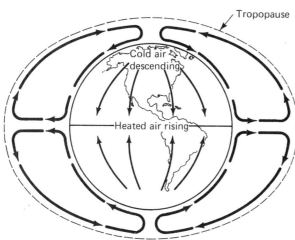

Figure 7-2

Global circulation on a nonrotating earth. A simple convection system produced by unequal heating of the atmosphere on a nonrotating earth.

In the 1920s a three-cell circulation in each hemisphere was proposed to carry on the task of maintaining the earth's heat balance. Although this model has been modified to fit recent upper-air observations, it is nonetheless still very useful. Figure 7-3 illustrates the three-cell model and the surface winds that result. Note that the surface flow resulting from the three-cell model has a much greater east–west component than a north–south component.

In the zone between the equator and roughly 30 degrees latitude the circulation closely resembles the convective model used by Hadley for the whole earth; hence, the name *Hadley cell* is generally applied to it. Here the surface flow is equatorward while the flow aloft is poleward. Near the equator the warm rising air that releases latent heat during the formation of cumulus towers is believed to provide the energy to drive this cell. These clouds also provide the rainfall that maintains the lush vegetation of the equatorial rain forests. As the upper flow in this cell moves poleward, it begins to subside in a zone between about 20 and 35 degrees latitude. Two factors are thought to contribute to the general subsidence found here. First, when this flow moves

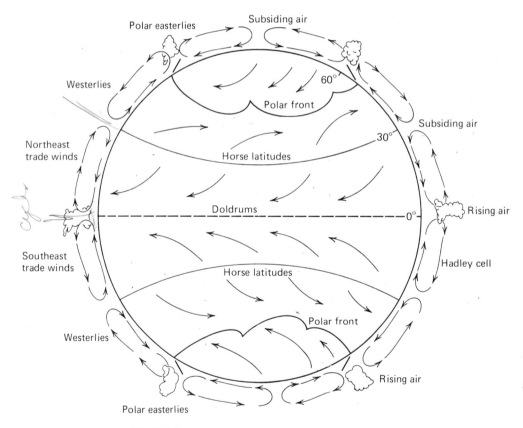

Figure 7-3

Idealized global circulation proposed for the three-cell circulation model.

away from the stormy equatorial region where the release of latent heat of condensation keeps the air warm and buoyant, radiation cooling would result in increased density of the air aloft. This could account for the subsidence. Satellites that monitor the radiation emitted in the upper troposphere have supplied data to support this mechanism for subsidence. Second, because the Coriolis force becomes stronger with increasing distance from the equator, the once poleward-directed winds are deflected into a nearly west-to-east flow by the time they reach 25 degrees latitude. This results in a restricted poleward flow of air. Stated another way, the Coriolis force produces a blocking effect that causes a general pileup of air (convergence) aloft. In either case, the result is general subsidence in the zone between 20 and 35 degrees latitude. This subsiding air is relatively dry, for it has released its moisture near the equator. In addition, the effect of adiabatic heating during descent further reduces the relative humidity of the air. Consequently, this subsidence zone is the site of the world's tropical deserts. Near the center of this zone of descending air the winds are generally weak and variable. Because of this, the region was popularly called the *horse latitudes*. The name is believed to have been coined by Spanish sailors, who, while crossing the Atlantic, were sometimes becalmed in these waters and reportedly were forced to throw horses overboard when they could no longer water or feed them. From the center of the horse latitudes the surface flow splits into a poleward branch and an equatorward branch. The equatorward flow is deflected by the Coriolis force and forms the reliable *trade winds*. In the Northern Hemisphere the trades are from the northeast where they provided the sail power to explore the New World; in the Southern Hemisphere the trades are from the southeast. The trades from both hemispheres meet near the equator in a region that has a weak pressure gradient. This region is called the *doldrums*. Here, light winds and humid conditions provide the monotonous weather that is the basis for the expression "down in the doldrums."

In the three-cell model the circulation between 30 and 60 degrees latitude is just opposite that of the Hadley cell. The net surface flow is poleward, and because of the Coriolis effect, the winds have a strong westerly component. These *prevailing westerlies* were known to Benjamin Franklin, perhaps the first American weather forecaster, who noted that storms migrated eastward across the colonies. Franklin also observed that the westerlies were much more sporadic and unreliable than the trades for sail power. We now know that it is the migration of cyclones and anticyclones across the midlatitudes that disrupts the general westerly flow at the surface. Again, referring to the three-cell model in Figure 7-3, we can see that the flow aloft in the midlatitudes is equatorward, from which the Coriolis force would produce an east wind. Since World War II, however, numerous observations have indicated that a general westerly flow exists aloft in the midlatitudes, as well as at the surface. Consequently, the center cell in this model does not completely fit observations. Because of this complication and the importance of the midlatitude circulation in maintaining the earth's heat balance, we will consider the westerlies in more detail in a later section.

Relatively little is known about the circulation in the high latitudes. It is generally believed that subsidence near the poles produces a surface flow that moves equatorward and is deflected into the *polar easterlies* of both hemispheres. As these cold polar winds move equatorward, they eventually encounter the warmer westerly flow of the midlatitudes. The region where these contrasting flows clash has been named the *polar front*. This important region will be considered in future discussions.

observed distribution of surface pressure and winds

As we would expect, the planetary circulation considered earlier is accompanied by a compatible distribution of surface pressure. Let us now consider the relationship between the mean surface winds and this pressure distribution. To simplify the discussion, we will first examine the idealized pressure distribution that would be expected if the earth's surface were uniform. Under this restriction, four latitudinally oriented belts of either high or low pressure emerge (Figure 7-4). Near the equator the converging air from both hemispheres is associated with the pressure zone known as the *equatorial low*, a region marked by abundant precipitation. Because this is the region where the trade winds meet, it is also referred to as the *intertropical convergence zone* (ITC). In the belts about 20 to 35 degrees on either side of the equator, where the westerlies and trade winds originate and go their separate ways, are located the *subtropical high*-pressure zones. These are regions of subsidence and divergent flow. Yet another surface pressure region is situ-

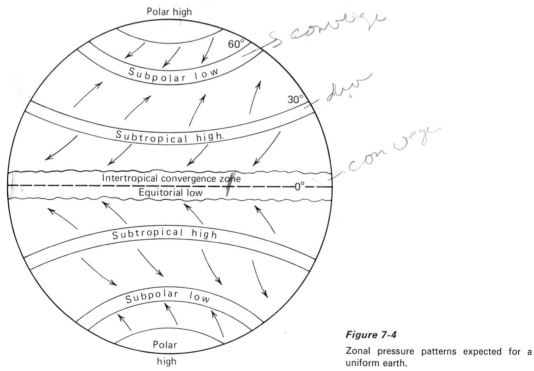

Figure 7-4

Zonal pressure patterns expected for a uniform earth.

152

ated at about 50 to 60 degrees latitude in a position corresponding to the polar front. Here the polar easterlies and westerlies meet to form a convergent zone known as the *subpolar low*. Finally, near the earth's poleward extremes, are the *polar highs* from which the polar easterlies originate.

Up to this point we have considered the surface pressure systems to be continuous belts around the earth. However, the only true zonal distribution of pressure exists in the region of the subpolar low in the Southern Hemisphere where the ocean is continuous. To a lesser extent the equatorial low is also zonal. At other latitudes, particularly in the Northern Hemisphere, where the bulk of the land exists, large seasonal temperature differences disrupt the idealized zonal pattern. Figure 7-5 shows the resulting mean pressure distribution for July and January. Notice on these pressure maps that, for the most part, the observed pressure regimes are cellular (or elongated) instead of zonal. The most prominent features on both maps are the subtropical highs. These systems are centered between 20 and 35 degrees latitude over all the larger oceans. It should also be noted that the subtropical highs are situated toward the east side of these oceans, particularly in the North and South Pacific and the North Atlantic. This fact greatly affects the west coast climates of the adjacent continents. When we compare Figures 7-5(*a*) and 7-5(*b*) we see that some of the pressure cells are more or less permanent features, like the subtropical highs, and can be seen on both the January and July charts. Others, however, are seasonal, such as the low situated over the southwestern United States in July, which can be seen on only one map. The seasonal change in pressure is much more noticeable in the Northern Hemisphere than in the Southern Hemisphere.

In the Southern Hemisphere relatively little pressure variation occurs from midsummer to midwinter, a fact we attribute to the dominance of water in that hemisphere. The most noticeable changes here are the seasonal 5-to 10-degree shifts of the subtropical highs that move with the position of the sun's vertical rays.

In the Northern Hemisphere numerous departures from the idealized zonal pattern shown in Figure 7-4 are evident. The main cause of these variations is the seasonal temperature fluctuations experienced over the land masses, especially those in the middle and higher latitudes. In January we find a very strong high pressure center situated over frigid Siberia and a somewhat weaker high pressure system existing over the chilled North American continent. These cold anticyclones are consistent with the greater density of this frigid air during this season. As these highs develop over the land, a weakening is observed in the subtropical anticyclones over the northern oceans. Although these subtropical highs do retain their identity during the winter, the circulation over the oceans is dominated by two intense cell-like cyclones, the *Aleutian* and *Icelandic lows*. These two semipermanent lows are statistical averages that represent the great number of cyclonic storms that migrate eastward across the globe and converge in these areas.

In the summer high surface temperatures over the continents generate lows that replace the winter highs. These are thermal lows in which warm

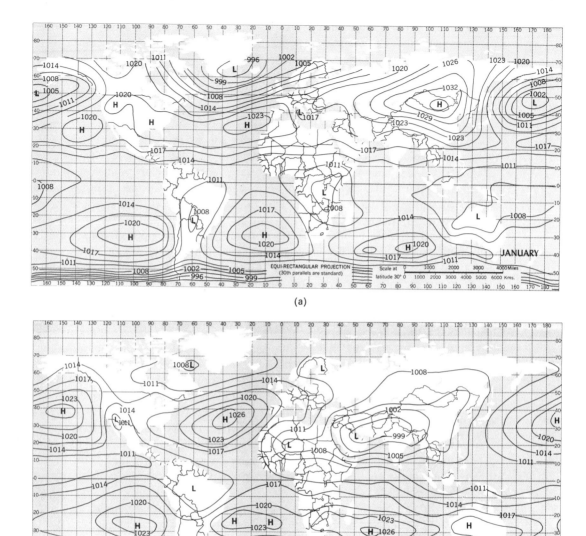

Figure 7-5

Average surface barometric pressure in millibars for (*a*) January and (*b*) July. (From Glenn T. Trewartha, *An Introduction to Climate,* © 1968, by McGraw-Hill Book Company, reprinted by permission of the publisher.)

outflow aloft induces inflow at the surface. The strongest of these low-pressure centers develops in northern India. It is this system, along with the northward migration of the ITC, which creates the well-known summer monsoon that is experienced in India. A weak thermal low is also generated

154

in the southwestern United States. In addition, notice in Figure 7-5 that during the summer months the subtropical highs in the Northern Hemisphere are more intense than during the winter months. These large pressure systems dominate the summer circulation over the oceans and aid the general influx of air onto the continents during that season.

Having examined the distribution of pressure in summer and winter, let us now turn our attention to the airflow associated with these pressure patterns. Figure 7-6 is a schematic representation showing the gross features of the surface circulation in July and January. Although these diagrams are highly idealized, they do depict the mean flow, and when they are used with Figure 7-5, they are useful in studying the effects of seasonal changes in pressure on the flow pattern. These charts clearly show the dominance of the subtropical anticyclones in generating the surface flow. Also notice the seasonal change in the circulation surrounding these subtropical highs.

Still another important seasonal change evident in Figure 7-6 is the migration of the intertropical convergence zone, which is indicated on these diagrams by a dashed line. Along with the migration of the ITC, we can detect the well-known monsoon circulation of Asia. In January the strong *Siberian high* produces flow off the Asian continent and across the equator toward the ITC. This outflow of cool, dry continental air produces the dry winter monsoon for much of South and Southeast Asia. With the onset of summer, the low that develops over northern India draws warm moist air from the adjacent oceans into Asia, producing the wet summer monsoon. Many other regions around the globe experience a somewhat similar seasonal wind shift caused by the seasonal fluctuation in the global pressure regimes. In fact, if you examine the flow over the United States, you will notice a much greater surge of warm moist air from the Gulf of Mexico during the summer when the high-pressure centers over the North Atlantic are most intense. Although the seasonal wind shift over the United States is sometimes referred of as a monsoon, it is not nearly as abrupt or dramatic a change as that experienced in Asia where even the most unobservant person cannot help but notice the change.

the westerlies

Prior to World War II upper-air observations were scarce. Since then data collected from aircraft and a network of radiosonde stations have aided in filling this gap. One of the most important pieces of information discovered was that at most latitudes, except near the equator where the Coriolis force is weak, the airflow in the middle and upper troposphere is westerly.

why westerlies?

Let us consider the reason for the predominance of westerly flow aloft. Recall from the gas law that cold air is more dense than warm air. Because of this, air pressure decreases more rapidly in a column of cool air than in a

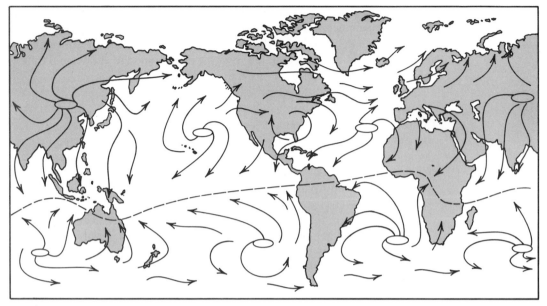

January

(a)

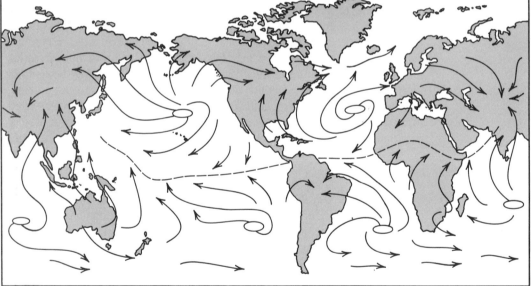

July

(b)

Figure 7-6

Schematic showing the gross features of the global circulation for (*a*) January and (*b*) July.

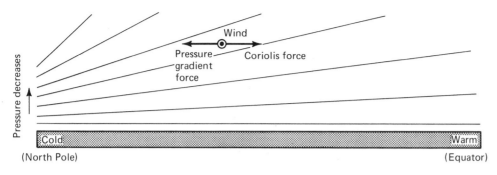

Figure 7-7

Cross section showing the northward-directed pressure gradient that is respon-sible for generating the westerlies of the middle latitudes.

column of warm air. Figure 7-7 shows the resulting pressure distribution with height. This figure is a vertical cross section through the Northern Hemisphere. The warm equatorial region is on the right side of the drawing and the cold polar region is on the left. The lines represent the normal pressure distribution with height and, to simplify matters, we have assumed the same surface pressure at all latitudes. Note that over the equator, where temperatures are highest, the pressure decreases more slowly than over the colder polar regions. Consequently, high pressure exists over the tropics and low pressure exists above the poles. Thus, the resulting pressure gradient is directed from south to north. Adding the effect of the Coriolis deflection, which directly opposes the pressure gradient force, we generate a wind with a strong westerly component. That is, a wind that is 90 degrees to the right of the northward-directed pressure gradient force. In our diagram the flow is into the page. Because the equator to pole temperature gradient shown in Figure 7-7 is typical over the globe, a westerly flow aloft should be expected, and on most occasions it is observed. It can also be seen that the pressure gradient increases with altitude and as a consequence so should wind speeds. This increase in wind speed only continues to the tropopause. Here the temperature gradients are reversed (warmer over the poles); hence, the west-erlies reach a maximum at the top of the troposphere and decrease upward into the stratosphere.

jet streams

It should be evident from our previous discussion that greater temperature contrasts at the surface will produce greater pressure gradients aloft, and hence faster upper-air winds as well. In the winter it would not be unusual to have a warm balmy day in southern Florida and near freezing temperatures in Georgia, only a few hundred kilometers to the north. Because of large wintertime temperature contrasts in the midlatitudes, like the one just men-tioned, we would expect faster westerly flow at that time of year. Observa-

157

tions substantiate such expections. In the flow directly above these regions of large temperature contrasts, very fast currents of air do indeed exist. These fast streams of air have been considered analogous to jets of water, and thus they are called *jet streams*.

Although predicted earlier, the existence of jet streams was first dramatically illustrated during World War II. United States bombers heading westward toward Japanese-occupied islands sometimes made little or no headway. Upon abandoning their missions the planes experienced tailwinds that sometimes exceeded 300 kilometers per hour during the return flight. Even today commercial aircraft use these strong tail winds to increase their ground speed when making eastward flights across the United States. Naturally, on westward flights attempts are made to fly away from these fast currents of air. On the average, these rather narrow bands of air have speeds of 125 kilometers per hour in the winter and roughly half that speed in the summer.

The cause of jet streams was alluded to in the previous section. Recall that large temperature contrasts over a short distance produce strong pressure gradients and strong winds. These large temperature contrasts occur along regions called fronts. In the midlatitudes a jet is found in association with the polar front. Remember that the surface position of the polar front is situated between the cool polar easterlies and warm westerlies. Because other jet streams have been discovered, this midlatitude jet is called the *polar-front jet*. Instead of flowing in a nearly straight west-to-east manner, the polar-front jet often meanders wildly, on occasions flowing almost due north and south. Sometimes it splits into two jets which may or may not rejoin. Like the polar front, this jet is not continuous around the globe. Because of the great variability of the path of the polar-front jet, it does not appear on charts depicting the mean circulation.

Several other jet streams are known to exist, but none has been studied in as much detail as the polar-front jet. An almost permanent jet exists over the subtropics centered at 25 degrees latitude. This westerly jet is located about 13 kilometers above the surface in the region where the Hadley cell and the middle latitude cell meet in Figure 7-3. The convergence of cool and warm air at this location generates a front aloft and hence the appropriate conditions for the development of a jet are met. There also exists a jet that flows from east to west at about 15 degrees north latitude over the region occupied by the Himalayas. This jet exists only in the summer when the ITC has shifted northward. Still other jets have been discovered above the troposphere. One such jet circles the Arctic Circle during the cold polar night at 25 to 30 kilometers above the surface.

waves in the westerlies

The west-to-east jet streams found in the upper troposphere are part of the total westerlies and together these winds follow the same course. In fact, the jet streams can more accurately be described as the fast cores of the overall westerly flow. Studies of upper-level wind charts reveal that the westerlies follow wavy paths that have rather long wavelengths. Much of our knowledge of these large-scale motions is attributed to C. G. Rossby who first explained the nature of these waves. The longest wave patterns (called *Rossby waves*) have wavelengths of 4000 to 6000 kilometers so that three to six waves will fit around the earth. Although the air flows eastward along this wavy path, these long waves tend to remain in the same position or move very slowly. In addition to Rossby waves, shorter waves also occur in the middle and upper troposphere. These shorter waves are often associated with cyclones at the surface and, like these storms, the waves travel eastward around the globe at rates of up to 15 degrees per day.

To get a better understanding of this wavy flow, let us examine the pattern on an upper-level weather chart. Figure 7-8 is a simplified 500-millibar height contour chart which is provided daily by the National Weather Service. Like most upper-air charts, this one shows the height at which a given pressure (500 millibars) is found, rather than giving the pressure at a given height, as do surface maps. But do not let that bother you, for there is a simple relationship between height contours and pressure. Places where the same pressure occurs at higher altitudes are experiencing higher pressures than places where the height contours indicate lower altitudes. Thus, higher contour readings indicate higher pressures and vice versa. Notice in Figure 7-8 that the height of the 500-millibar level decreases poleward as well. This fact is in agreement

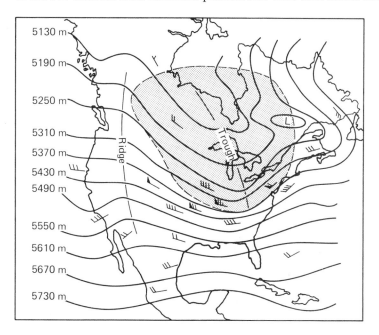

Figure 7-8

Simplified 500-millibar height contour chart. Note the position of the trough (low) and ridge (high). The position of the jet stream core is found by reading the wind speeds using Appendix B.

with the pressure gradient we concluded must exist in the upper troposphere in order to produce westerly winds there. Recall that the winds aloft are nearly geostrophic. Hence, the flow is nearly parallel to the contours, having wind speeds proportional to the contour spacing. Notice the large sweeping wavy pattern of the contour lines, which generally outline the flow pattern as well. Although this chart is well below the altitude of the jet stream core, the position of the jet at 500 millibars can be estimated from the indicated wind speeds.

Now let us return to the wind's function of transporting heat from the equator toward the poles. In Chapter 2 we showed that the equator receives more solar radiation than it radiates to space, while the poles experience the reverse situation. Thus, the equator has excess heat while the poles experience a deficit. Although the flow near the equator is somewhat meridional (north to south), at most other latitudes the flow is zonal (west to east). The reason for the zonal flow, as we have seen, is the Coriolis force. The question we will now consider is "How can wind with a west-to-east flow transfer heat from south to north?"

In Figure 7-8 the shaded area represents cold air which is bounded by the polar front to the south. We can also see that the polar front is displaced with the wavy flow of the jet. It should be remembered, however, that the surface winds and temperature gradients will be somewhat different from those aloft. Notice in Figure 7-8 that where the jet bends equatorward an elongate low or *trough* is produced that allows colder air to move southward. Conversely, a poleward bend in the jet produces a *ridge* of high pressure that draws warmer air poleward. Also notice that along one limb of a meander warm air is directed poleward and along the other limb of the meander cold air is directed equatorward. Thus, the wave flow of the westerlies provides an important mechanism for heat transfer across the midlatitudes. In addition, vortices (cyclones and anticyclones) help in this redistribution of energy. Imagine the counterclockwise circulation around a cyclone in the Northern Hemisphere; the east side draws warm air northward while the west side pulls cool air southward. Although the amount of air carried southward is roughly equal to the amount of air carried northward, more heat is transported northward than is transported southward.

Laboratory experiments using rotating fluids to duplicate the earth's circulation support the existence of waves and eddies to carry on the task of heat transfer in the middle latitudes. In these studies, called *dishpan experiments*, a large circular pan is heated around the outer edge to represent the equator while the center is cooled to duplicate the poles. Colored particles are added so that the flow can be easily observed and photographed. When the pan is heated, but not rotated, a very simple convection cell forms to redistribute the heat. This cell is similar to the Hadley cell we considered earlier. However, when the pan is rotated, the simple circulation breaks down and the flow develops a wavy pattern with eddies imbedded between meanders, as seen in Figure 7-9(a). These experiments indicate that changing the rate of rotation as well as varying the temperature gradient largely determines the

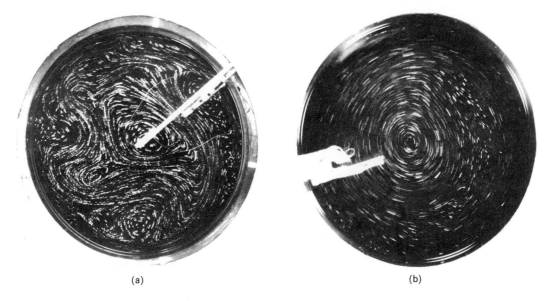

(a)   (b)

Figure 7-9

Photographs obtained from a "dishpan" experiment simulating (*a*) wavy and
(*b*) flat flow of the general circulation. (Courtesy of D. H. Fultz, University of
Chicago Hydrodynamics Laboratory.)

flow pattern produced. These studies have added greatly to our understanding
of global circulation. It is now known that the wavy flow aloft largely determines surface pressure patterns. During periods when the wavy flow aloft is
relatively flat (small amplitude waves), very little cyclonic activity is generated
at the surface. By contrast, when the flow acquires large amplitude waves
having short wavelengths, vigorous cyclonic storms are experienced. This
important relationship between the flow aloft and cyclonic storms will be
considered in more detail in Chapter 9.

local winds

Having examined the large macroscale circulation for the earth, let us now
turn our attention to some mesoscale winds. Remember that all winds are
produced for the same reason: temperature differences that arise because of
unequal heating of the earth's surface. In turn, these differences generate
pressure differences. Local winds are simply small-scale winds produced by a
locally generated pressure gradient. Although many winds are given local
names, some are actually part of the general circulation just described. For
example, the "norther" of Texas is a cold southward flow produced by the
circulation around anticyclones that invade the United States from Canada
in the winter. Since these are not locally generated winds, they cannot be considered true local winds. Others, like those described in the following discussion, are truly mesoscale and are caused either by topographic effects or
variations in surface composition in the immediate area.

land and sea breezes

The daily temperature contrast between the land and the sea, and the pressure pattern that generates a *sea breeze*, was discussed in the preceding chapter. Recall that the land is heated more intensely during the daylight hours than is an adjacent body of water. As a result, the air above the land surface heats and expands, creating an area of low pressure. A sea breeze then develops, blowing cooler air off the water and onto the land. At night, the reverse may take place; the land cools more rapidly than the sea and a *land breeze* develops.

The sea breeze has a significant moderating influence on the temperatures in coastal areas. Shortly after the breeze begins land temperatures may drop by as much as 5°C to 10°C. However, the cooling effect of these breezes generally reaches a maximum of only 100 kilometers inland in the tropics and often only half that distance in the middle latitudes. These cool sea breezes generally begin shortly before noon and reach their greatest intensity, about 10 to 20 kilometers per hour, in the midafternoon. Smaller-scale sea breezes can also develop along the shores of large lakes. People who live in a city near the Great Lakes, such as Chicago, recognize the lake effect, especially in the summer. These people are reminded daily by weather reports of the cool temperatures near the lake as compared to warmer outlying areas. In many places sea breezes also affect the amount of cloud cover and rainfall (Figure 7-10). The peninsula of Florida, for example, experiences a summer precipitation maximum that is caused partly by the convergence of sea breezes from both the Atlantic and Gulf coasts.

Figure 7-10

Sea breeze moving from the Atlantic onto the peninsula of Florida is partly responsible for the cloudy conditions over the land as shown in this Gemini V photograph. (Courtesy of NASA.)

The scale of land and sea breezes depends upon the location and the time of year. Tropical areas where intense solar heating is continuous throughout the year experience more frequent and stronger sea breezes than do midlatitude locations. The most intense sea breezes develop along tropical coastlines adjacent to cool ocean currents. In the middle latitudes sea breezes are most common during the warmest months of the year, but their counterpart, the land breeze, is often missing since the land does not always cool below the ocean temperature. In the higher middle latitudes the frequent migration of pressure systems generally dominates the circulation; hence, the land and sea breezes are less noticeable in these areas.

mountain and valley breezes

A diurnal, or daily, wind similar to land and sea breezes occurs in most mountainous regions. Here, during the daylight hours the air along the slopes of the mountains is heated more intensely than the air at the same elevation over the valley floor. This warm air glides up along the slope and generates a *valley breeze*. The occurrence of these daytime upslope breezes can often be identified by the cumulus clouds that develop over adjacent mountain peaks. The late afternoon thundershowers so common on warm summer days in the mountains can also be attributed to this phenomenon. After sunset the pattern is reversed. Rapid radiation heat loss along the mountain slopes results in cool air drainage into the valley below and causes the so-called *mountain breeze*. The same type of cool air drainage can occur in regions that have very little slope. The result is that the coldest pockets of air are usually found in the lowest spots. Consequently, low areas are the first to experience radiation fog and are also the most likely spots for frost damage to crops.

Like many other winds, mountain and valley breezes have seasonal preferences. Although valley breezes are most common during the warm season when solar heating is most intense, mountain breezes tend to be more dominant in the cold season.

chinook (foehn) winds

Warm, dry winds sometimes move down the east slopes of the Rockies, where they are called *chinooks*, and the Alps, where they are called *foehns*. Such winds are often created when a pressure system such as a cyclone is situated on the leeward side of the mountains where it pulls air over these imposing barriers. As the air descends the leeward slopes of the mountain, it is heated substantially by compression. Since condensation may have occurred as the air ascended the windward side releasing latent heat, the air descending the leeward side will be warmer and drier than it was at a similar elevation on the windward side. Although the temperature of these winds is generally less than 10°C, which is not particularly high, they generally occur in the winter and spring when the affected area may be experiencing below freezing temperatures. Thus, by comparison, these dry, warm winds often bring a drastic change. Within a few minutes after the arrival of a chinook, the temperature will usually climb 20°C. When the ground has a snow cover, these winds are known to melt it in short order. The word chinook literally means "snow-eater."

The chinook is viewed by some as beneficial to ranchers east of the Rockies, for their grasslands are kept clear of snow during much of the winter, but this benefit is offset by the loss of moisture that the snow would bequeath to the land had it remained until the spring melt.

Another chinooklike wind that occurs in the United States is the *Santa Ana*. Found in southern California, these hot desiccating winds greatly increase the threat of fire in this already dry area.

katabatic winds

In the winter season areas adjacent to highlands may experience a local wind called a *katabatic* or *fall wind*. These winds originate when cold air, situated over a highland area such as the ice sheets of Greenland or Antarctica, is set in motion. Under the influence of gravity, the cold air cascades over the rim of a highland like a waterfall. Although the air is heated adiabatically, as are chinooks, the initial temperatures are so low that the wind arrives in the lowlands still colder and more dense than the air it displaces. In fact, it is a requirement that this air be colder than the air it invades since it is the air's greater density that causes it to descend. Occasionally, as this frigid air descends, it is channeled into narrow valleys where it acquires very high velocities capable of great destruction.

A few of the better known katabatic winds have been given local names. The most famous is the *mistral* that blows from the Alps over France toward the Mediterranean Sea. Another is the *bora* that originates in the mountains of Yugoslavia and blows to the Adriatic Sea.

global distribution of precipitation

Even a casual glance at Figure 7-11(*a*) reveals the complex nature of the global distribution of precipitation. Nevertheless, the gross features of the earth's precipitation regimes shown in this figure can be explained by using our knowledge of the global wind and pressure systems. In general, regions influenced by high pressure, with its associated subsidence and divergent winds, will experience rather dry conditions. On the other hand, regions under the influence of low pressure and its converging winds and ascending air will receive ample precipitation. If the wind–pressure regimes were the only control of precipitation, the pattern shown in Figure 7-11(*a*) would be much simpler. However, the inherent nature of the air involved is also an important factor in determining the potential for precipitation. Since cold air has a very low capacity for moisture as compared to warm air, we would expect a latitudinal variation in precipitation, with low latitudes receiving the greatest amounts of precipitation and high latitudes receiving the least amounts. An

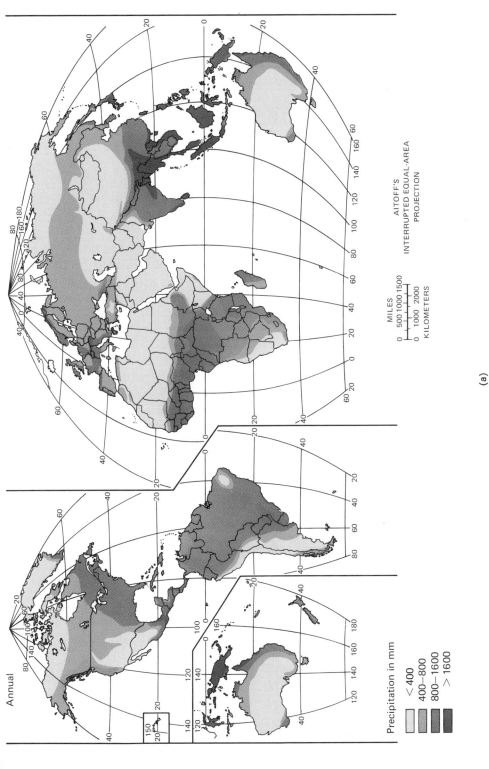

Annual

Precipitation in mm

　　< 400
　400—800
　800—1600
　　> 1600

MILES
0　500 1000 1500
0　1000　2000
KILOMETERS

AITOFF'S
INTERRUPTED EQUAL-AREA
PROJECTION

(a)

Figure 7-11

Global distribution of precipitation. (*a*) Annual, (*b*) July, and (*c*) January. (From Joseph E. Van Riper, *Man's Physical World.* © 1971 by McGraw-Hill Book Company, reprinted by permission of the publisher.)

Lutgens/Tarbuck, Fig. 7-11

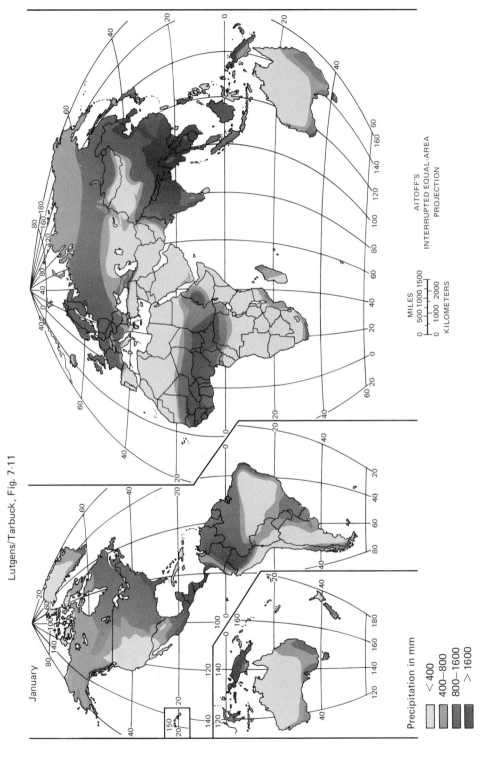

January

AITOFF'S
INTERRUPTED EQUAL-AREA
PROJECTION

MILES
500 1000 1500
0   1000   2000
KILOMETERS

Precipitation in mm
< 400
400–800
800–1600
> 1600

(b)

Lutgens/Tarbuck, Fig. 7-11

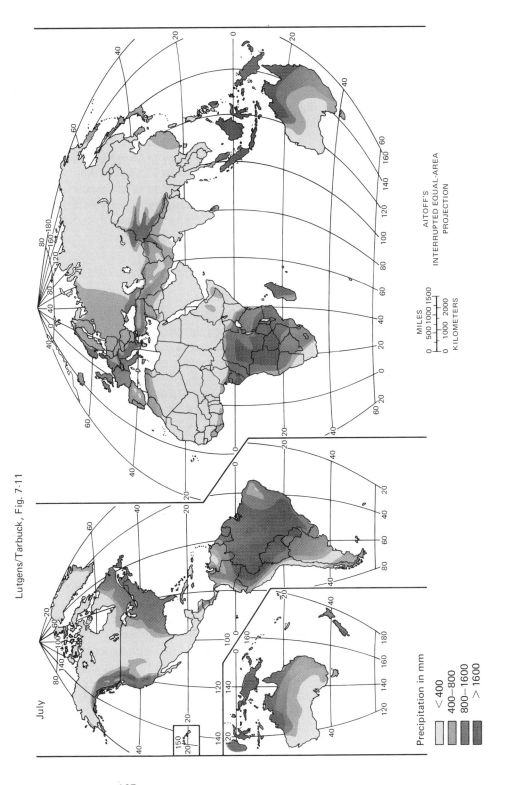

July

Precipitation in mm

< 400
400—800
800—1600
> 1600

MILES
0   500 1000 1500
0   1000 2000
KILOMETERS

AITOFF'S
INTERRUPTED EQUAL-AREA
PROJECTION

(c)

examination of Figure 7-11(*a*) does indeed reveal heavy precipitation in equatorial regions and meager precipitation near the land masses poleward of 60 degrees latitude. However, a noticeably arid region is also found in the subtropics. This situation can be explained by examining the global wind–pressure regimes.

In addition to the latitudinal variations in precipitation, the distribution of land and water complicates the precipitation pattern found over the earth. Large land masses in the middle latitudes generally experience decreased precipitation toward their interiors. For example, central North America and central Eurasia receive considerably less precipitation than coastal regions at the same latitude. Further, the effects of mountain barriers alter the idealized precipitation regimes we would expect from the global wind systems. Windward mountain slopes receive abundant precipitation while leeward slopes and adjacent lowlands are usually deficient in moisture.

zonal distribution of precipitation

Let us first examine the zonal distribution of precipitation that we would expect on a uniform earth and then add the variations caused by the influences of land and water. Recall from our earlier discussion that on a uniform earth four major pressure zones emerge in each hemisphere. These zones include the equatorial low, the subtropical high, the subpolar low, and the polar high. Also, remember that these pressure belts show a marked seasonal shift toward the summer hemisphere. The idealized precipitation regimes expected from these pressure systems are shown in Figure 7-12 in which we can see that the equatorial regime is centered over the equatorial low throughout most of the year. Hence, in this region where the trade winds converge (ITC), heavy precipitation is experienced in all seasons. Poleward of the equatorial low in each hemisphere lies the belt of subtropical high pressure. In these regions subsidence contributes to the rather dry conditions found here throughout the year. Between the wet equatorial regime and the dry subtropical regime lies a zone that is influenced by both pressure systems. Because the pressure systems migrate with the sun, these transitional regions receive most of their precipitation in the summer when they are under the influence of the ITC. They experience a dry season in the winter when the subtropical high moves equatorward.

The midlatitudes receive most of their precipitation from traveling cyclones that frequent this region and generate the subpolar low. This region is also the site of the polar front, the convergent zone between cold polar air and the warmer westerlies. It is along the polar front that cyclones often form (Figure 7-13). Since the position of the polar front migrates freely between approximately 30 and 70 degrees latitude, most midlatitude areas receive ample precipitation. However, the mean position of this zone also moves with the sun on a seasonal basis, such that a narrow belt between 30 and 40 degrees latitude experiences a marked seasonal fluctuation in precipitation. In

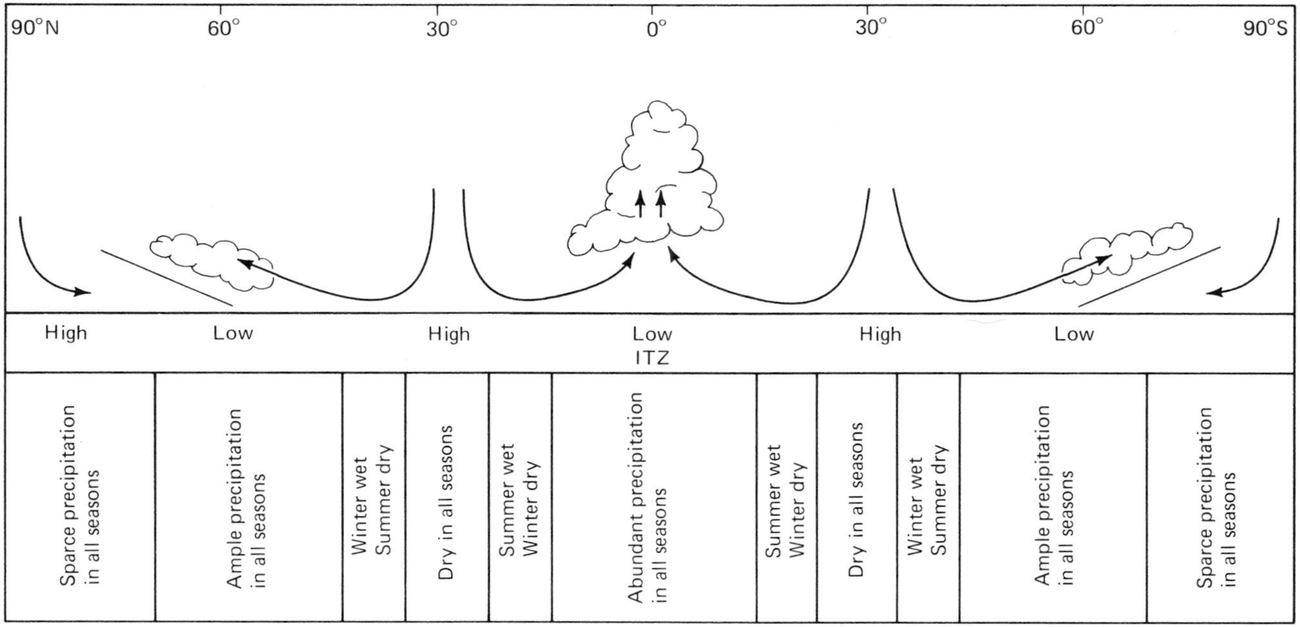

Figure 7-12
Schematic illustration of zonal precipitation patterns.

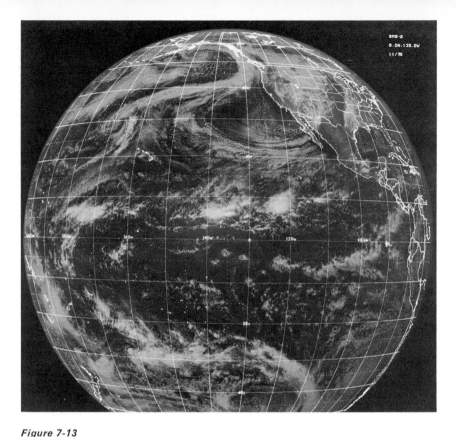

Figure 7-13

Satellite photographs of the earth's cloud cover reveal the global wind systems. The curved cloud patterns found poleward of 30 degrees latitude in both hemispheres are associated with traveling cyclones. The ITC over the Pacific is visible as a band of clouds located at 5 degrees north latitude. (Courtesy of NOAA.)

the winter this zone experiences numerous cyclones with associated precipitation as the position of the polar front moves equatorward. During the summer, however, the circulation of this region is dominated by subsidence associated with the subtropical high, and hence rather dry conditions prevail. Compare the January and July precipitation data for the west coast of North America by referring to Figure 7-11(*b*) and (*c*).

The polar regions are dominated by cold air that holds very little moisture. Throughout the year this region experiences only meager amounts of precipitation. Even in the summer when temperatures rise, these areas of ice and snow are dominated by high pressure that blocks the movement of the few cyclones that do travel poleward.

distribution of precipitation over the continents

The zonal pattern outlined in the previous section roughly approximates the major aspects of global precipitation. Abundant precipitation is found in the equatorial and midlatitude regions, while substantial portions of the subtropics and polar areas are relatively dry. However, numerous exceptions to this idealized zonal pattern can also be found in Figure 7-11(*a*). For example, several arid areas are found in the midlatitudes. The desert of Patagonia, located along the southeast sector of South America, is one example. For the most part, midlatitude deserts like that of Patagonia are found on the leeward (rain shadow) side of a mountain barrier or in the interior of a continent cut off from a source of moisture. Most of the other departures from the true zonal scheme result from the effects of the distribution of continents and oceans on the global circulation.

The most notable breakdown in the zonal distribution of precipitation occurs in the subtropics. Here we find not only many of the world's largest deserts but also regions of abundant rainfall. This pattern results because the high-pressure centers that dominate the circulation in these latitudes have different characteristics on their eastern and western flanks. On the eastern side of these oceanic highs subsidence is most pronounced and a strong temperature inversion is encountered very near the surface and results in stable atmospheric conditions. The upwelling of cold water along the west coasts of the adjacent continents cools the air from below and adds to the stability on the east sides of these highs. Since these anticyclones tend to crowd the east side of an ocean, we find that the west sides of the continents adjacent to these subtropical highs are very arid. For example, centered at approximately 25 degrees latitude north or south on the west side of their respective continents, we find the Sahara Desert of North Africa, the Namib of Southwest Africa, the Atacama of South America, the deserts of the Baja Peninsula of North America, and the Great Desert of Australia. On the western side of these highs, however, subsidence is less pronounced and convergence with associated rising air appears to be more prevalent. In addition, as this air travels over a large expanse of warm water, it acquires moisture through evaporation which acts to enhance its instability. Consequently, the eastern regions of subtropical continents generally receive abundant precipitation all year round. Southern Florida is a good example.

When we consider the influence of land and water on the distribution of precipitation, the general pattern illustrated in Figure 7-14 emerges. This figure is a highly idealized precipitation scheme for a hypothetical land mass in the Northern Hemishere. By rotating this diagram 180 degrees around the line representing the equator, the pattern for the Southern Hemisphere is established. Although this diagram contains the most salient features of the global precipitation pattern, frequent reference should also be made to Figure 7-11 as you read the following discussion. To make this diagram more useful, Figure 7-15 provides precipitation data for locations in selected precipitation regimes.

First, notice in Figure 7-14 that the precipitation regime for the western section of the hypothetical continent is the same as the zonal pattern depicted earlier for a uniform earth. The precipitation data for San Francisco and Mazatlán illustrate the marked seasonal variations found in regimes 3 and 5, respectively. Recall that it is the seasonal migration of the pressure systems

171

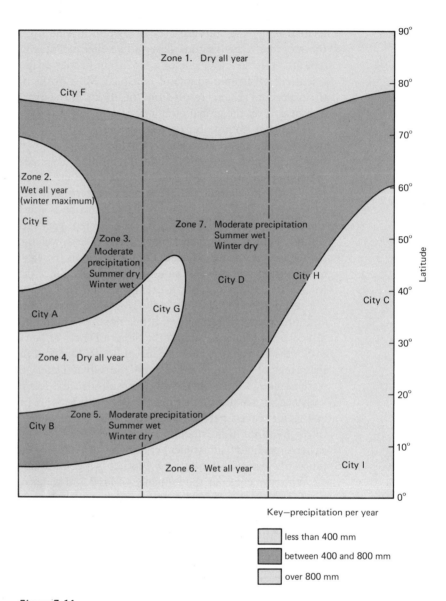

Figure 7-14
Idealized precipitation regimes for a hypothetical land mass in the Northern Hemisphere. Refer also to Figure 7-15.

that causes these fluctuations. Second, notice that precipitation regime 2 shows a strong maximum in the winter, as illustrated by the data for Juneau. This seasonal variation results because cyclonic storms are more prevalent in the winter.

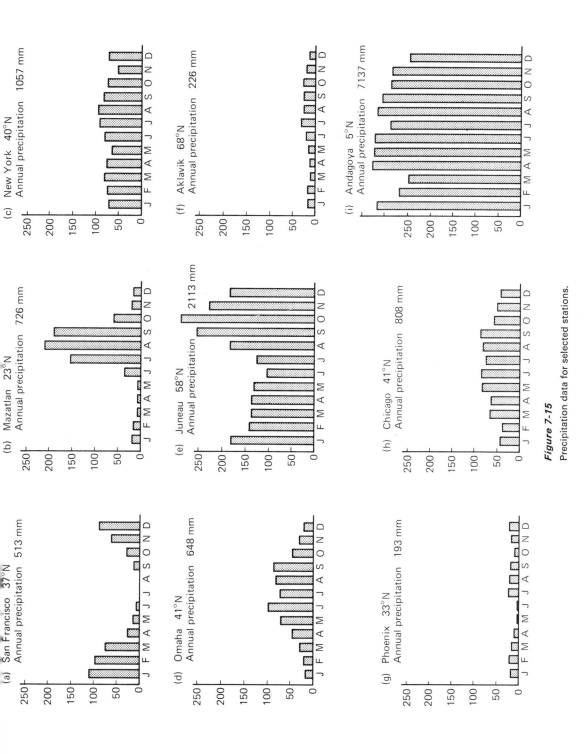

Figure 7-15

Precipitation data for selected stations.

The eastern segment of the continent shows a marked contrast to the zonal pattern described earlier. Only the dry polar regime is similar in size and position to the western seaboard. The most noticeable departure from the zonal pattern is the absence of the arid subtropical region on the east coast. As was stated earlier, this difference is caused by the predominance of convergence and lifting on the west side of the oceanic anticyclones and subsidence on the east side. Also notice that as one moves inland from the east coast, a decrease in precipitation is experienced. This, however, does not hold true for mountainous regions; even relatively small ranges like the Appalachians are able to extract more than their share of precipitation.

Another variation is seen with latitude. As we move poleward along the eastern section, we note a decrease in precipitation. This is the decrease you would expect with cooler air temperatures and corresponding lower moisture capacities. But we do not find a similar decrease in precipitation in the higher middle latitudes along the west coast. Since this region lies in the westerly wind belt, the west coasts of midlatitude regions such as Canada and Norway receive abundant moisture. In fact, some of the rainiest regions on earth are found in such settings. By contrast, the east coasts of continents in the mid-latitudes experience temperature and moisture regimes more typical of continental locations. This is especially true in the winter when the flow is off the land. Let us now consider the interior of the continents which shows a somewhat different precipitation pattern, especially in the middle latitudes. Since cyclones are more common in the winter, we would expect that all midlatitude locations would also experience maximum precipitation then. However, the precipitation regimes of the interior are affected to a degree by the so-called monsoon circulation that offsets this. The cold winter temperatures and dominance of flow off the land in the winter makes this the dry season. This holds true, for example, for our own Great Plains. Although cyclonic storms frequent these regions in the winter, they do not become precipitation producers until they have drawn in abundant moist air from the Gulf of Mexico. Thus, most of the winter precipitation falls in the eastern one-third of the United States. The lack of winter precipitation in the Plains states as compared to the East can be seen in the data for Omaha and New York City. In the summer the inflow of warm moist air from the ocean is aided by the thermal low that develops over the land and causes a general increase in precipitation over the midcontinent. The dominance of summer monsoon-type precipitation is very evident when we compare the patterns for India in summer and winter by using Figure 7-11(*b*) and (*c*). In the United States areas in the southeastern portion of the midlatitudes begin their rainy season in the early spring, but more northerly and westerly locations do not reach their peak precipitation periods until summer or early fall.

REVIEW   **1.** Distinguish between macroscale, mesoscale, and microscale winds. Give an example of each.

2. If you were to view a weather map of the entire world for any single day of the year, would it be likely that the global pattern of winds would be visible? Explain your answer.

3. Briefly describe the idealized global circulation as it was proposed by George Hadley. Did subsequent observations confirm Hadley's proposal?

4. What factors are believed to cause air to subside in the latitude zone 20 degrees to 35 degrees?

5. In the idealized three-cell model of atmospheric circulation most of the United States is situated in which belt of prevailing winds?

6. Briefly explain each of the following statements that relate to the global distribution of surface pressure:
   a. The only true zonal distribution of pressure exists in the region of the subpolar low in the Southern Hemisphere.
   b. The subtropical highs are more continuous in the winter hemisphere.
   c. The subpolar low in the Northern Hemisphere is represented by individual oceanic cells that are strongly seasonal in character.
   d. There is a strong high-pressure cell over Eurasia in the winter.

7. Why is the flow aloft predominantly westerly?

8. At what time of year should we expect the fastest westerly flow? Explain.

9. Describe the situation in which jet streams were first encountered.

10. Describe the manner in which pressure distribution is shown on upper-air charts. How are high-and low-pressure areas depicted on these charts?

11. What were the "dishpan experiments" and what have we learned from them?

12. Why is the well-known Texas "norther" not a true local (mesoscale) wind?

13. The most intense sea breezes develop along tropical coasts adjacent to cool ocean currents. Explain.

14. What are katabatic (fall) winds? Name two examples.

15. Relying only on your knowledge of the zonal distribution of precipitation, describe the seasonal pattern of precipitation at the following representative locations: equator, 15 degrees north latitude, Tropic of Cancer, 40 degrees north latitude, 80 degrees north latitude.

16. What factors, other than global wind and pressure systems, exert an influence upon the world distribution of precipitation?

VOCABULARY REVIEW

macroscale circulation	prevailing westerlies
synoptic scale	polar easterlies
mesoscale winds	polar front
microscale winds	equatorial low
convection cell	intertropical convergence zone (ITC)
Hadley cell	subtropical high
horse latitudes	subpolar low
trade winds	polar high
doldrums	Aleutian low

global circulation

Icelandic low

Siberian high

jet stream

polar-front jet

Rossby waves

trough

ridge

dishpan experiments

sea breeze

land breeze

valley breeze

mountain breeze

chinook

foehn

Santa Ana

katabatic winds

fall winds

mistral

bora

Chapter Eight

air masses

Most of us living in the middle latitudes have experienced hot, "sticky" summer heat waves and/or frigid winter cold waves. In the first case, after several days of sultry weather the spell may come to a dramatic end that is marked by thundershowers and is followed by several days of cool relief. In the second case, thick stratus clouds and snow may replace the clear skies that had prevailed and temperatures may climb to values that seem mild by comparison to what had preceded them. In both examples, what was experienced was a period of generally uniform weather conditions followed by a relatively short period of change and the subsequent reestablishment of a new set of weather conditions that remains for perhaps several days before changing again.

The weather patterns just described are the result of the movements of very large bodies of air, called air masses. An *air mass*, as the term implies, is an immense body of air, usually 1600 kilometers or more across, and perhaps several kilometers thick, which is characterized by homogeneous physical properties (in particular temperature and moisture content) at any given altitude. When this air moves out of its region of origin, it will carry these temperatures and moisture conditions elsewhere, eventually affecting a large portion of a continent.

The horizontal uniformity of an air mass is not complete because it may extend through 20 degrees or more of latitude and cover hundreds of thousands to millions of square kilometers. Consequently, small differences in temperature and humidity from one point to another at the same level are to

177

be expected. However, the differences observed within an air mass are small in comparison to the rapid rates of change that are experienced across the boundaries between air masses. Since it may take several days for an air mass to traverse an area, the region under its influence will probably experience generally constant weather conditions, a situation called *air-mass weather*. Certainly, there may be some day-to-day variations, but the events will be very unlike those in an adjacent air mass.

The air mass concept is an important one because it is closely related to the study of atmospheric disturbances. Most of the disturbances in the middle latitudes originate along the boundary zones that separate different air masses.

source regions

Where do air masses form? What factors determine the nature and degree of uniformity of an air mass? These two basic questions are very closely related, for where an air mass forms vitally affects the properties that characterize it.

Areas where air masses originate are termed *source regions*. Since the atmosphere is heated chiefly from below and gains its moisture by evaporation from the earth's surface, the nature of the source region largely determines the initial characteristics of an air mass. An ideal source region must meet two essential criteria. First, it must be an extensive and physically uniform area. A region having highly irregular topography or one that has a surface consisting of both water and land is not satisfactory. The second criterion is that the area be characterized by a general stagnation of atmospheric circulation so that air will stay over the region long enough to come to some measure of equilibrium with the surface. In general, this means regions dominated by stationary or slow-moving anticyclones with their extensive areas of calms or light winds. Regions under the influence of cyclones are not likely to produce air masses because such systems are characterized by converging surface winds. The winds in lows are constantly bringing air with unlike temperature and humidity properties into the area. Since the time involved is not long enough to eliminate these differences, steep temperature gradients result, and air-mass formation is precluded.

Figure 8-1 shows the source regions that produce the air masses that most often influence North America. The waters of the Gulf of Mexico and Caribbean Sea and similar regions in the Pacific west of Mexico yield warm air masses, as does the land area consisting of the southwestern United States and northern Mexico. In contrast, the North Pacific, the North Atlantic, and the snow and ice-covered areas comprising northern North America and the adjacent Arctic Ocean are major source regions for cold air masses.

Notice that the major source regions are not found in the middle latitudes but instead are confined to tropical and polar locations. Since the middle latitudes are the site where cold and warm air masses clash, often because the converging winds of a traveling cyclonic system draws them together, this zone lacks the homogeneous conditions that are necessary for a source

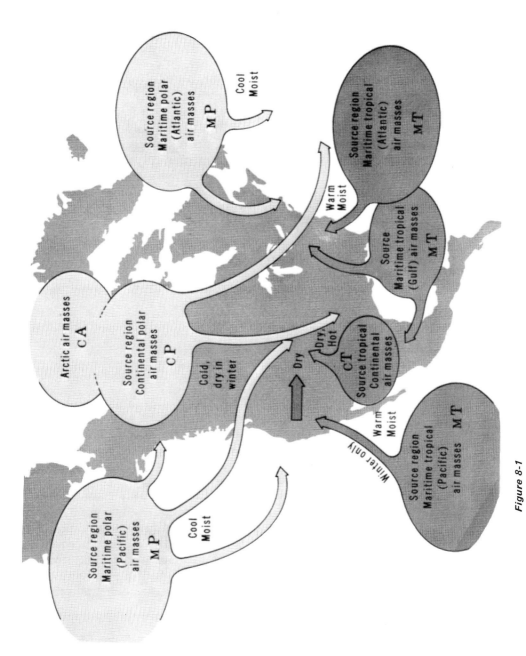

Figure 8-1

Air-mass source regions for North America. (Courtesy of Ward's Natural Science Establishment, Inc., Rochester, N.Y.)

region. Almost every air mass found in the middle latitudes had its source elsewhere.

classifying air masses

The classification of an air mass depends upon the latitude of the source region and the nature of the surface in the area of origin—ocean or continent. The latitude of the source region gives an indication of the temperature conditions within the air mass, and the nature of the surface below strongly influences the moisture content of the air.

Air masses are identified by two-letter codes. With reference to latitude (temperature), air masses are placed into one of four categories: *polar* (P), *arctic* (A), *tropical* (T), and *equatorial* (E). The differences between polar and arctic and between tropical and equatorial are generally very small and simply serve to indicate the degree of coldness or warmness of the respective air masses. The lower-case letter m (for maritime) or the lower-case letter c (for continental) is used to designate the nature of the surface in the source region and hence the humidity characteristics of the air mass. *Maritime air masses* have a high water vapor content indicating that condensation is likely. *Continental air masses* have a relatively low water vapor content suggesting that the condensation will be meager.

When this classification scheme is applied, the following air masses may be identified:

cA	continental	arctic
cP	continental	polar
cT	continental	tropical
mT	maritime	tropical
mP	maritime	polar
mE	maritime	equatorial

Notice that the list includes neither mA (maritime arctic) nor cE (continental equatorial). These air masses are not listed because they seldom, if ever, form. Although arctic air masses form over the Arctic Ocean, this water body is ice-covered throughout the year. Consequently, the air masses that originate here consistently have the moisture characteristics associated with a continental source region. The region of the trade winds, however, produces warm and humid air masses almost exclusively. Continental equatorial air masses do not generally form because this area is dominated by the oceans. In the latitude belt that includes the region 10 degrees on either side of the equator, more than 75 percent of the surface is ocean. Furthermore, the land areas adjacent to the equator are warm and wet tropical rain forests. Hence, if an air mass were to originate over such locales, its moisture content would be relatively high.

air-mass modification

After an air mass forms it normally migrates from the area where it acquired its distinctive properties to a region that has different surface characteristics. Once the air mass moves from its source region, it not only modifies the weather of the area it is traversing, but it is also gradually modified by the

surface over which it is moving. Warming or cooling from below, the addition or subtraction of moisture, and vertical movements all act to bring
about changes in an air mass. The amount of modification may be relatively
small, or as the following example illustrates, the changes may be profound
enough to completely alter the original identity of the air mass.

When cA or cP air moves over the ocean in winter, it undergoes considerable change. Evaporation from the water surface rapidly transfers large
quantities of moisture to the once dry continental air. In addition, since the
underlying water is warmer than the air above, the air is also heated from
below. This leads to instability and vertically ascending currents that rapidly
transport heat and moisture to higher levels. In a matter of days, cold, dry,
and stable continental air is transformed into an unstable mP air mass.

When an air mass is colder than the surface over which it is passing, as
was the case in the preceding example, the lower-case letter k is added after
the air-mass symbol. If, however, an air mass is warmer than the underlying
surface, the lower-case letter w is added. It should be remembered that the k
or w suffix does not mean that the air mass itself is cold or warm. It means
only that the air is relatively cold or warm in comparison to the underlying
surface over which it is traveling. For example, an mT air mass from the
Gulf of Mexico is usually classified as mTk as it moves over the southeastern
states in summer. Although the air mass is warm, it is still cooler than the
highly heated continent over which it is passing.

The k or w designation gives an indication of the stability of an air mass,
and hence the weather that might be expected. An air mass that is colder than
the surface is obviously going to be warmed in its lower layers. This results
in an increased lapse rate and greater instability which favor the ascent of the
heated lower air and create the possibility of cloud formation and precipitation. Indeed, a k air mass is often characterized by cumuliform clouds and,
should precipitation occur, it will be of the shower or thunderstorm variety.
Further, visibility is generally good (except in rain) because of the stirring
and overturning of the air. Conversely, when an air mass is warmer than the
surface over which it is moving, its lower layers are chilled. A surface inversion often develops which increases the stability of the air mass. This condition does not favor the ascent of air, and hence it opposes cloud formation
and precipitation. Any clouds that do form will be stratiform and precipitation, if any, will be drizzle or light rain. Further, because of the lack of vertical
movements, smoke and dust often become concentrated in the lower layers
of the air mass and cause poor visibility. During certain times of the year
fogs, especially of the advection type, may also be common in some regions.

In addition to modifications resulting from temperature differences
between an air mass and the surface below, vertical movements induced by
cyclones and anticyclones or topography may also affect the stability of an
air mass. Such modifications are often called mechanical or dynamic and are
generally independent of the changes caused by surface cooling or heating.
For example, significant modification can result when an air mass is drawn
into a low. Here convergence and lifting dominate and the air mass is

181

rendered more unstable. Conversely, the subsidence associated with anti-cyclones acts to stabilize an air mass. Similar alterations in stability occur when an air mass is lifted over highlands or descends the leeward side of a mountain barrier. In the first case, the air's stability is reduced; in the second case, the air becomes more stable.

properties of north american air masses

Since air masses are continually passing over us, the day-to-day weather that we experience depends primarily upon the temperature, stability, and mois-ture content of these large bodies of air. In the following section we will briefly examine the properties of the principal North American air masses. In addition, Table 8-1 serves as a summary.

continental polar (cP) and continental arctic (cA) air masses

Continental polar and continental arctic air masses are, as their classifica-tion implies, cold and dry. Continental polar air originates over the snow-covered interior regions of Canada and Alaska, poleward of the fiftieth parallel; continental arctic air forms farther north over the Arctic basin and the Greenland ice cap (Figure 8-1). Continental arctic air is distinguished from cP air by its generally lower temperatures, although at times the differ-ences may be slight. In fact, some meteorologists do not differentiate between cP and cA.

During the winter both air masses are bitterly cold and very dry. Since the winter nights are long, the earth radiates heat that is, for the most part, not replenished by incoming solar energy. Hence, because of the prolonged earth radiation, the surface reaches very low temperatures and the air near the ground is gradually chilled to heights of perhaps 1 kilometer or more. The result is a strong and persistent temperature inversion with the lowest temperatures near the ground. Marked stability is therefore the rule. Since the air is very cold and the surface below is frozen, the specific humidity of these air masses is of necessity very low, ranging from perhaps 0.1 gram per kilogram in cA up to 1.5 gram per kilogram in some cP air.

As wintertime cP or cA air moves outward from its source region, it car-ries its cold, dry weather to the United States, generally entering between the Great Lakes and the Rockies. Because there are no major barriers to their movement between their source regions and the Gulf of Mexico, cP and cA air masses can sweep rapidly and with relative ease far southward into the United States. Consequently, the winter cold waves experienced in much of the central and eastern United States are closely associated with such polar outbreaks. One such cold wave, which was among the most intense and widespread in this century, is depicted on the weather map in Figure 8-2. Commonly, the last freeze in spring and the first in autumn can be correlated with outbreaks of polar or arctic air.

Although cP air masses are not usually associated with heavy precipita-tion, those that cross the Great Lakes often bring snow to the leeward shores.

182

Table 8-1

Weather Characteristics of North American Air Masses

Air Mass	Source Region	Temperature and Moisture Characteristics in Source Region	Stability in Source Region	Associated Weather
cA	Arctic basin and Greenland ice cap	Bitterly cold and very dry in winter	Stable	Cold waves in winter
cP	Interior Canada and Alaska	—Very cold and dry in winter —Cool and dry in summer	Stable entire year	a. Cold waves in winter b. Modified to cPk in winter over Great Lakes bringing "lake-effect" snow to leeward shores
mP	North Pacific	Mild (cool) and humid entire year	—Unstable in winter —Stable in summer	a. Low clouds and showers in winter b. Heavy orographic precipitation on windward side of western mountains in winter c. Low stratus and fog along coast in summer; modified to cP inland
mP	Northwestern Atlantic	—Cold and humid in winter —Cool and humid in summer	—Unstable in winter —Stable in summer	a. Occasional "northeaster" in winter b. Occasional periods of clear, cool weather in summer
cT	Northern interior Mexico and southwestern U.S. (summer only)	Hot and dry	Unstable	a. Hot, dry, and clear, rarely influencing areas outside source region b. Occasional drought to southern Great Plains
mT	Gulf of Mexico, Caribbean Sea, western Atlantic	Warm and humid entire year	Unstable entire year	a. In winter it usually becomes mTw moving northward and brings occasional widespread precipitation or advection fog b. In summer, hot and humid conditions, frequent cumulus development and showers or thunderstorms
mT	Subtropical Pacific	Warm and humid entire year	Stable entire year	a. In winter it brings fog, drizzle, and occasional moderate precipitation to N.W. Mexico and S.W. United States b. In summer this air mass occasionally reaches the western United States and is a source of moisture for infrequent convectional thunderstorms

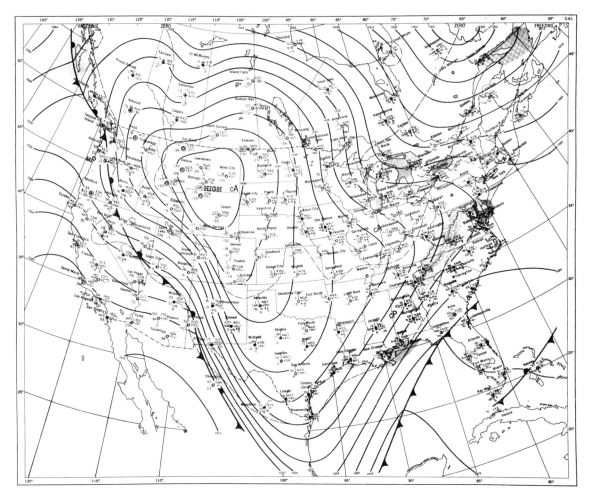

Figure 8-2

An intense winter cold wave depicted on the January 10, 1962 surface weather map. This outbreak of continental arctic air brought subfreezing temperatures as far south as the Gulf of Mexico. As shown here, these frigid air masses are generally associated with high pressure in the winter. (Courtesy of NOAA.)

As Figure 8-3 illustrates, during its journey the air acquires heat and moisture from the relatively warm lake surface. By the time it reaches the southern and eastern shore, this cPk air is very moist and unstable and heavy *lake-effect snow* is characteristic. As the modified cP air continues toward the rugged terrain of the Appalachians, the forced ascent often yields a general overcast and locally heavy snow on the western side of the mountains. However, to the east of the Appalachians the clouds dissipate and the snow ends because of the latent heat released during condensation and the adiabatic warming that occurs as the air descends the leeward slopes of the mountains.

Since cA air is present principally in the winter, only cP air has any influence upon our summer weather and this effect is considerably reduced when

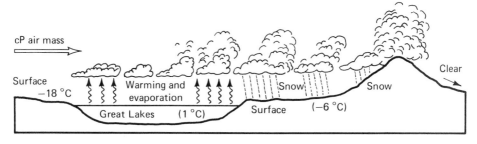

Figure 8-3

As continental polar air crosses the Great Lakes in winter, it acquires moisture and is made less stable because of warming from below. "Lake-effect" snow showers on the lee side of the lakes is often the consequence of this air-mass modification.

compared to winter. During the summer months the properties of the source region for cP air are very different from those during winter. Instead of being chilled by the ground, the air is warmed from below as the long days and higher sun angle warm the snow-free land surface. Although summer cP air is warmer and has a higher moisture content than its wintertime counterpart, the air is still cool and relatively dry in comparison to areas farther south. In the northern portions of the eastern and central United States summer heat waves are often ended by the southward advance of cP air, which for a day or two brings cooling relief and bright pleasant weather.

maritime polar (mP) air masses

Maritime polar air masses form over oceans at high latitudes. As the classification would indicate, mP air is cool to cold and it is humid, but when it is compared to cP and cA air masses in winter, mP air is relatively mild because of the higher temperatures of the ocean surface as contrasted to the colder continents. Two regions are important sources for mP air that influences the United States: the North Pacific and the northwestern Atlantic from Newfoundland to Cape Cod (Figure 8-1). Because of the general west-to-east circulation in the middle latitudes, mP air masses from the North Pacific source region have a much more profound influence on American weather than mP air masses generated in the northwestern Atlantic. While air masses that form in the Atlantic generally move eastward toward Europe, mP air from the North Pacific largely controls the weather along the West Coast of North America, especially in the winter.

During the winter mP air masses from the Pacific usually begin as cP air in Siberia. Although air rarely stagnates over this area, the source region is extensive enough to allow the air moving across it to acquire its characteristic properties. As the air moves eastward over the relatively warm water, there is active evaporation and heating in the lower levels. Consequently, what was once a very cold, dry, and stable air mass is changed into one that is mild and humid near the surface and relatively unstable. As this mP air arrives at

185

the West Coast of North America, it is often accompanied by low clouds and shower activity. When the mP air advances inland against the western mountains, orographic uplift produces heavy rain or snow on the windward slopes of the mountains.

As is the case for mP air from the Pacific, air masses forming in the northwestern Atlantic source region were originally cP air masses that moved from the continent and were transformed over the ocean. However, unlike the air masses from the North Pacific, mP air from the Atlantic seldom affects the weather of North America. Nevertheless, on occasion the general west-to-east flow is temporarily reversed, for example, when the northeastern United States is on the northern or northwestern edge of a passing low. When this occurs, the cyclonic winds draw mP air into the region. Its influence is generally confined to the area east of the Appalachians and north of Cape Hatteras. The weather associated with an invasion of mP air from the Atlantic is known locally as a *northeaster*. Strong northeast winds, freezing or near freezing temperatures, high relative humidity, and the possibility of precipitation make this weather phenomenon a most unwelcome event.

During the summer months the influence of mP air masses from the North Pacific on American weather changes. At this time of year the water is cooler than the surrounding continents. In addition, the Pacific anticyclone lies off the West Coast of the United States (Figure 7-5). Consequently, there is an almost continuous southward flow of air having moderate temperatures. Although the air near the surface may often be conditionally unstable, the presence of the Pacific high means that there is subsidence and stability aloft. Hence, low stratus clouds and summer fogs characterize much of the West Coast. Once summer mP air from the Pacific moves inland, it is heated at the surface over the hot and dry interior. The heating and resulting turbulence act to reduce the relative humidity in the lower layers and the clouds dissipate. By the time it has traversed the Rocky Mountains, mP air cannot be distinguished from a cP air mass.

While mP air masses from the Atlantic may produce an occasional unwelcome "northeaster" during the winter, summertime incursions of this air mass often bring very plesant weather. Like the Pacific source region, the northwestern Atlantic is dominated by high pressure during the summer (Figure 7-5). Hence, the upper air is stable because of subsidence and the lower air is essentially stable because of the chilling effect of the cold water. As the circulation on the south side of the anticyclone carries this stable and relatively dry mP air into New England, and occasionally as far south as Virginia, the region enjoys clear, cool weather and good visibility.

maritime tropical (mT) air masses

Maritime tropical air masses affecting North America most often originate over the warm waters of the Gulf of Mexico, the Caribbean Sea, or the adjacent western Atlantic Ocean (Figure 8-1). The tropical Pacific is also a source region for mT air. However, the land area affected by this latter source is very small in comparison to the size of the region on the continent influenced by air masses produced in the Gulf of Mexico and adjacent waters.

As expected, mT air masses are warm to hot and they are humid. In addition, they are often unstable. It is through invasions of mT air masses that the subtropics export much heat and moisture to the less endowed areas to the north. Consequently, these air masses are important to the weather whenever they are present because they are capable of releasing considerable precipitation.

Maritime tropical air masses from the Gulf–Caribbean–Atlantic source region greatly affect the weather of the United States east of the Rocky Mountains. Although the source region is dominated by the North Atlantic subtropical high, the air masses it produces are not stable but are neutral or unstable. The reason lies in the fact that the source region is located on the weak western edge of the anticyclone where pronounced subsidence is absent.

During the winter when cP air dominates in the central and eastern United States mT air does not often enter this part of the country. When an invasion does occur, the lower portions of the air mass are chilled and stabilized as it moves northward. Its classification is changed to mTw. As a result, the formation of convective showers is precluded. Widespread precipitation does occur, however, when northward moving mT is pulled into a traveling cyclone and forced to ascend. In fact, much of the wintertime precipitation over the eastern and central states results when mT air from the Gulf is lifted along fronts in traveling cyclones. When northward moving mT air does not become part of a cyclonic storm, the chilling of the air mass by the cold ground may produce dense advection fogs.

During the summer mT air masses from the Gulf, Caribbean, and adjacent Atlantic cover a much wider area of North America and are present for a much greater percentage of the time than during the winter. As a result, they exert a strong and often dominating influence over the summer weather of the United States east of the Rocky Mountains. The reason is the general sea to land (monsoonal) airflow over the eastern portion of North America during the warm months which brings more frequent incursions of mT air that penetrate much deeper into the continent than during the winter months. Consequently, these air masses are largely responsible for the hot and humid conditions that prevail over the eastern and central United States.

Initially, summertime mT air from the Gulf is unstable. As it moves inland over the warmer land, it becomes an mTk air mass as daytime heating of the surface layers further increases the air's instability. Since the relative humidity is high, only slight lifting is necessary to bring about active convection, cumulus development, and thunderstorm or shower activity. This is, indeed, a common warm weather phenomenon associated with mT air.

It should also be noted here that air masses from the Gulf–Caribbean–Atlantic source region are the primary source of much, if not most, of the precipitation received in the eastern two-thirds of the United States. Pacific air masses contribute little to the water supply east of the Rockies because the western mountains effectively "drain" the moisture from the air by numerous episodes of orographic uplift. Figure 8-4, which shows the distribution of average annual precipitation for the eastern two-thirds of the United States by using *isohyets* (lines connecting places having equal rainfall), illustrates this nicely. Aside from the complexity of the pattern in mountainous areas, the pattern of isohyets shows the greatest rainfall in the Gulf region and a decrease in precipitation with increasing distance from the mT source region.

Compared with mT air from the Gulf of Mexico, mT air masses from the Pacific source region have much less of an impact upon North American weather. In winter only northwestern Mexico and the extreme southwestern United States are influenced by air from the tropical Pacific. Since the source region lies along the eastern side of the Pacific anticyclone, subsidence aloft produces upper-level stability. When the air mass moves northward, cooling at the surface also causes the lower layers to become more stable, often resulting in fog or drizzle. If lifted along a front or forced over mountains, moderate precipitation results.

For many years the summertime influence of air masses from the tropical Pacific source region upon the weather of the southwestern United States and northern Mexico was thought to be minimal. There was a widely held belief that the source of moisture for the infrequent summer thunderstorms that occur in the region could be traced to occasional westward thrusts of mT air from the Gulf of Mexico, but the Gulf of Mexico is no longer believed to be a significant supplier of moisture for the area west of the Continental Divide. Rather, it has been demonstrated that the tropical North Pacific west of central Mexico is a more important source of moisture for this area. In summer the mT air moves northward from its Pacific source region up the Gulf of California and into the interior of the western United States. This movement, which is confined largely to July and August, is essentially monsoonal in character. That is, the inflow of moist air is a response to the thermally produced low over the heated land mass.

continental tropical (cT) air masses

Since North America narrows as it extends southward through Mexico, the continent has no extensive source region for continental tropical air masses. Only in summer do northern interior Mexico and the adjacent parts of the arid southwestern United States produce hot, dry cT air. Because of the intense daytime heating at the surface, there is a steep lapse rate and turbulence to considerable heights. However, although the air is unstable, it generally remains clear because of the extremely low humidity. Consequently, the prevailing weather is hot, with an almost complete lack of rainfall. Large daily temperature ranges are the rule. Although cT air masses are generally confined to the source region, they occasionally move into the southern Great Plains. If the cT air persists for long, drought may be the result.

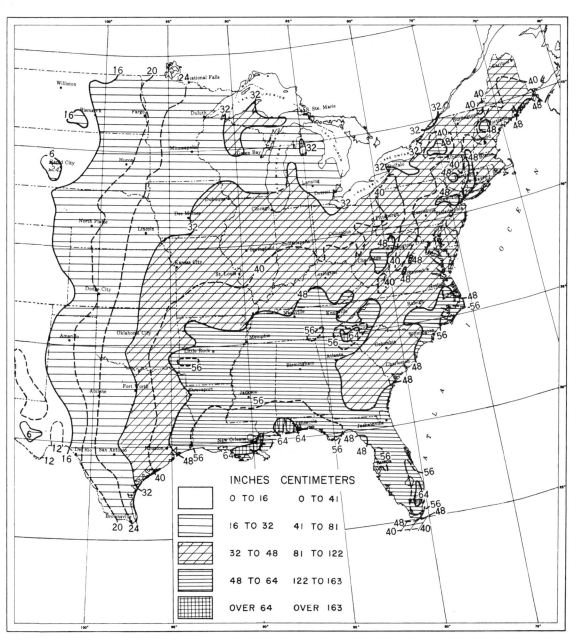

Figure 8-4

Average annual precipitation for the eastern two-thirds of the United States. Note the general decrease in yearly precipitation totals with increasing distance from the Gulf of Mexico, the source region for mT air masses, Isohyets are labeled in inches. (Courtesy of Environmental Data Service, NOAA.)

189

1. Define the terms air mass and air-mass weather.

2. For an area to be an air-mass source region, what two criteria must it meet?

3. Why are regions that have a cyclonic circulation generally not conducive to air-mass formation?

4. Why do air masses not usually form in the middle latitudes?

5. On what basis are air masses classified? Compare the temperature and moisture characteristics of the following air masses: cP, mP, mT, cT.

6. Why were mA and cE left out of the air-mass classification scheme?

7. What do the lower-case letters k and w indicate about an air mass? List the general weather conditions associated with k and w air masses.

8. How might vertical movements induced by a pressure system or topography act to modify an air mass?

9. What two air masses are most important to the weather of the United States east of the Rocky Mountains? Explain your choice.

10. What air mass dominates the weather of the Pacific Coast more than any other?

11. Why do cA and cP air masses often sweep so far south into the United States?

12. Describe all of the modifications that occur as a cP air mass traverses the Great Lakes and Appalachian Mountains.

13. Why do mP air masses from the North Atlantic source region seldom affect the eastern United States?

14. What air mass and source region provide the greatest amount of moisture to the eastern and central United States?

15. For each of the statements below indicate which air mass is most likely involved and from what source region it came:
a. summer drought in the southern Great Plains
b. wintertime advection fog in the Midwest
c. heavy winter precipitation in the western mountains
d. summertime convectional showers in the Midwest and East
e. a "northeaster"

VOCABULARY
REVIEW

air mass	equatorial (E) air mass
air-mass weather	maritime (m) air mass
source region	continental (c) air mass
polar (P) air mass	lake-effect snow
arctic (A) air mass	northeaster
tropical (T) air mass	isohyet

Chapter Nine

weATHer
PATTerns

In the preceding chapters we examined the basic elements of the atmosphere and the dynamics of the motion therein. Let us now consider these diverse phenomena in a unified framework that will serve as the basis for understanding the day-to-day weather in the middle latitudes. For our purposes, the middle latitudes refer to the region between southern Florida and Alaska. The primary weather producer here is called the *middle-latitude cyclone* or *wave cyclone*.

As early as the 1800s it was known that cyclones (lows) were the bearers of precipitation and severe weather. Thus, the barometer was established as the main tool in "forecasting" day-to-day weather changes. These early methods of weather prediction largely ignored the importance of air-mass interactions in the formation of cyclones. Consequently, it was not possible to determine the conditions under which cyclone development was favorable. The first encompassing model of the wave cyclone was constructed by a group of Norwegian scientists during World War I. At that time the Norwegians were cut off from weather reports, especially those from the Atlantic. To counter this deficiency, a tight network of weather stations was established throughout the country. In the years that followed several Norwegian-trained meteorologists made great advances in broadening man's understanding of the weather. Included in this early group were J. Bjerknes, V. Bjerknes, H. Solberg, and T. Bergeron. In 1918 J. Bjerknes published his theory of cyclone formation in an article entitled "On the Structure of Moving Cyclones." This theory, which became known as the *polar front theory*, provides us with a useful model of the wave cyclone. Although some changes had to be made in the original version because of subsequent discoveries, the main tenets of this theory remain an integral part of meteorological thought.

In the Norwegian model the wave cyclone develops in conjunction with a frontal zone known as the polar front. Recall that the polar front separates cold polar air from warm subtropical air. During the cool months the polar front is generally well defined and forms an essentially continuous band around the earth that can be recognized on upper-air charts. Near the surface this frontal zone is often broken into distinct segments that are separated by regions that experience a more gradual temperature change. Other fronts capable of generating cyclones have also been discovered. For example, it was learned that fronts can form between continental arctic and continental polar air masses, as well as between continental polar and maritime polar air masses. Since wave cyclones form in conjunction with fronts, we will first consider the nature of fronts and the weather associated with their movement and then apply what we have learned to the cyclone model.

fronts

Fronts are defined as boundary surfaces separating air masses of different densities, one warmer and often higher in moisture content than the other. Thus, ideally, fronts can form between any two contrasting air masses. Although small changes in density occur through a given air mass, these are small in comparison to those associated with frontal zones. Figure 9-1 shows the temperature distribution across a "typical" front. Considering the vast size of the air masses involved, these 15- to 200-kilometer wide bands of discontinuity are relatively narrow. On the scale of a weather map they are generally narrow enough to be satisfactorily represented by a broad line.

Above the ground the frontal surface slopes at a low angle so that warmer air overlies cooler air, as shown in Figure 9-1. In the ideal case, the air

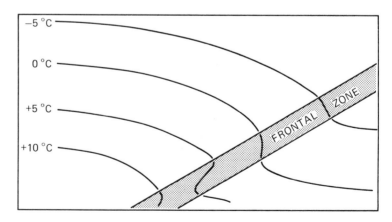

Figure 9-1

Temperature distribution across a frontal zone.

masses on both sides of the front move in the same direction and at the same speed. Under this condition, the front acts as a barrier with which the air masses must move but cannot penetrate. Generally, however, the pressure field across a front is such that air on one side is moving faster in the direction perpendicular to the front than the air mass on the other side of the front. Thus, one air mass actively advances into another and "clashes" with it. Because of this, the Norwegian meteorologists visualized these zones of interaction as being analogous to battle lines and tagged them fronts. As one air mass moves into another, some mixing does occur along the frontal surface, but for the most part, the air masses retain their identity as one air mass is displaced upward over the other. No matter which air mass is advancing, it is always the warmer, lighter air that is forced aloft, while the cooler, heavier air acts as the wedge upon which lifting takes place. The term *overrunning* is generally applied to warm air gliding up a retreating cold air mass.

warm fronts

When the surface position of a front moves so that warm air occupies territory formally covered by cooler air, it is called a *warm front*. On a weather map the surface position of a warm front is denoted by a line with semicircles extending into the cooler air. East of the Rockies warm tropical air often enters the United States from the Gulf of Mexico and overruns receding cool air. As the cold wedge retreats, friction slows the advance of the surface position of the front more so than its position aloft, so that the boundary separating these air masses acquires a small slope (Figure 9-2). The average

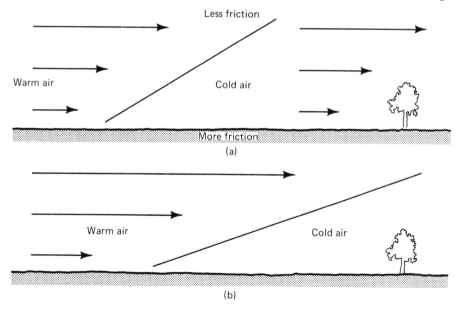

Figure 9-2
Changing slope of a warm front through time.

193

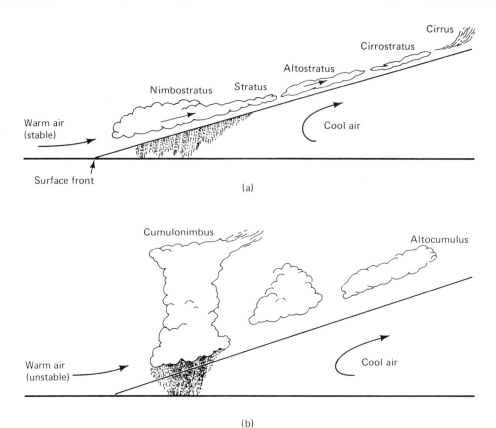

Warm air
(stable)

Cool air

Surface front

Nimbostratus Stratus Altostratus Cirrostratus Cirrus

(a)

Cumulonimbus Altocumulus

Warm air
(unstable)

Cool air

(b)

Figure 9-3

(a) Warm front with stable air and associated stratiform clouds. Precipitation is moderate and occurs within a few hundred kilometers of the surface front. (b) Warm front with unstable air and cumuliform clouds. Precipitation is heavy near the surface front.

slope of a warm front is about 1:200, which means that if you went a distance of 200 kilometers ahead of the surface location of a warm front, you would find the frontal surface at a height of 1 kilometer overhead.

As warm air ascends the retreating wedge of cold air, it cools by adiabatic expansion to produce clouds and frequently precipitation. Typically, the sequence of clouds shown in Figure 9-3 precedes a warm front. The first sign of the approach of a warm front is the appearance of cirrus clouds overhead. These high clouds form 1000 kilometers or more ahead of the surface front where the overrunning warm air has ascended high up the wedge of cold air. Another indication of the approach of a warm front is provided by aircraft contrails. On a clear day, when these condensation trails persist for several hours, you can be fairly certain that relatively warm moist air is ascending overhead. As the front nears, cirrus clouds grade into cirrostratus which blend into denser sheets of altostratus. About 300 kilometers ahead of the front thicker stratus and nimbostratus clouds appear and rain or snow commences. Usually, warm fronts produce several hours of moderate to gentle precipitation over a large region. This is in agreement with the

194

rather gentle slope of a warm front that does not generally encourage convective activity. However, warm fronts are occasionally associated with cumulonimbus clouds and violent thunderstorms. This results when the overrunning air is inherently unstable and the front is rather sharp. When these conditions exist, cirrus clouds are generally followed by cirrocumulus clouds giving us the familiar "mackerel sky" that warned sailors of an impending storm, as indicated by the following:

Mackerel scales and mares' tails
Make lofty ships carry low sails.

At the other extreme, a warm front associated with a rather dry air mass could pass unnoticed by those of us at the surface.

As we can see from Figure 9-3, the precipitation associated with a warm front occurs ahead of the surface position of the front. Some of the rain that falls through the cool air below evaporates. As a result, the air directly below the cloud base often becomes saturated and a stratus cloud deck develops. These clouds occasionally grow downward very rapidly, which can cause problems for pilots of small aircraft that require visual landings. One minute pilots may encounter visibility that is adequate and the next they find themselves in the middle of a cloud deck that has the landing strip "socked in." Further, we can see in Figure 9-1 that the temperature increases vertically through a front in the region where the front is situated near the surface. During the winter this temperature inversion is responsible for producing freezing rain and sleet in the sub-freezing air mass ahead of a warm front. This temperature distribution is also responsible for the general stability of the lower atmosphere in this region as is reflected by the stratiform clouds located there. You have probably frequently experienced these stable conditions with their associated fog and drizzle. Such weather is typical of a warm front during the cool season.

With the passage of a warm front, a rather gradual increase in temperature occurs. As you would expect, the increase is most apparent when a large temperature contrast exists between the adjacent air masses. Furthermore, a wind shift from the east to the southwest is generally noticeable. The reason for this shift will be evident later. The moisture content and stability of the encroaching warm air mass largely determine the time period required for clear skies to return. During the summer cumulus and occasionally cumulonimbus clouds will be embedded in the warm unstable air mass that follows the front. These clouds can produce precipitation, but it is usually randomly organized and restricted in extent.

cold fronts

When cold air is actively advancing into a region that is occupied by warmer air, the zone of discontinuity is called a *cold front*. As is the case with warm fronts, friction tends to slow the surface position of a cold front more so than its position aloft. However, because of the relative positions of the adjacent air masses, the cold front steepens as it moves (Figure 9-4). On the average, cold fronts are about twice as steep as warm fronts, having a slope of perhaps 1:100. Occasionally, the position of the cold front aloft may even move ahead of its surface location (Figure 9-4). When this happens, cold air will overlie warm air. This steepens the lapse rate along the front, thus greatly enhancing instability. In addition, cold fronts advance more rapidly than warm fronts. These two differences, rate of movement and steepness of slope, largely account for the more violent nature of cold front weather as compared to the weather generally accompanying a warm front. The displacement of air along a cold front is often rapid enough to permit the released latent heat to appreciably increase the air's buoyancy. This frequently

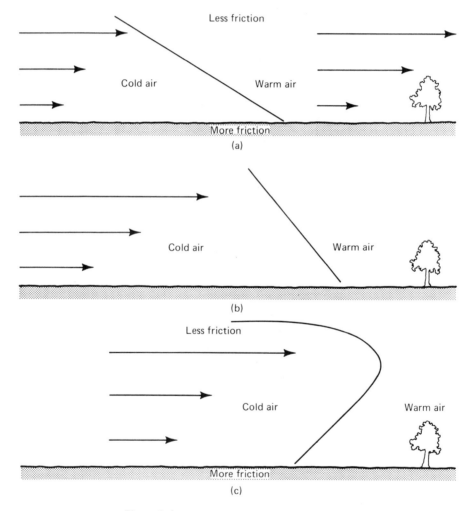

Figure 9-4
Changing slope of a cold front through time.

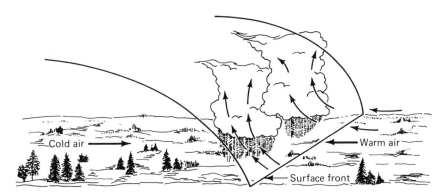

Figure 9-5

Fast-moving cold front and cumulonimbus clouds. Often thunderstorms occur if the warm air is unstable.

results in the sudden downpours and vigorous gusts of wind associated with mature cumulonimbus clouds (Figure 9-5). Since cold fronts produce roughly the same amount of lifting as a warm front but over a shorter distance, the intensity of precipitation is greater, but the duration is shorter.

The arrival of a cold front is sometimes preceded by altocumulus clouds. As the front approaches, generally from the west or northwest, towering clouds can often be seen in the distance. Near the front a dark band of ominous clouds foretells of the ensuing weather. A marked temperature drop and a wind shift from the south to west or northwest usually accompany frontal passage. The violent weather and the sharp temperature contrast along the cold front are indicated by the symbol used to depict them on a weather map. The symbol consists of a line with triangle-shaped points extending into the warmer air mass.

The weather behind a cold front is dominated by a subsiding and relatively cold air mass. Hence, clearing conditions prevail after the front passes. Although general subsidence causes some adiabatic heating, this has a minor effect on surface temperatures. In the winter the clear skies associated with these cold outbreaks further reduce surface temperatures because of more rapid radiation cooling at night. If the continental polar air mass, which most frequently accompanies a cold front, moves into a relatively warm and humid area, surface heating can produce shallow convection which, in turn, may generate low cumulus or stratocumulus clouds behind the front.

stationary fronts

Occasionally, the flow on both sides of a front is almost parallel to the position of the front. The surface position of the front does not move and it is therefore named a *stationary front*. On a weather map stationary fronts are shown with triangular points on one side of the front and semicircles on the other. Occasionally, some overrunning occurs along a stationary front and in this situation precipitation of the gentle warm front type is likely.

197

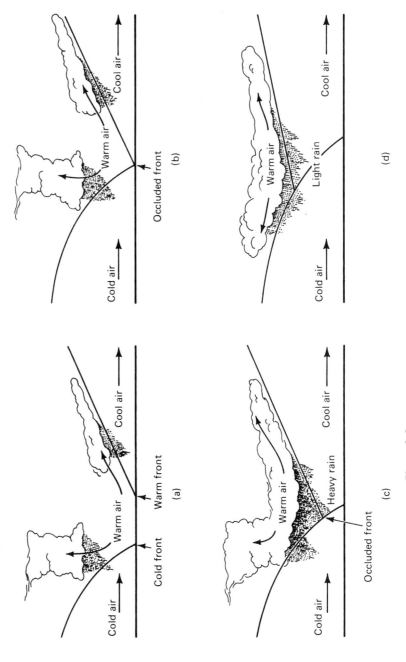

Figure 9-6

Stages in the formation and eventual dissipation of an occluded front.

occluded fronts

Another commonly occurring front is the *occluded front*. Here, an active cold front overtakes a warm front, as shown in Figure 9-6. As the advancing cold air wedges the warm front upward, a new front emerges between the advancing cold air and the air over which the warm front is gliding. The weather of an occluded front is generally very complex. Most of the precipitation is associated with the warm air being forced aloft, but when conditions are suitable, the newly formed front is capable of initiating precipitation of its own.

In the occluded front shown in Figure 9-7(*a*) the air behind the cold front is colder than the air underlying the warm front it is overtaking. This is the most common occluded front to form east of the Rockies and is termed a *cold-type occluded front*. It is also possible for the air behind the active cold front to be warmer than the air underlying the warm front. These *warm-type occluded fronts* frequently occur along the Pacific coast where somewhat milder maritime polar air invades polar air that had its origin over the continent. Notice in Figure 9-7(*a*) and (*b*) that in the warm-type occluded front the warm front aloft, and hence the precipitation, often precedes the arrival of the surface front but that the situation is reversed for the cold-type occluded front. It should also be noted that cold-type occluded fronts may very frequently resemble cold fronts in terms of the type of weather they generate.

A word of caution is in order concerning the weather associated with the various fronts. Although the preceding discussion should aid you in

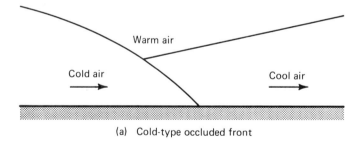

(a)   Cold-type occluded front

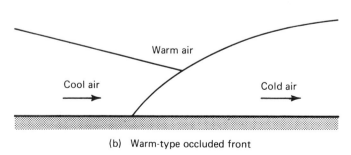

(b)   Warm-type occluded front

Figure 9-7

Occluded fronts of the (*a*) cold-type and (*b*) warm-type.

recognizing the weather patterns associated with fronts, you should remember that these descriptions are rather idealized. The weather generated along any individual front may or may not conform fully to our picture. Fronts, like all aspects of nature, never lend themselves to classification as nicely as we would like.

wave cyclone   The Norwegian model of the middle-latitude or wave cyclone was created primarily from near surface observations. As data from the middle and upper troposphere and from satellite photos became available, some modifications were necessary. Yet this model is still an accepted working tool used in the interpretation of the weather. The model provides a visual picture of the dynamic atmosphere as it generates a storm. If you keep this model in mind when you consider changes in the weather, the observed changes will no longer come as a surprise. You should begin to see some order in what had appeared to be disorder, and at best you might even occasionally "predict" the impending weather.

life cycle of a wave cyclone

According to the wave cyclone model, cyclones form along fronts and proceed through a somewhat predictable life cycle. This cycle can last for a few hours or for several days, depending upon whether or not conditions for development are favorable. Figure 9-8 is a schematic representation of the stages in the development of a "typical" wave cyclone. As the figure shows, cyclones originate along a front where air masses of different densities (temperatures) are moving parallel to the front in opposite directions. In the classic model this would be continental polar air associated with the polar easterlies north of the front and maritime tropical air of the westerlies south of the front. The result of this opposing air flow is the development of cyclonic shear, that is, a net counterclockwise rotation. To better visualize this effect, place a pencil between the palms of your hands. Now move your right hand ahead of your left hand and notice that your pencil rotates in a counterclockwise fashion. It is possible for cyclonic shear to develop in other ways that may also initiate a wave cyclone. In any event, under the correct conditions, the frontal surface will take on a wave shape. These waves are analogous to the waves that are produced on the surface of a water body by moving air, except that the scale is different. The waves generated between two contrasting air masses are usually several hundred kilometers long. Some waves tend to dampen out while others become unstable and grow in amplitude. The latter ones change in shape with time much like a gentle ocean swell does as it moves into shallow water and becomes a tall breaking wave (see Figure 9-8).

Once a small wave forms, warm air invades this weak spot along the front and extends itself poleward and the surrounding cold air moves equatorward. This change causes a readjustment in the pressure field that results

200

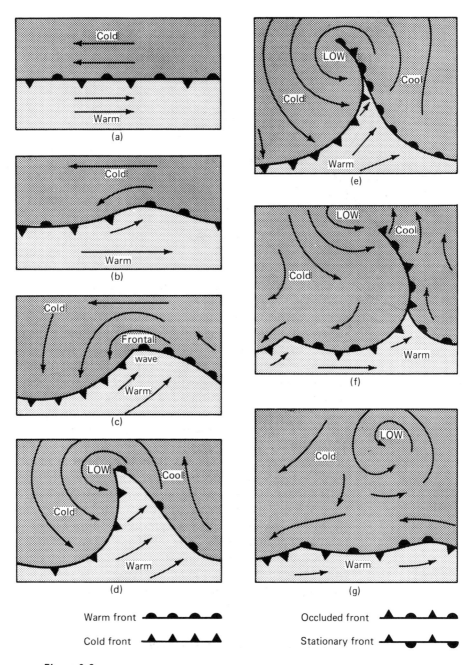

Figure 9-8
Stages in the life cycle of a middle-latitude cyclone as proposed by J. Bjerknes.

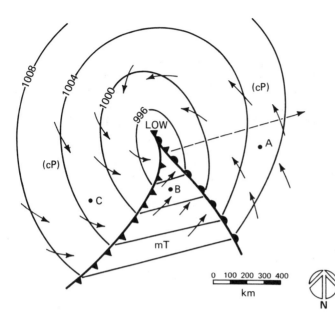

Figure 9-9

Idealized circulation of a mature cyclone. Study this figure to determine the wind shifts that would be expected for a city as the storm moved so that the city's relative position shifted from location *A* to *B* and then from *B* to *C*.

in almost circular isobars, with the low pressure centered at the apex of the wave. The resulting cyclonic circulation is shown in Figure 9-9. Once the cyclonic circulation develops, we would expect general convergence to result in vertical lifting, especially where warm air is overrunning colder air. We can see in Figure 9-9 that the air in the warm sector is flowing from the southwest toward colder air flowing from the southeast. Since the warm air is moving faster than the cold air in a direction perpendicular to this front, we can conclude that warm air is invading a region formally occupied by cold air; therefore, this must be a warm front. Similar reasoning indicates that in the rear of the cyclonic disturbance cold air is underrunning the air of the warm sector, generating a cold front there. Generally, the position of the cold front advances faster than the warm front and begins to close the warm sector, as shown in Figure 9-8. This process, called *occlusion*, results in the formation of an occluded front with the displaced warm sector located aloft. The cyclone enters maturity (maximum intensity) when it reaches this stage in its development. A steep pressure gradient and strong winds develop as lifting continues. Eventually all of the warm sector is forced aloft and cold air surrounds the cyclone at low levels. Once the sloping discontinuity (front) between the air masses no longer exists, the pressure gradient weakens. At this point the cyclone has exhausted its source of energy, and hence the storm comes to an end.

idealized weather of a wave cyclone

As stated earlier, the cyclone model provides a useful tool for examining the weather patterns of the middle latitudes. Figure 9-10 illustrates the distribution of clouds and thus the regions of possible precipitation associated

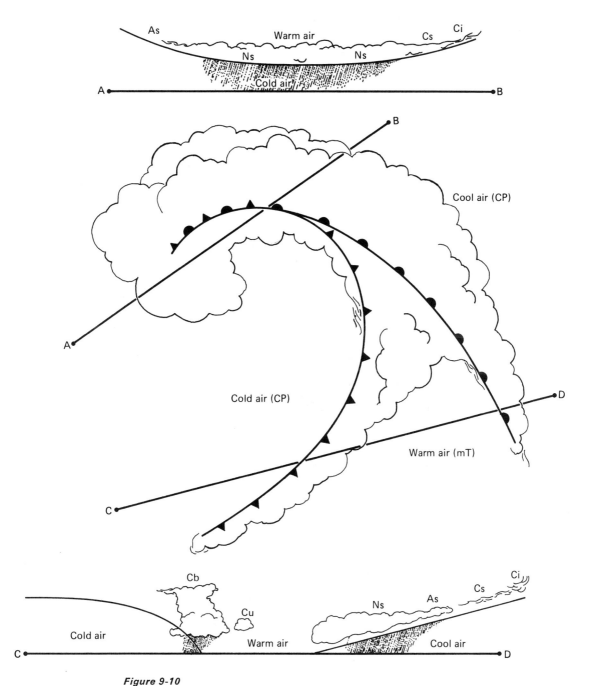

Figure 9-10

Cloud patterns usually associated with a mature middle-latitude cyclone. Upper and lower sections are vertical cross sections along A–B and C–D, respectively; the middle section represents a map view.

with a mature wave cyclone. Compare this drawing to the satellite photo of a cyclone shown in Figure 9-11. Guided by the westerlies aloft, cyclones generally move eastward across the United States so that we can expect the first signs of their arrival in the west. However, often in the region of the Mississippi valley, cyclones begin a more northeasterly trajectory and occasionally move directly northward. Typically, a midlatitude cyclone requires 2 to 4 days to pass over a given region. During that relatively short time period rather abrupt changes in atmospheric conditions can be experienced. This is particularly true in the spring of the year when the largest temperature contrasts occur across the midlatitudes.

Using Figure 9-10 as a guide, we will now consider these weather producers and the conditions we should expect from them as they pass an area in the spring of the year. To facilitate our discussion, profiles are provided along lines *A–B* and *C–D*. Imagine the change in weather as you move along profile *C–D*. Here the sighting of high cirrus clouds would be the first sign of the approaching cyclone. These high clouds can precede the surface front by 1000 kilometers or more and they will generally be accompanied by falling pressure. As the warm front advances, a lowering and a thickening of the cloud deck are noticed. Within 12 to 24 hours after the first sighting of cirrus clouds, light precipitation usually commences. As the front nears, the rate of precipitation increases, a rise in temperature is noticed, and winds begin to change from an easterly to a southerly flow. With the passage

Figure 9-11

Satellite view of a mature cyclone situated over the eastern one-third of the United States. It is easy to see why we often refer to the cloud pattern of a cyclone as having a "toboggan" shape. Photograph taken on June 12, 1975. (Courtesy of NOAA.)

of the warm front, the area is under the influence of the maritime tropical air mass of the warm sector. Generally, the region affected by this sector of the cyclone experiences warm temperatures, southerly winds, and generally clear skies, although fair weather cumulus or altocumulus are not uncommon here. The rather pleasant weather of the warm sector passes quickly in the spring of the year and is replaced by gusty winds and precipitation generated along the cold front. The approach of a rapidly advancing cold front is marked by a wall of rolling black clouds. Severe weather accompanied by heavy precipitation, hail, and an occasional tornado is common at this time of year. In addition, intense squall line thunderstorms of short duration frequently precede the cold front. On many occasions the squall activity is more severe than that associated with the cold front itself. The passage of the cold front is easily detected by a wind shift, the southerly flow is replaced by winds from the west to northwest, and there is a pronounced drop in temperature. Also, rising pressure hints of the subsiding cool, dry air behind the front. Once the front passes, the skies clear quickly as the cooler air invades the region. A day or two of almost cloudless deep blue skies can usually be expected, unless another cyclone is edging into the region.

A very different set of weather conditions will prevail in regions that encounter that portion of the cyclone that contains the occluded front as shown along profile A–B. Here the temperatures remain cool during the passage of the storm; however, a continual drop in pressure and increasingly overcast conditions strongly hint at the approach of the low-pressure center. This sector of the cyclone most often generates snow or icing storms during the cool months. Further, the occluded front often moves more slowly than the other fronts; hence the wishbone-shaped frontal structure shown in Figure 9-10 rotates in a counterclockwise manner so that the occluded front appears to "bend over backward." This effect adds to the misery of the region influenced by the occluded front since it remains over the area longer than the other fronts. Also, the storm reaches its greatest intensity during occlusion; consequently, the area affected by the developing occluded front can expect to receive the brunt of the storm's fury.

A few observations on the discussion above can serve to illustrate the value in knowing the cyclone model. In particular, shifts in wind direction are very useful in predicting the impending weather. Notice that with the passage of both the warm and cold fronts, the wind arrows changed position in a clockwise direction, for example, with the passage of the warm front, the wind changed from easterly to southerly. From nautical terminology the word *veering* is applied to a wind shift in a clockwise manner. Since clearing conditions occur with the passage of both fronts, veering winds are indicators that the weather will improve. On the other hand, the area located north of the cyclone will experience winds that shift in a counterclockwise direction, as can be seen in Figure 9-9. Winds that shift direction in a counterclockwise direction are said to be backing. With the approach of a wave cyclone, *backing* winds indicate cool temperatures and continued foul weather.

The preceding discussion provides only a brief sketch of the weather usually experienced with the passage of a wave cyclone. After we examine the daily weather map in Chapter 11, we will use this aid to look more closely at cyclonic weather as we consider a "real" cyclone.

energy mechanisms for a wave cyclone

The energy for cyclonic circulation comes from the *potential energy* (energy of position) that exists whenever cold and warm air are in juxtaposition, as for example, along a frontal surface. This potential energy is transferred into *kinetic energy* (energy of motion) as the air masses involved move in response to density differences. The cold air mass will move under the warmer air so that the discontinuity between them becomes horizontal instead of sloping, as is the case along a front. This means of transferring potential energy into energy of motion might be more easily visualized if you imagine a tank that has a vertical partition down the middle to separate oil from water. Like air of different densities, the juxtaposition of these two fluids of different densities provides potential energy. When the partition separating these fluids is removed, the water will flow under the less dense oil until a horizontal surface separates them. In this way potential energy is transferred into motion (kinetic energy). A cyclone is simply an area where the potential energy of a dome of cold air is released during subsidence and concentrated as kinetic energy. In other words, a cyclone is a region of strong winds.

You might expect that anywhere cold and warm air are in contact, the cold air would quickly underrun the warm air and thereby generate winds, but this does not readily occur because forces other than gravity determine the positions of the air masses involved. The forces that control the movement of air, in particular the Coriolis force, tend to hold a mass of cold air in a dome shape, thus preserving the sloping frontal boundary. It is only when the flow near the front becomes disturbed that cold air is able to sink and thus release the energy for cyclonic development. Once the sinking begins, more and more potential energy is converted into organized cyclonic flow. This energy is available to the storm until the occlusion process is completed, at which time all of the warm air has been displaced upward and over the cold air mass. At that point, without a continued source of energy, friction quickly brings the rotating system to a halt.

Before we leave this subject, we should also point out that the flow aloft is believed to play an important role in cyclonic development. We will consider this relationship shortly. Furthermore, some of the energy for cyclonic development is derived from the release of latent heat of condensation. Recall that latent heat provides the buoyancy for the development of convectional clouds. This vertical motion, in turn, encourages cyclonic inflow at the surface. However, this energy mechanism provides a somewhat smaller quantity of energy for cyclonic circulation in comparison to the sources that were described earlier. The preceding statement, however, is not true for tropical cyclones, such as hurricanes, in which latent heat is the major source of energy.

cyclogenesis   We saw in the Norwegian model that *cyclogenesis* (cyclone formation) occurs where a frontal surface is distorted into a wave-shaped discontinuity. Several surface factors are thought to produce a wave in a frontal zone. Topographic effects such as mountains, temperature contrasts as between sea and land, or ocean current influences can disrupt the general zonal flow sufficiently to produce a wave along a front. It is also believed that the disturbance of one storm center may initiate cyclogenesis elsewhere. But, perhaps more often than not, the initiation of a cyclonic system can be attributed to flow aloft.

When the earliest studies of cyclones were made, little data was available on the nature of the airflow in the middle and upper troposphere. Since then a close relationship between surface disturbances and the flow aloft has been established. Whenever the flow aloft is relatively straight, that is, from west to east, very little cyclonic activity occurs at the surface. However, when the upper air begins to meander widely in a north-to-south manner, thereby producing high amplitude waves consisting of alternating troughs and ridges, surface cyclonic activity intensifies. Further, when surface cyclones form, they almost invariably occur just ahead of an upper-air trough. Figure 9-12 illustrates the relative positions of a wave cyclone and the upper-level flow. The dashed lines represent 500-millibar contours and the solid lines show the pressure at the surface. The wavy pattern of the contours corresponds roughly to the airflow, which, as you recall, is almost geostrophic. Also recall

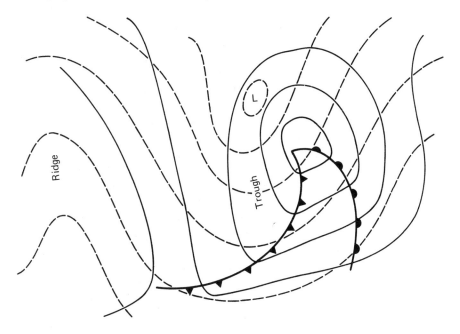

Figure 9-12

Schematic drawing showing the relative positions of upper-level waves and a surface cyclone. Solid lines represent the surface isobars and fronts, and the dashed lines represent the upper-level contour lines.

that above the surface troughs (lows) contain cold air and ridges (highs) contain warm air. Note in Figure 9-12 that the surface cyclone is found just ahead of the upper trough with the jet axis situated nearly overhead. The close association of the wave cyclone and polar jet should not come as a surprise in view of what we know about wind speeds aloft and temperature gradients. Also notice in Figure 9-12 that the surface cyclone is found directly below the location where the wavy flow aloft changes from a counter-clockwise (cyclonic) flow to a clockwise (anticyclonic) flow. Studies of fluid motion reveal that this is a region of net divergence. Earlier we learned that divergence aloft is a requirement for the maintenance of a cyclone with its surface convergence. For similar reasons, a surface high (anticyclone) is located below the region in the wavy flow where net convergence is expected. This is also the region of the cold dome where general subsidence provides the energy for cyclogenesis. Recall that the energy for a cyclone comes from the redistribution of cold and warm air masses. The subsiding cold air underruns the warm air which, as a consequence, is displaced upward. The flow aloft is able to provide the necessary convergence and divergence needed to maintain cyclonic flow. The general upward motion of warm air ahead of the trough is aided by divergence aloft, while the subsidence of the cold dome behind the cyclone is associated with upper-air convergence. In this larger sense we are justified in stating that the flow aloft provides the energy that is needed to maintain the surface cyclonic flow.

Thus far we have shown that a rather close association exists between the surface circulation and the flow aloft. This is an indication, but not proof, that surface disturbances are initiated by activity aloft. Other evidence, however, does show that cyclones can be generated by the wavy flow of the westerlies. Cyclones have been known to form without the prior existence of fronts. These storms are initiated by an upper-level trough and grow toward the surface. Once these surface cyclones form, they often "pull" contrasting air masses together, thereby generating fronts. From this point on these systems usually develop in much the same way as the wave cyclones discussed earlier. It should be noted that contrasting air masses must be available for concentration into a front before development can progress. Also, not all upper troughs generate surface cyclones.

From the preceding discussion it appears that wave cyclones can form in one of two ways. They can form along a discontinuity such as the polar front where the flow is disturbed most often by the flow aloft. In these situations, when the flow aloft becomes wavy, it tends to kink the front giving it its characteristic wave shape. In addition, wave forms can develop from a low-pressure center aloft that grows downward and then creates fronts. In either case, it seems apparent that the flow aloft plays a major role in initiating and maintaining cyclonic flow at the surface. We will also show later that wave cyclones are steered by the wavy motion aloft. Because of the significant role that the upper-level flow has on cyclogenesis, it should be evident that any attempt at weather prediction must take this flow into account.

traveling cyclones and anticyclones

In the previous section we indicated that waves in the westerlies were perhaps the most important cause for cyclonic development. The flow aloft is also important in determining how rapidly these pressure systems advance and the direction they will follow. Compared with the flow aloft (500-millibar level), cyclones generally travel at somewhat less than half the speed. Normally, they advance at a rate of from 20 to 50 kilometers per hour, so that distances of roughly 240 to 1200 kilometers are traversed each day. The faster speeds correspond with the colder months when temperature gradients are the greatest.

One of the most exacting tasks in weather forecasting is predicting the path of cyclonic storms. As stated earlier, the flow aloft tends to steer developing pressure systems. Let us examine an example of this steering effect by noting how changes in the upper flow correspond to changes in the direction of the path taken by a cyclone. Figure 9-13 illustrates the changing position of a wave cyclone over a period of a few days. Notice that on March 21 the 500-millibar contours are relatively flat and that for the following 2 days the cyclone moves in a rather straight easterly direction. By March 23 the 500-millibar contours make a sharp bend northward on the east side of a trough situated over Wyoming. Notice that on the next day the path of the cyclone makes a similar northward migration. Although this is an admittedly oversimplified example, it does illustrate the steering effect of upper-level flow. It should be remembered that we examined the effect of upper-air flow on cyclonic movement after the fact. In order to make useful predictions of future positions of cyclones, an accurate appraisal of changes in the westerly flow aloft is required. For this reason, predicting the behavior of the wavy flow aloft makes up an important part of modern weather forecasting.

As evidenced by the preceding example, wave cyclones have a tendency to migrate first in an easterly direction and then travel a more northeastward path (on some occasions developing storms move southeastward prior to migrating northward). The northeastward trajectory of cyclones frequently causes them to merge with one of the two semipermanent low-pressure systems (the Aleutian and Icelandic lows) that are centered at roughly 60 degrees north latitude. It is probably more accurate to state that the great number of traveling cyclones that merge in these regions show up in the mean picture as semipermanent low-pressure centers.

The cyclones that reach the West Coast originate in the Pacific along the polar front. These storms usually occur as members of a "family" of cyclones in various stages of development, as shown in Figure 9-14. Most of these systems move northeastward toward the Gulf of Alaska where they merge with the Aleutian low. During the winter months these storms develop farther south in the Pacific and often reach the coast of the contiguous 48 states, occasionally moving as far south as Southern California. These systems provide the rather short winter rainy season that affects California and the western states in general. Most of the Pacific storms do not cross the Rockies

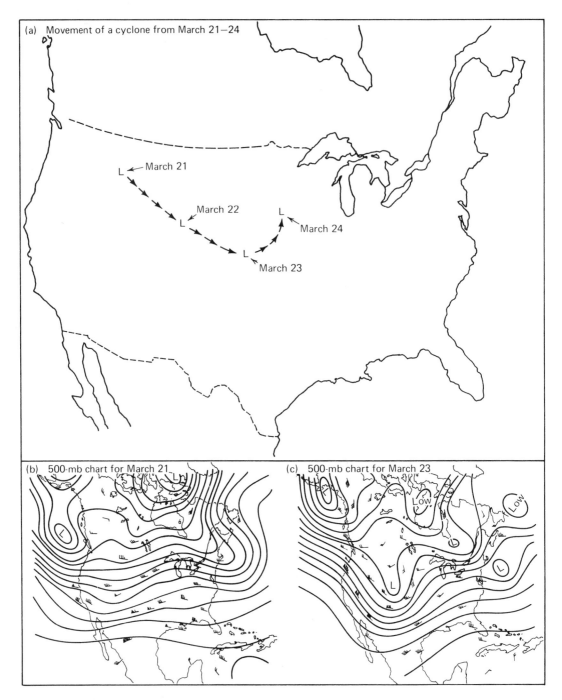

Figure 9-13

Steering of cyclonic storms. (*a*) Notice that the cyclone (low) moved almost in a straight southeastward direction during the 2 days from March 21 to March 23. On the morning of March 23 it abruptly turned northward. This change in direction corresponds to the change from rather flat contours (*b*) on the upper-air chart for March 21 to curved contours (*c*) on the chart for March 23.

Figure 9-14

Two well-developed cyclones are apparent in this satellite image from their characteristic toboggan-shaped cloud pattern. Located in the area of only scattered clouds between this so-called family of cyclones is an anticyclone. (Courtesy of NOAA.)

intact, but many of them redevelop on the lee side of these mountains. A favorite site for redevelopment is Colorado, but other common sites of formation exist as far south as Texas and as far north as Alberta. The cyclones that form in Canada tend to move southward toward the Great Lakes and then turn northeastward and move out into the Atlantic. Cyclones that redevelop over the Plains states generally migrate eastward until they reach the central United States where a northeastward or even northward trajectory is followed. Most of these storms also traverse the Great Lakes region, making this the most storm-ridden region in the country. Not all of the cyclones that affect the United States originate in the Pacific. Some form over the Plains states and are associated with an influx of maritime tropical air from the Gulf of Mexico. Another area where cyclogenesis occurs is east of the southern Appalachians. These storms tend to move northward with the Gulf Stream and eventually merge with the Icelandic low.

Although the paths of cyclones tend to be northeastward, anticyclones more frequently move southeastward. However, some anticyclones are imbedded between members of a family of cyclones and consequently travel northeastward with the group (see Figure 9-14). After the last member of a family of cyclones passes through the United States, the cold anticyclone situated behind the cold front breaks out and moves southward. This produces the cold waves that are experienced deep into the southern states during the winter. These anticyclones generally continue to move toward the subtropical high-pressure center situated over the Atlantic. Because anticyclones are associated with clear conditions and are therefore "weatherless," much less study has been conducted on their development and movement. However, because anticyclones occasionally become stagnant and remain over a region for several days, they are becoming ever more important to man as he tries to cope with air pollution. Recall that it is the stable and rather calm conditions associated with high pressure that contribute to pollution episodes. Large stagnant anticyclones are also important because they often block the eastward migration of cyclonic centers. This effect can keep one section of the country dry for a week or longer while another area is continually under the influence of a cyclonic storm.

thermal lows

Not all low-pressure systems move, and not all low-pressure systems are associated with fronts. As indicated earlier, the name cyclone merely implies a low-pressure system that may or may not produce severe weather. Two large-scale lows that exist without fronts are tropical cyclones (hurricanes) and thermal lows. Tropical cyclones are discussed in Chapter 10.

Thermal lows are found in two major regions of the world: in the area near the Persian Gulf and in the American Southwest. Both of these regions are generally dominated by the circulation of the semipermanent subtropical highs that were considered earlier in the discussion of global circulation. The general subsidence that prevails in these systems provides the clear skies which, in turn, allow the intense surface heating we associate with these areas around the globe. However, during the hottest summer months these regions develop lows that dominate the surface circulation. These lows form because the intense surface heating causes expansion of the overlying air column; hence the name thermal lows. This effect generates a general outflow aloft that encourages surface inflow. The formation of this circulation is very similar to the development of the sea breeze that we considered earlier. In the United States a thermal low center develops each summer over southwestern Arizona. It is believed that this low contributes to a low-level influx of relatively moist air from the Gulf of California during the hottest months of the year. This would then account for the general increase in precipitation this region experiences in the hot summer months as compared to the cooler spring season. For example, Phoenix, Arizona receives only meager precipitation all year long, but, on the average, it receives ten times more in July than it does in May.

REVIEW

1. If you were located 400 kilometers ahead of the surface position of a warm front, how high would the frontal surface be above you?

2. Why is cold front weather usually more violent than warm front weather?

3. Explain the basis for the following weather proverb:

> Rain long foretold, long last;
> Short notice, soon past.

4. Distinguish between cold-type and warm-type occluded fronts.

5. Although the formation of an occluded front represents the period of maximum intensity for a wave cyclone, it also marks the beginning of the end of the system. Explain why this is the case.

6. For each of the weather elements listed below, describe the changes that an observer experiences when a wave cyclone passes with its center north of the observer.
   a. wind direction
   b. pressure tendency
   c. cloud type
   d. cloud cover
   e. precipitation
   f. temperature

7. Describe the weather conditions an observer would experience if the center of a wave cyclone passed to the south.

8. Distinguish between veering and backing winds.

9. Briefly explain how the flow aloft initiates and maintains cyclones at the surface.

10. Explain why predicting the behavior of upper-level flow is an important task in modern weather forecasting.

11. Why does a city such as Phoenix, Arizona, experience a precipitation maximum in midsummer?

VOCABULARY REVIEW

middle-latitude cyclone	cold-type occluded front
wave cyclone	warm-type occluded front
polar front theory	occlusion
front	veering wind shift
overrunning	backing wind shift
warm front	potential energy
cold front	kinetic energy
stationary front	cyclogenesis
occluded front	thermal low

Chapter Ten

severe weather

As with many other natural catastrophies, hurricanes and tornadoes attract a great deal of attention. Although these phenomena are relatively rare, they have a fascination that ordinary weather events cannot provide. Furthermore, because of the death and destruction that these storms leave in their wake, they have been and continue to be an important focus of atmospheric research.

what is in a name?

In Chapter 9 we examined the middle-latitude cyclones that play such an important role in causing day-to-day weather changes. However, the use of the term cyclone is very often confusing. To many people the term implies only an intense storm such as a hurricane or a tornado. For example, when a hurricane unleashes its fury upon India or Bangladesh, it is usually reported in the media as a cyclone (the local term denoting a hurricane in that part of the world). In a similar manner, tornadoes are occasionally referred to as cyclones. Recall that in the *Wizard of Oz* Dorothy's house was carried to the land of Oz by a cyclone. Although hurricanes and tornadoes are, in fact, cyclones, the vast majority of cyclones are not hurricanes and tornadoes. The term cyclone simply refers to the circulation around any low-pressure center, no matter how large or intense it is.

Tornadoes and hurricanes are both smaller and more violent than middle-latitude cyclones. While middle-latitude cyclones may have a diameter of 1600 kilometers or more, hurricanes average only 600 kilometers across, and tornadoes, with a diameter of just $\frac{1}{4}$ kilometer, are much too small to show up on a weather map.

The thunderstorm is a much more familiar weather event, and it hardly needs to be distinguished from tornadoes, hurricanes, and middle-latitude cyclones. Unlike the flow of air about these latter storms, the circulation associated with thunderstorms is characterized by strong up and down movements. Winds in the vicinity of a thunderstorm do not follow the inward spiral of a cyclone, but they are typically variable and gusty.

Although thunderstorms form "on their own" away from cyclonic storms, they also form in conjunction with cyclones. For instance, thunderstorms are frequently spawned along the cold front of a middle-latitude cyclone where on rare occasions a tornado may descend from the thunderstorm's cumulonimbus tower. Hurricanes have also been known to generate widespread thunderstorm activity. Hence, thunderstorms can be related in some manner to all three types of cyclones mentioned in this section.

**thunder-
storms**

Most everyone has observed a small-scale phenomenon that is a consequence of the vertical motion of warm, unstable air. Perhaps you have seen a small dust devil form over an open field on a hot day and whirl its dusty load to great heights, or, perhaps, you have seen a bird glide effortlessly skyward upon an invisible thermal of hot air. These examples illustrate a much more dynamic thermal instability that occurs during the development of a *thunderstorm*. At any given time over the earth's surface nearly 2000 thunderstorms are in progress, mostly in tropical regions. Thunderstorm activity is associated with massive cumulonimbus clouds that generate torrential downpours, thunder, lightning, and occasionally hail.

stages in the development of a thunderstorm

All thunderstorms require warm, moist air, which, when lifted, will release sufficient latent heat to provide the buoyancy necessary to maintain its upward flight. Although this instability and associated buoyancy are triggered by a number of different processes, all thunderstorms have a similar life cycle.

Because instability is enhanced by high surface temperatures, thunderstorms are most common in the afternoon and early evening. However, surface heating is generally not sufficient in itself to cause the growth of towering cumulonimbus clouds. A solitary cell of rising hot air produced by surface heating alone could, at best, produce a small cumulus cloud. Mixing between the moist air of the infant cloud and the cool, dry air aloft causes evaporation that dissipates the cloud in 10 to 15 minutes. The development of a 12,000-meter or, on rare occasions, an 18,000-meter cumulonimbus tower requires a continuous supply of moist air. Each new surge of warm air rises higher than the last, adding to the height of the cloud (Figure 10-1). This phase in the development of a thunderstorm, called the *cumulus stage*, is

Figure 10-1

Cumulonimbus clouds in various stages of development. (Courtesy of R. R. Braham, Cloud Physics Laboratory, The University of Chicago.)

dominated by updrafts (Figure 10-2). These updrafts must occasionally reach speeds of 160 kilometers per hour to accommodate the large hailstones they are capable of carrying upward.

Once the cloud passes beyond the freezing level, the Bergeron process begins producing precipitation. Usually within half an hour of its inception, the accumulation of precipitation in the cloud is too great for the updrafts to support. The falling precipitation causes drag on the air and initiates a downdraft. The creation of the downdraft is further aided by the influx of cool, dry air surrounding the cloud, a process termed *entrainment*. This process intensifies the downdraft because the air added during entrainment is cool and therefore heavy, and possibly of greater importance, it is dry. Hence, it causes some of the falling precipitation to evaporate (a cooling process), thereby cooling the air within the downdraft. As the resulting downdraft leaves the bottom of the cloud, precipitation is released, marking the beginning of the cloud's *mature stage*. At the surface the cool downdraft spreads laterally and can be felt before the actual precipitation reaches the ground. The sharp cool gusts at the surface are indicative of the downdrafts aloft. During the mature stage updrafts exist side by side with downdrafts and continue to enlarge the cloud. When the cloud grows to the top of the unstable region often located at the base of the warmer stratosphere, the updrafts spread laterally and produce the characteristic anvil top (Figure 10-3). Generally, ice-laden cirrus clouds make up the top and are spread downwind by rapid winds aloft. The mature stage is the most active period of a thunderstorm. Gusty winds, lightning, heavy precipitation, and sometimes hail are experienced.

216

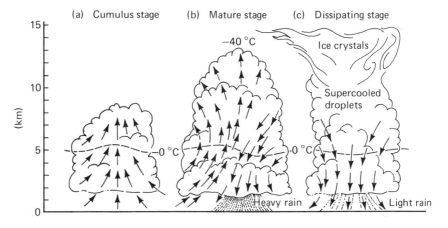

Figure 10-2

Stages in the development of a thunderstorm. During the cumulus stage strong updrafts act to build the storm. The mature stage is marked by heavy precipitation and cool downdrafts in part of the storm. When the warm updrafts disappear completely, precipitation becomes light and the cloud begins to evaporate.

Once a downdraft begins, the vacating air encourages more entrainment of the cool, dry air surrounding the cell. Eventually, downdrafts dominate throughout the cloud and initiate the *dissipating stage* (Figure 10-2). The cooling effect of falling precipitation and the influx of colder air aloft mark

Figure 10-3

Satellite view near the Florida coast of several well-developed cumulonimbus clouds exhibiting anvil tops. (Courtesy of NASA.)

the end of the thunderstorm activity. Without a supply of moisture, the cloud will soon evaporate. The life span of a single cumulonimbus cell within a thunderstorm complex is only about an hour, but as the storm moves, fresh supplies of warm, water-laden air generate new cells to replace those that are dissipating.

To summarize, the stages in the development of a thunderstorm are: (1) the cumulus stage in which updrafts dominate throughout the cloud and growth from a cumulus to a cumulonimbus cloud occurs, (2) the mature stage characterized by violent weather when downdrafts are found side by side with updrafts, and (3) the dissipating stage dominated by downdrafts and entrainment causing evaporation of the structure.

generating a thunderstorm

Any factor that destabilizes the air aids in generating a thunderstorm. Since many processes do affect the stability of air, it should come as no surprise that thunderstorms can be triggered in a number of ways. For convenience of discussion, we will divide thunderstorms into three types: (1) isolated thunderstorms produced within tropical air masses, (2) thunderstorms produced by forceful lifting, either frontal or orographic, and (3) thunderstorms produced along squall lines.

In the United States air-mass thunderstorms generally occur in warm, moist (maritime tropical) air that originates over the Gulf of Mexico and moves northward. Recall that air of this type contains most of its moisture in the lower portion and can be rendered unstable upon uplifting. Because this air becomes unstable most often in the spring and summer when it is warmed sufficiently from below, air-mass thunderstorms are most frequent then. They also have a strong preference for midafternoon when the surface temperatures are the highest. Some do occur after sunset when the growth of immature cells becomes restimulated by cloud top cooling. Because local differences in surface heating aid in the growth of air-mass thunderstorms, they generally occur as scattered, isolated cells instead of being organized in narrow bands as are frontal thunderstorms.

Mountainous regions like the Rockies in the West and the Appalachians in the East experience a greater number of air-mass thunderstorms than do the Plains states. The air near the mountain slope is heated more intensely than air at the same elevation over the adjacent lowlands. This causes a general upslope movement during the daytime that generates thunderstorm cells. These cells remain almost stationary above their source region, the slopes below.

Although the growth of thunderstorms is aided by high surface temperatures, most air-mass thunderstorms are not generated solely by surface heating, mountainous areas being the possible exception. Recall that the thunderstorms that form over the peninsula of Florida are believed to be the result of converging winds from the ocean on both sides. Similarly, the air-mass thunderstorms that occur over the eastern two-thirds of the United States occur as part of the general convergence associated with a passing midlatitude cyclone. Near the equator they form in association with the convergence along the equatorial low.

A few places in the United States, western Kansas and Nebraska in particular, experience a thunderstorm maximum at night. This may be the result of general cooling and downflow in the adjacent mountains to the west. As the air flows out of the mountains during the night, it produces a general convergence over the plains which triggers these storms.

Frontal and orographic thunderstorms result from lifting of potentially unstable air. This upgliding furnishes the trigger that releases large reserves of latent heat that provide the needed vertical buoyancy. In orographic thunderstorms hilly or mountainous terrain provides the lift. Frontal lifting occurs when potentially unstable air is forced over a wedge of colder, denser air. Because advancing cold air provides a blunter nose for uplift than do the more gently inclined warm fronts, most frontal thunderstorms are associated with cold fronts.

Many of the most awesome thunderstorms are associated with *squall lines*. These rather narrow belts of thunderstorms are often forerunners of cold fronts, being as much as 300 kilometers ahead. They can, however, spawn in the warm sector of a midlatitude cyclone independently of the cold front. Squall-line disturbances generally move eastward across the United States at 40 kilometers per hour. When it is associated with a cold front, the squall line is generated by the cold downdrafts that descend from the towering cumulonimbus clouds along the front (Figure 10-4). Upon reaching the surface, the cool air surges along the ground primarily in the direction in which the storm is advancing, producing a pseudo-cold front. The zone where

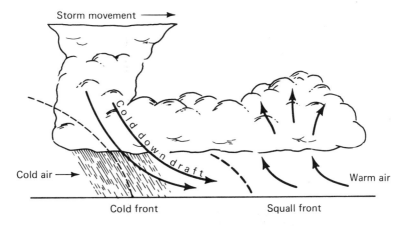

Figure 10-4

A squall front ahead of an advancing cold front is produced by cold downdrafts. Often the weather is more severe along a squall line than along the trailing cold front that has generated it.

the downdraft meets the warm air is called the squall line. Here the cooler, denser downdraft underruns the warmer air and forces it to ascend, thus generating the clouds of the squall line. Once formed, a squall line can become self-propagating. The dissipating cells of the initial squall disturbance provide new downdrafts that trigger instability in the warm air ahead advancing the storm's position. Since a self-propagating squall line can move faster than the initiating cold front, it occasionally precedes the front by several hours.

The approach of a squall line is often preceded by a mammatus sky—dark cloud rolls that have downward pouches. A sharp veer in the wind direction, cooler temperatures, and gusty conditions capable of inflicting heavy destruction are also common to squall disturbances. In addition, the most violent of all windstorms, the tornado, is an occasional companion.

thunder and lightning

By international agreement a storm is classified as a thunderstorm only after thunder is heard. Because thunder is produced by lightning, lightning must also be present (Figure 10-5). *Lightning* is very similar to the electrical shock you may have experienced upon touching a metal object on a very dry day. Only the intensity is different.

During the development of a large cumulonimbus cloud a separation of charge occurs. This simply means that part of the cloud obtains an excess negative charge while another part of the cloud acquires an excess positive charge. The object of lightning is to equalize these electrical differences by

Figure 10-5
Lightning. (Courtesy of NOAA.)

producing a negative flow of current from the region of excess negative charge to the region with excess positive charge, or vice versa. Because air is a poor conductor of electricity (good insulator), the electrical potential (charge difference) must be very high before lightning will occur, on the order of 3000 volts per meter.

The origin of charge separation in clouds, although not fully understood, must hinge on rapid vertical movements within, since lightning occurs primarily in the violent mature stage of a cumulonimbus cloud. Because the formation of these tall clouds is primarily a summertime phenomenon in the midlatitudes, it also explains why lightning is seldom observed there in the winter. Furthermore, lightning rarely occurs before the growing cloud penetrates the 5-kilometer level where sufficient cooling begins to generate some ice crystals. It is believed by some cloud physicists that charge separation occurs during the formation of ice pellets. Experimentation shows that as droplets begin to freeze, positively charged ions are concentrated in the colder regions of the droplets while negatively charged ions are concentrated in the warmer regions. Thus, as the droplets freeze from the outside in, they develop a positively charged ice shell and a negatively charged interior. As the interior begins to freeze, it expands and shatters the outside shell. The small positively charged ice fragments are carried upward by turbulence and the relatively heavy droplets eventually carry their negative charge toward the cloud base. The result is that the upper part of the cloud is left with a positive charge and the lower portion of the cloud maintains an overall negative charge with small positively charged pockets (Figure 10-6). As the cloud moves, the negatively charged cloud base alters the charge at the earth's surface directly below by repelling negatively charged particles. Thus, the earth's surface beneath the cloud acquires a net positive charge. These charge differences build to millions and even hundreds of millions of volts before a lightning stroke acts to discharge the negative region of the cloud by striking the positive area of the earth below, or more frequently the positively charged portion of that cloud, or a nearby cloud.

Cloud-to-earth strokes are of most interest and have been studied in detail. High-speed cameras have greatly aided in these studies. They show that the lightning we see as a single flash is really several very rapid strokes between the cloud and the earth (Figure 10-6). We will call the total discharge, which lasts for only a few tenths of a second and appears as a bright streak, the *flash*. The individual components that make up each flash are called *strokes*. Each stroke is separated by roughly 50 milliseconds and there are usually three to four strokes per flash. When a lightning flash appears to flicker, it is because your eyes discern the individual strokes that make up this discharge. Further, each stroke consists of a downward propagating leader that is immediately followed by a very luminous return stroke.

Each stroke is believed to begin when the electrical field near the cloud base frees electrons in the air immediately below, thereby ionizing the air. Once ionized, the air becomes a conductive path having a radius of roughly 10 centimeters and a length of 50 meters. This path is called a *leader*. During

221

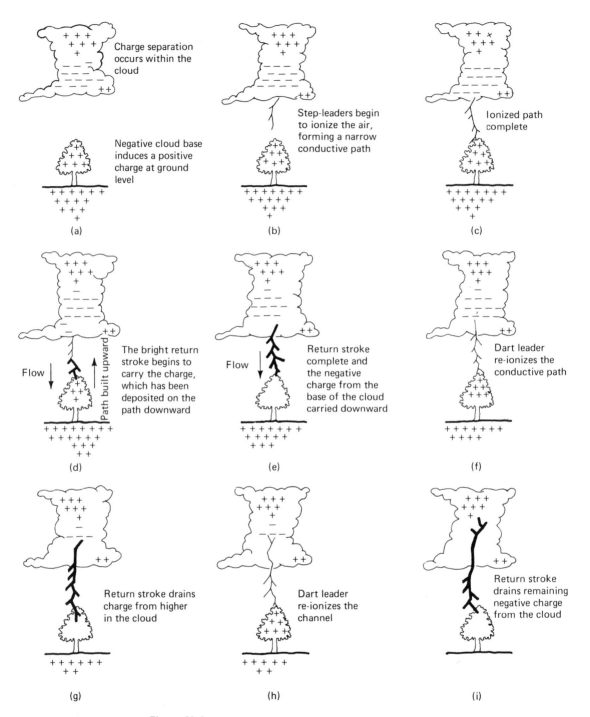

Figure 10-6

Discharge of a cloud via cloud-to-ground lightning. Examine this drawing carefully while reading the text.

this electrical breakdown the mobile electrons in the cloud base begin to flow down this channel. This flow increases the electrical potential at the head of the leader, which causes a further extension of the conductive path through further ionization. Because this initial path extends itself earthward in short, nearly invisible bursts, it is called a *step leader*. Once this channel nears the ground, the electrical field at the surface ionizes the remaining section of the path. When the channel is complete, a wave of ionizing potential called the *return stroke* moves upward along the channel and nearly fully ionizes it. As the wave front advances, the negative charge that was deposited on the channel is effectively lowered to the earth. It is this intense return stroke that illuminates the conductive path and discharges the lowest kilometer or so of the cloud. During this phase tens of coulombs of negative charge are lowered to the ground.*

The first stroke is usually followed by additional strokes which apparently drain charges from higher areas within the cloud. Each subsequent stroke begins with a *dart leader* that once again ionizes the channel and carries the cloud potential toward the earth. The dart leader is continuous and less branched than the step leader. When the current between strokes has ceased for periods greater than one-tenth of a second, further strokes will be preceded by a stepped leader whose path is different from that of the initial stroke. The total time of each flash consisting of 3 to 4 strokes is about 0.2 second.

The electrical discharge of lightning heats the air and causes it to expand explosively. This expansion produces the sound waves we hear as *thunder*. Since lightning and thunder occur simultaneously, it is possible to estimate the distance to the stroke. Lightning is seen instantaneously, but the rather slow sound waves, which travel approximately 330 meters per second, reach us a little later. If thunder is heard 5 seconds after the lightning is seen, the lightning occurred about 1650 meters away (approximately 1 mile).

The thunder we hear as a rumble is produced along a rather long lightning path located at some distance from the observer. The sound that originates along the path nearest the observer arrives before the sound that originated farthest away. This lengthens the duration of the thunder. Reflection of the sound waves further delays their arrival and adds to this effect. When lightning occurs more than 20 kilometers away, thunder is rarely heard. This type of lightning, popularly called heat lightning, is no different from the lightning we associate with thunder.

tornadoes

Tornadoes are local storms of short duration that may be ranked high among nature's most destructive forces. The lack of forewarning, the incredible fury of its winds, and the near total destruction of the stricken area have led many to liken its passage to a bombing raid during war.

Tornadoes, sometimes called twisters or cyclones, are intense centers of

---

*A coulomb is a unit of electrical charge equal to the quantity of charge transferred in 1 second by a steady current of 1 ampere.

Figure 10-7

Funnel cloud. (Photograph by Tom Carter. Courtesy of the *Peoria Journal Star.*)

low pressure having a whirlpool-like structure of winds rotating around a central cavity where centrifugal force produces a partial vacuum. Pressures within the center of the tornado may be 100 millibars less than immediately outside the storm. Because of such a tremendous pressure gradient, some meteorologists have estimated that wind speeds may reach 650 kilometers per hour or more. Air sucked into the vortex of the storm is rapidly lifted and cooled adiabatically. The resulting condensation creates the pale and ominous appearing funnel cloud, which may darken as it moves across the ground picking up dust and debris (Figure 10-7).

Few people have had the eye of a funnel cloud pass over them and lived to tell about it. However, Will Keller, a farmer living near Greensburg, Kansas, is one who did.

> On the afternoon of June 22, 1928, between three and four o'clock, I noticed an umbrella-shaped cloud in the west and southwest and from its appearance suspected there was a tornado in it. The air had that peculiar oppressiveness which nearly always precedes the coming of a tornado.
>
> I saw at once my suspicions were correct. Hanging from the greenish black base of the cloud were three tornadoes. One was perilously near and apparently headed directly for my place . . . .
>
> Two of the tornadoes were some distance away and looked like great ropes dangling from the parent cloud, but the one nearest was shaped more like a funnel, with ragged clouds surrounding it. It appeared larger than the others and occupied the central position, with great cumulus clouds over it.
>
> Steadily the cloud came on, the end gradually rising above the ground. I probably stood there only a few seconds, but was so impressed with the

sight it seemed like a long time. At last the great shaggy end of the funnel hung directly overhead. Everything was still as death. There was a strong, gassy odor, and it seemed as though I could not breathe. There was a screaming, hissing sound coming directly from the end of the funnel. I looked up, and to my astonishment I saw right into the heart of the tornado. There was a circular opening in the center of the funnel, about fifty to one hundred feet in diameter and extending straight upward for a distance of at least half a mile, as best I could judge under the circumstances. The walls of this opening were rotating clouds and the whole was brilliantly lighted with constant flashes of lightning which zigzagged from side to side . . . .

Around the rim of the great vortex small tornadoes were constantly forming and breaking away. These looked like tails as they writhed their way around the funnel. It was these that made the hissing sound. I noticed the rotation of the great whirl was anticlockwise, but some of the small twisters rotated clockwise . . . . The tornado was not traveling at a great speed. I had plenty of time to get a good view of the whole thing, inside and out.*

Other than the fact that tornadoes more often occur singly rather than in multiples and sometimes do not have lightning, Mr. Keller's account describes a typical tornado.

Although tornadoes probably generate the greatest wind speeds on earth, no meteorological installation has ever withstood a direct hit. Hence, most knowledge concerning tornadoes comes from examining the aftereffects of the storm. Although it might be possible to construct a shelter and set of instruments capable of withstanding the fury of a tornado, it would not be worthwhile. Since tornadoes are highly localized and randomly distributed, the probability of placing a set of instruments in the right place would be infinitesimally small. For example, the probability of a tornado striking a given point in the area most frequently subject to tornadoes is about once in 250 years.

the development of tornadoes

Although many theories about tornado formation have been advanced, none has won general acceptance. The general atmospheric conditions that are most likely to develop into tornado activity, however, are known. Tornadoes are most often spawned along the cold front of a middle-latitude cyclone, in conjunction with cumulonimbus clouds and severe thunderstorms. Throughout the spring the air masses associated with middle-latitude cyclones are most likely to have greatly contrasting conditions. Continental polar air from the Canadian arctic may still be very cold and dry, while maritime tropical air from the Gulf of Mexico is very warm and moisture laden. The greater the contrast, the more intense the storm. Since these two contrasting air

_____

*From Alonzo A. Justice, "Seeing the Inside of a Tornado," *Monthly Weather Review*, LVIII, 5 (1939), 205–206.

masses are most likely to meet in the central United States, it is not surprising that this region generates more tornadoes than any other area of the country, and, in fact, the world. Figure 10-8, which depicts tornado incidence by state and area, readily substantiates this fact.

About 700 tornadoes are reported each year in the United States and they have been known to occur during every month of the year. April through June is the period of greatest tornado frequency in the United States, while the number is lowest during December and January. In February, when the incidence of tornadoes begins to increase, the center of maximum frequency lies over the central Gulf states. During March this center moves eastward to the southeast Atlantic states, where tornado frequency reaches a peak in April. During May and June the center of maximum frequency moves through the southern plains and then to the northern plains and Great Lakes area. The reason for this drift is the increasing penetration of warm, moist air while contrasting cool, dry air still surges in from the north and northwest. Thus, when the Gulf states are substantially under the influence of warm air after May, there is no cold air intrusion to speak of, and tornado frequency drops. This is the case across the country after June. Winter cooling permits fewer and fewer encounters between warm and cold air masses, and tornado frequency returns to its lowest level by December.

The average funnel cloud has a diameter of between 150 and 600 meters,

TORNADO INCIDENCE BY STATE AND AREA, 1953–1976

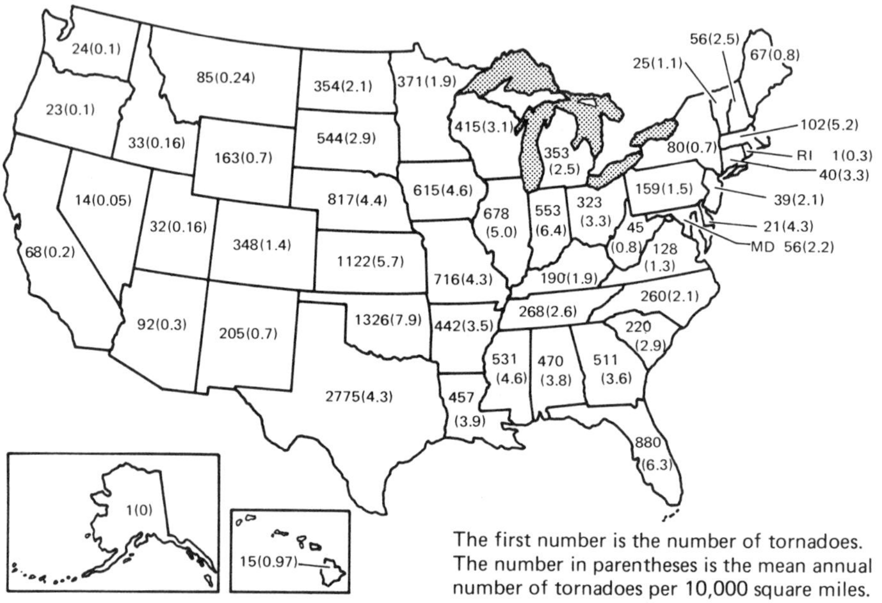

The first number is the number of tornadoes.
The number in parentheses is the mean annual
number of tornadoes per 10,000 square miles.

Figure 10-8

Tornado incidence by state and area, 1953–1976. Since the sizes of states vary widely, the lower figure is a more accurate indication of tornado activity. (Data courtesy of NOAA.)

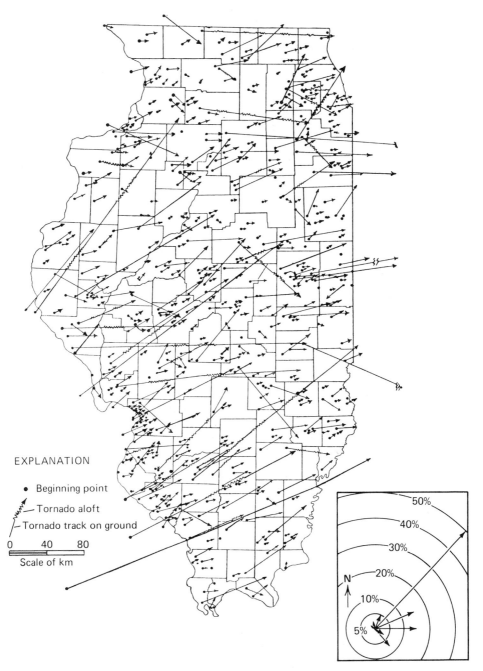

EXPLANATION

• Beginning point

⌇— Tornado aloft

— Tornado track on ground

0   40   80
Scale of km

Figure 10-9

Paths of Illinois tornadoes (1916–1969). Since most tornadoes occur slightly
ahead of a cold front, in the zone of southwest winds, they tend to move toward
the northeast. Tornadoes in Illinois verify this. Over 80 percent exhibited directions of movement toward the northeast through east. (After John W. Wilson
and Stanley A. Changnon, Jr., *Illinois Tornadoes*, Illinois State Water Survey
Circular 103, 1971, pp. 10, 24.)

travels across the landscape at 45 kilometers per hour, and cuts a path about 26 kilometers long. Since many tornadoes occur slightly ahead of a cold front, in the zone of southwest winds most move toward the northeast. The Illinois example demonstrates this nicely (Figure 10-9). Figure 10-9 also shows that many tornadoes do not fit the description of the "average" tornado. Many have had paths a great deal longer than 26 kilometers and have traveled not at 45 kilometers per hour, but at speeds in excess of 100 kilometers per hour (Figure 10-10). Furthermore, there have been tornadoes that have had diameters approaching 2 kilometers, more than four times the average size.

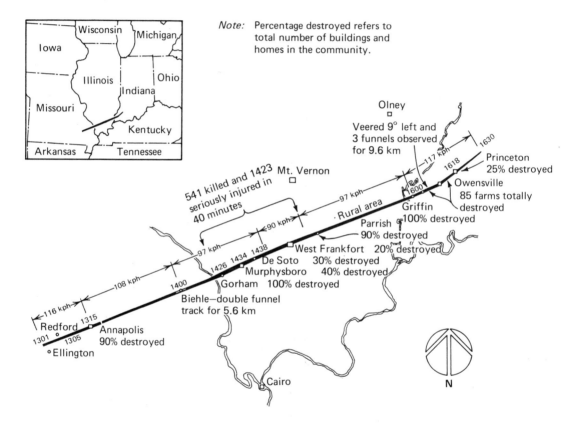

Figure 10-10

"One tornado among the more than 13,000 which have occurred in the United States since 1915 easily ranks above all others as the single most devastating storm of this type. Shortly after its occurrence on 18 March 1925, the famed Tri-State tornado was recognized as the worst on record, and it still ranks as the nation's greatest tornado disaster. The tornado remained on the ground for 219 miles. The resulting losses included 695 dead, 2027 injured, and damages equal to $43 million in 1970 dollars. This represents the greatest death toll ever inflicted by a tornado and one of the largest damage totals." (Map and description from J. W. Wilson and S. A. Changnon, Jr., *Illinois Tornadoes*, Illinois State Water Survey Circular 103, 1971, p. 32.)

Figure 10-11

The force of the wind during a tornado in Clarendon, Texas, in 1970 was enough to drive a wooden stick through a 4-centimeter metal pipe. (Courtesy of NOAA.)

tornado destruction

Funnels have accomplished many seemingly impossible tasks, such as driving a piece of straw through a thick wooden plank and uprooting huge trees (Figure 10-11). In 1931 a tornado actually carried an 83-ton railroad coach and its 117 passengers 24 meters through the air and dropped them in a ditch.

Most tornado losses are associated with a few storms that strike urban areas or devastate entire small communities. The near-total destruction wrought by such storms is linked to the combined effects of the exceedingly strong winds and the partial vacuum in the center of the storm. The winds may rip apart everything in the path of the storm, and the abrupt pressure decrease may cause some buildings to literally explode, especially if the windows are closed, because of the much lower pressure outside the structure. Thus, contrary to one's natural inclination, if a tornado seems imminent, it is a wise precaution to open doors and windows before seeking shelter. Only buildings of reinforced concrete and rigidly constructed structural steel generally escape serious structural damage from severe tornadoes. Windows, roofs, and siding are always vulnerable.

Tornadoes take many lives each year (Figure 10-12), sometimes hundreds in a single day. When tornadoes struck an area stretching from Canada to Georgia on April 3, 1974, the death toll exceeded 300, the worst in half a century. If there is some question about the causes of tornadoes, there certainly is no question about the destructive effects of these violent storms. A severe tornado leaves the affected area stunned and disorganized and may

229

Upper figure is number of deaths.
Middle figure is mean annual number of deaths.
Lower figure is number of deaths per 10,000
square miles.

Figure 10-12

Deaths from tornadoes, 1953–1976. The greatest potential for casualties from
tornadoes is not necessarily where the greatest number of tornadoes occur, but
where there is a combination of high tornado incidence and a dense concentration of population. (Data courtesy of NOAA.)

require a response of the magnitude demanded in war. A quotation from the
government publication *Tornado* puts it this way:

> Their time on earth is short, and their destructive paths are rather small.
> But the march of these shortlived, local storms through populated areas
> leaves a path of terrible destruction. In seconds, a tornado can transform
> a thriving street into a ruin, and hope into despair.

predicting tornadoes

Since severe thunderstorms and tornadoes are small and short-lived phenomena, they are the most difficult weather features to forecast precisely
given our present knowledge, theory, equipment, and techniques. Two
researchers at the National Severe Storms Laboratory describe the difficulty
of tornado forecasting this way:

> They are less specific than we would like, because our observations are too
> sparse to describe atmospheric variability on the scale producing the
> tornado or thunderstorm phenomena. A severe thunderstorm may extend

230

Figure 10-13

Sometimes tornadoes appear as hook-shaped echoes on a radar screen. (Courtesy of NOAA.)

10 to 25 miles, and exist for about 6 hours, while the distance between primary surface weather stations is about 100 miles and between upper-air stations more than 200 miles. Observations are made hourly at the surface stations (more often under special conditions) but usually at only 12-hour intervals at the upper-air stations. Even if our knowledge were otherwise adequate to the task, the current weather observing system limits us to indicating the probability of thunderstorms and accompanying tornadoes in regions much larger than the storms.*

Such forecasts, called *tornado watches*, generally refer to situations that are expected to begin from 1 to 7 hours after the forecast is issued in areas covering about 65,000 square kilometers. About 40 percent of these predictions are correct, that is, one or more tornadoes occurred somewhere in the region. The incorrect predictions are about evenly divided between cases when no tornadoes were sighted and cases when tornadoes occurred outside, but near the forecast areas.

While a tornado watch is designed to alert people to the possibility of tornadoes, a *tornado warning* is issued when a tornado has actually been sighted in an area or is indicated by radar. The Midwest is covered by a network of large radar sets that are capable of tracking cumulonimbus clouds as far away as 300 kilometers. When severe weather threatens, the radar screens look for very intense echoes, for they are associated with heavy precipitation, and the greater likelihood of hail, strong winds, and tornadoes. In addition, the echo from a tornadic storm within about 100 kilometers of the radar may display a hook-shaped appendage, as shown in Figure 10-13.

---

*Robert Davies-Jones and Edwin Kessler, *Weather and Climate Modification*, edited by Wilmot N. Hess (New York: John Wiley and Sons, Inc., 1974), p. 566.

If the direction and approximate speed of the storm are known, an estimate of its most probable path can be made. Since tornadoes often move erratically, the warning area is fan-shaped downwind from the point where the tornado has been spotted. Since the late 1960s warnings of from several minutes to an hour or more have been given for almost all major tornadoes. It is believed that these warnings have substantially reduced the number of deaths that might otherwise have occurred. As one might expect, the loss of life is commonly greatest in the first community struck by a tornado because downwind locations have the benefit of a longer warning time.

As was noted earlier, the probability of one place being struck by a tornado, even in the area of greatest frequency, is very slight. Nevertheless, although the probabilities may be small, tornadoes have provided many mathematical exceptions. For example, Oklahoma City has been struck 26 times since 1892, and the small town of Codell, Kansas was hit 3 years in a row–1916, 1917, and 1918–and each time on the same date, May 20! Needless to say, tornado watches and warnings should never be taken lightly.

hurricanes

The whirling tropical cyclones that on occasion have wind speeds reaching 320 kilometers per hour are known in the United States as *hurricanes*—the greatest storms on earth. Out at sea they can generate 15-meter waves capable of inflicting destruction hundreds of kilometers from their source. Should a hurricane smash into land, gale force winds coupled with extensive flooding can impose great loss of life and millions of dollars in damages. These awesome storms form in all tropical waters (except those of the South Atlantic) between the latitudes of 5 degrees and 20 degrees (Figure 10-14) and are known in each locale by a unique name. In the western Pacific they are called *typhoons*, in Australia they are called *willy-willys*, and in the Indian Ocean they are called *cyclones*. In the following discussion these storms will all be referred to as hurricanes. The North Pacific has the greatest number of these storms, averaging 20 per year. Fortunately for those living in the coastal regions of the southern and eastern United States, fewer than

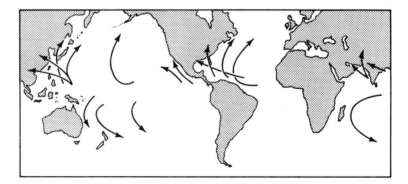

Figure 10-14
Regions of hurricane formation and paths of movement.

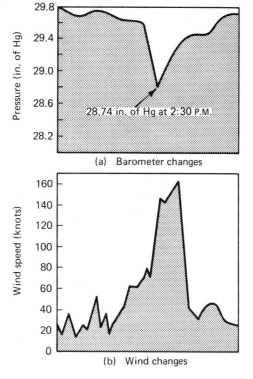

(a) Barometer changes

(b) Wind changes

Figure 10-15

Changes in (*a*) pressure and (*b*) wind speed during the passage of a hurricane at San Juan, Puerto Rico.

five hurricanes, on the average, develop each year in the warm sector of the North Atlantic.

Although hurricanes are most noted for their destruction, some parts of the world, especially eastern Asia, rely on them for much of their precipitation. Consequently, while a resort owner in Florida dreads the coming of the hurricane season, a farmer in Japan welcomes its arrival.

profile of a hurricane

Although numerous tropical storms develop each year, only a few reach hurricane status, which by international agreement requires wind speeds in excess of 119 kilometers per hour and a rotary circulation. Hurricanes average 600 kilometers in diameter and often extend 12,000 meters above the ocean surface. From the outer edge of the hurricane to the center the barometric pressure has on occasion dropped 60 millibars, from 1010 to 950 millibars (Figure 10-15). The lowest pressure ever recorded in the United States was 892.31 millibars, which was measured during a hurricane in September, 1935. A steep pressure gradient like that shown in Figure 10-16 generates the rapid, inward spiraling winds of a hurricane. As the inward rush nears the core of the storm, it is whirled upward (Figure 10-17). Upon ascending, the air condenses, generating the cumulonimbus clouds that constitute the doughnut-shaped inner structure of the hurricane. Surrounding this core are curved

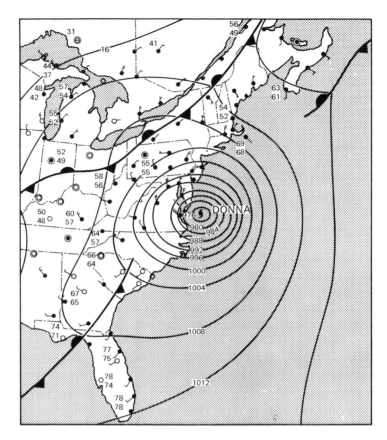

Figure 10-16

Hurricane Donna on the surface weather map, 1200 Greenwich mean time, September 12, 1960.

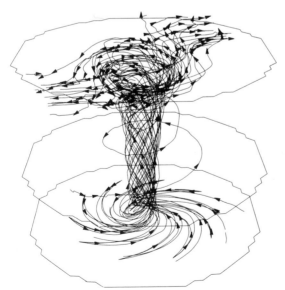

Figure 10-17

Circulation in a well-developed hurricane. Cyclonic winds are experienced at the surface. At the core of the storm the air is whirled in great spirals upward. Outward flow dominates near the top of the storm and seems to be aided by upper-level airflow.

Figure 10-18

This satellite view of a hurricane clearly
depicts the calm eye of the storm and the
bands of clouds that appear to whirl like
a pinwheel. (Courtesy of NOAA.)

bands of cumulonimbus clouds that trail away in a spiral fashion. Near the top of the hurricane the airflow is outward, carrying the rising air away from the storm center, thereby providing room for more inward flow at the surface.

At the center of the storm is the spectacular *eye* (Figure 10-18). Averaging 20 kilometers in diameter, this zone of calm and scattered cloud cover is unique to the hurricane. Separating periods of torrential rains, the eye, with its blue sky and intermittent sunlight against a background of circling clouds that are 12 kilometers high, is an awesome sight. The air within the eye slowly descends and heats by compression, making it the warmest part of the storm.

hurricane formation

A hurricane can be described as a heat engine that is fueled by the energy liberated during the condensation of water vapor (latent heat). The enormous amount of energy involved is evident when we consider that it is equal to the total amount of electricity consumed in the United States over a 6-month period. The release of latent heat warms the air and provides buoyancy for its upward flight. The result is to reduce the pressure near the surface that encourages a more rapid inward flow of air. To get this engine started, a large quantity of warm, moisture-laden air is required and a continual supply is needed to keep it going.

Hurricanes develop most often in the late summer when the water has reached 27°C or more and thus is capable of providing the warm, moist air required. This fact is thought to account for the fact that the coolest tropical ocean, the South Atlantic, does not experience hurricanes. For the same reason, hurricane formation is confined to the warm sectors of the oceans, which are located not more than 20 degrees on either side of the equator. Hurricanes are not known to form within 5 degrees of the equator, presumably because at low latitudes the Coriolis force is too weak to initiate necessary rotary motion.

Although the exact mechanism of formation is not completely understood, it is known that smaller tropical storms initiate the process. These initial disturbances are regions of low-level convergence and lifting. Many tropical disturbances like these occur each year and move westward across the warm oceans, but only a few develop into full-fledged hurricanes. It is believed that the upper-level airflow acts to further intensify selected storms by "pumping out" the rising air as it reaches the top of the storm, thus encouraging the influx of warm, moist air at the surface, which ascends and releases latent heat to fuel the storm. It seems that if the air is "pumped out" at the top faster than it is being replaced at the surface, the storm intensifies. However, if the rising air is not removed, the convergence at the surface will "fill" the storm center, equalizing the pressure differences, and the storm will die.

Whenever a hurricane moves onto land, it loses its punch rapidly, for its source of warm, water-laden air is cut off. Also, the added frictional effect of land causes the wind to move more directly into the center of the low, helping to eliminate the large pressure differences.

North Atlantic hurricanes develop in the trade winds, which generally move these storms from east to west at about 25 kilometers per hour. Then, almost without exception, hurricanes curve poleward and are deflected into the westerlies, which increase their forward motion up to a maximum of 100 kilometers per hour. Some move toward the mainland, but their irregular paths make prediction of their movement difficult (Figure 10-19).

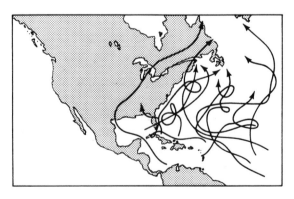

Figure 10-19

Paths of hurricanes in the North Atlantic. These represent a few of the many irregular paths hurricanes have been known to take.

A location only a few hundred kilometers from a hurricane—just one day's striking distance away—may experience clear skies and virtually no wind. Before the age of weather satellites such a situation made it difficult to warn people of impending stroms. Since the successful launching of TIROS 1* in April, 1960, which inaugurated the era of weather observation by satellites, meteorologists have been able to identify and track tropical storms even before they become hurricanes. Once a storm develops cyclonic flow, and the spiraling bands of clouds so typical of a hurricane, it receives continual monitoring.

As a typical storm approaches land, wind speeds increase, first to gale intensity of 60 kilometers per hour and eventually to over 160 kilometers per hour at the eye wall of the storm. With the increased wind speed comes torrential rain, which drops from 15 to 30 centimeters of water as it passes overhead. Flooding usually inflicts more death and destruction than wind. The greatest flooding in coastal areas is caused by the rising sea level that accompanies the low pressure of the storm. It may raise the ocean a meter, and the resulting surge of water may reach 5 meters (15 feet) above sea level as it moves into shallow waters. We can easily visualize the damage this surge of water could inflict on low-lying coastal towns. In the delta areas of Bangladesh, for example, the land is generally less than 2 meters above sea level. When a storm surge superimposed upon normal high tide inundated that area on November 13, 1970, the official death toll was 200,000; unofficial estimates ran to 500,000. This was one of the worst disasters of modern times. The worst United States death toll occurred in 1900 when 6000 persons living in the low-lying areas of Galveston, Texas, lost their lives. As one survivor explained the situation, the water level in his residence rose from a few inches to above his waist before he could turn to move. In the United States early-warning systems have reduced the death toll substantially, but the total property damage is on the upswing, mainly because of increased man-made developments in coastal areas (Figure 10-20).

One of the most destructive hurricanes to unleash its fury on the United States was Agnes. In 1972 this storm inflicted damage in excess of $3 billion and at least 118 persons lost their lives. Most of the destruction was attributed to flooding caused by an inordinate amount of rainfall. It is estimated that Agnes dropped 25.5 cubic miles of rainwater (28.1 trillion gallons) over the eastern United States. Agnes was born on June 15, 1972, as a tropical depression located in the entrance to the Gulf of Mexico between the tip of Yucatan and Cuba. Within a day she matured to become the first tropical storm of the season and was thus named Agnes (see the next section on naming hurricanes). By the following day the size of the storm increased and Agnes became a full-fledged hurricane with sustained wind speeds of 120 kilometers per hour and gusts of 150 kilometers per hour. Moving nearly straight northward, Agnes passed over the panhandle of Florida and into Georgia. Much of the destruction in these states was caused by tornadoes and wind

*Tiros stands for Television and Infra-Red Observation Satellite.

Figure 10-20

Hurricane Camille placed the fishing trawler, Wayde Klein, in the front yard of an unsuspecting resident of Biloxi, Mississippi, in August, 1969. (Courtesy NOAA.)

storms spawned by the hurricane. In Georgia Agnes' wind speed decreased and she was reduced to tropical storm status. Although some crop damage occurred in Georgia because of flooding, most farmers welcomed the rain since dry conditions prevailed prior to the storm. In fact, the value of the rains here far exceeded the losses. Unfortunately, this was not true for the northeastern portion of the United States, especially in Pennsylvania where record rainfalls were recorded. Harrisburg received 12.55 inches in 24 hours (previous record was 7.66 inches received in June, 1889) while western Schuylkill County received a record 19.0 inches in 24 hours. The widespread nature of the flooding resulted in President Nixon's declaring the entire state a disaster area. Then, ever so slowly, Agnes moved toward the northeast, leaving behind a drenched and battered landscape.

Methods devised to halt hurricane destruction cannot rely on tracking alone. They must also be designed to disarm the storm. Project Stormfury is an attempt by the National Oceanic and Atmospheric Administration and the U.S. Navy to do just that. Recall that the energy for hurricanes comes from the release of latent heat in the region of the eye wall, thus creating the steep pressure gradient that drives the winds. The object of hurricane modification is to disrupt the pressure gradient by seeding portions of the hurricane that are not in the area of maximum wind velocity. Theoretically, at least, this would cause the supercooled droplets to freeze, releasing latent heat of fusion, which would then stimulate additional cloud growth in several regions away from the eye wall. The artificial development of these secondary centers of low pressure would reduce the pressure gradient between them and the eye wall of the storm and thus reduce the wind speeds. Experimental tests on a few selected hurricanes indicate that this mechanism is an effective method of reducing some of the fury of a hurricane. Tests are being conducted each hurricane season to refine and improve the technique.

the naming of hurricanes

To many people, one of the interesting aspects of the warning system of the National Hurricane Center is the way in which these severe tropical storms are named. Most of us are familiar with the fact that women's names were used exclusively to describe each tropical storm that attained hurricane stature. However, many do not realize that feminine names were not always the rule. Furthermore, beginning with the 1978 hurricane season we had to become accustomed to hearing not only about storms named Aletta, Carlotta, and Emilia, but also to hurricanes named Bud, Daniel, and Hector. The naming of hurricanes solely after women had come to an end.

The naming of great storms stretches back at least to the early 1800s, and the Australians were naming tropical storms after women before the turn of the twentieth century. In the Caribbean, the practice for many years was to name hurricanes for the saint's day on which they occurred. This system was not used in the United States; rather our method involved a cumbersome latitude-longitude identification. However, the advent of high-speed communications and the confusion that arose when more than one tropical cyclone existed in the same area forced a change. For a time letters of the alphabet were used, and during World War II the designations were transmitted using the phonetic alphabet—Able, Baker, Charlie, and so forth. In

Table 10-1
Proposed Names, Atlantic Hurricanes, 1979–83

1979—Ana, Bob, Claudette, David, Elena, Frederic, Gloria, Henri, Isabel, Juan, Kate, Larry, Mindy, Nicolas, Odette, Peter, Rose, Sam, Teresa, Victor, Wanda.
1980—Allen, Bonnie, Charley, Danielle, Earl, Frances, Georges, Hermine, Ivan, Jeanne, Karl, Lisa, Mitch, Nicole, Otto, Paula, Richard, Shary, Tomas, Virginie, Walter.
1981—Arlene, Bret, Carla, Dennis, Emily, Floyd, Gert, Harvey, Irene, Jose, Katrina, Lenny, Maria, Nate, Ophelia, Philippe, Rita, Stan, Tammy, Vince, Wilma.
1982—Alberto, Beryl, Chris, Debby, Ernesto, Florence, Gilbert, Helene, Isaac, Joan, Keith, Leslie, Michael, Nadine, Oscar, Patty, Rafael, Sandy, Tony, Valerie, William.
1983—Alicia, Barry, Chantal, Dean, Erin, Felix, Gabrielle, Hugo, Iris, Jerry, Karen, Luis, Marilyn, Noel, Opal, Pablo, Roxanne, Sebastien, Tanya, Van, Wendy.

1951 the use of the phonetic alphabet to name hurricanes became official United States policy. However, in 1952, a new international alphabet was introduced, but this resulted in confusion as various organizations used different names for the same storms. Beginning with the 1953 hurricane season, United States agencies agreed to use common lists of female names.

The practice of using feminine names continued until 1978 when a list containing both male and female names was adopted for tropical cyclones in the eastern Pacific. In the same year a United States proposal that both male and female names be adopted for Atlantic hurricanes, beginning with the 1979 season, was accepted by the hurricane committee of the World Meteorological Organization. (Table 10-1) As was the case in the past, the new lists do not contain names that begin with the letters Q, U, X, Y, and Z because of the scarcity of names beginning with those letters.

REVIEW

1. If you hear that a cyclone is approaching, should you immediately seek shelter?
2. What is the only fact of which you can be certain when you hear that a cyclone is approaching?
3. Compare the wind speeds and the sizes of middle-latitude cyclones, tornadoes, and hurricanes.
4. How do thunderstorms differ from the storms mentioned in question 3? In what ways are thunderstorms related to each one?
5. What is the primary requirement for the formation of thunderstorms?
6. Where would you expect thunderstorms to be most common on earth? In the United States?
7. At what time of day do most thunderstorms occur? Why?
8. Explain the formation of a squall line that has formed in association with a cold front.
9. How is thunder produced? How far away is a lightning stroke if the thunder is heard 15 seconds after the lightning is seen?
10. What is heat lightning?
11. Why do tornadoes have such high wind speeds?
12. The winds of a tornado have never been measured directly. Why not?
13. What general atmospheric conditions are most conducive to the formation of tornadoes?
14. When is the "tornado season"? Can you explain why it occurs when it does? Why does the area of greatest tornado frequency migrate?
15. Why should you open the doors and windows of a house if you expect a tornado to strike?
16. Distinguish between a tornado watch and a tornado warning. Why cannot tornadoes be predicted accurately?
17. Why do some areas of the world welcome the arrival of the hurricane season?
18. After examining Figure 10-15, write a generalization relating wind speed and pressure during the passage of a hurricane.

19. Using Figure 10-16, determine the pressure gradient (in millibars per 100 kilometers) from the 1004-millibar isobar to the center of the hurricane, a distance of approximately 400 kilometers.

20. Tropical storms that form near the equator do not acquire a rotary motion as cyclones of higher latitudes do. Why?

21. Why is upper-level airflow thought to be important to the formation of hurricanes?

22. A hurricane has slower wind speeds than a tornado, but a hurricane inflicts more total damage. How might this be explained?

23. The number of deaths in the United States attributable to hurricanes has continually declined over the last 50 years, but the number of tornado deaths has increased. Write an explanation to account for this.

24. When a storm warning is issued in some parts of the United States, some people board up their windows. Are they expecting a hurricane or a tornado? Explain.

VOCABULARY
REVIEW

thunderstorm	return stroke
cumulus stage	dart leader
entrainment	thunder
mature stage	tornado
dissipating stage	tornado watch
squall line	tornado warning
lightning	hurricane
flash	typhoon
stroke	willy-willy
leader	cyclone
step leader	

Chapter Eleven

weather analysis

Modern society's ever increasing demand for more accurate weather forecasts is evident to most of us. The spectrum of needs for weather predictions ranges from the general public's desire to know if the weekend's weather will permit an outing at the beach to NASA's desire for clear skies on a particular launch date. Such diverse industries as airlines and fruit growers depend heavily upon accurate weather forecasts. In addition, the designs of buildings, smokestacks, and many industrial facilities rely on a sound knowledge of the atmosphere. We are no longer satisfied with short-range predictions, but instead are demanding accurate long-range predictions. A question such as, "Will the Northeast experience an unseasonably cold winter?" has become common. All of these demands are placing a greater burden on the National Weather Service for better and longer-range forecasts.

To produce even a short-range forecast is an enormous task involving numerous steps. Included in these are collecting, transmitting, and compiling weather data on a global scale. These data must then be analyzed so that an accurate assessment of the current conditions can be made. From the current weather patterns a number of methods are used to determine the future state of the atmosphere, a task generally called *weather forecasting*. Although the goal of this chapter is to provide insight into the job of weather forecasting, many of the procedures used in modern weather prediction are beyond the scope of the text. Consequently, we can provide only a brief overview of this important aspect of the weather business.

Before the weather can be accurately predicted, the forecaster must have a firm grasp of the current atmospheric conditions. This enormous task, called *weather analysis*, involves collecting, transmitting, and compiling millions of pieces of observational data. Because of the ever changing nature of the atmosphere, this job must be accomplished as rapidly as possible. In this regard modern high-speed computers have greatly aided the weather analyst. In addition to collecting an overwhelming quantity of data, the analyst must display it in a form that can be easily comprehended by the forecaster. This is accomplished by placing the information on a number of *synoptic weather charts* (Figure 11-1). They are called synoptic, which means coincident in time, because they display the weather conditions of the atmosphere at a given moment. These weather charts can be thought of as symbolic representations of the state of the atmosphere. Thus, to the trained eye a weather chart is a picture of the atmosphere that depicts its conditions, including motion.

A vast network of weather stations is required to produce a weather chart that will encompass enough to be useful for short-range forecasts. On a global scale, the *World Meteorological Organization*, which consists of over 130 nations, is responsible for gathering the needed data and for producing some general prognostic charts. About 10,000 surface stations located on both land and on ships at sea report the atmospheric conditions four times each day at 000, 0600, 1200, and 1800 Greenwich mean time. In addition, satellite photos and radiosonde data are used to determine conditions aloft. This network is becoming more complete each year; nevertheless, even today, vast sections of the globe, especially the large portions of the oceans outside the shipping lanes, are inadequately monitored.

Once the information is collected, it is transmitted to three World Meteorological Centers located in Melbourne, Australia, Moscow, USSR, and Washington, D.C. From these world centers the compiled data are sent to the National Meteorological Center of each participating country. In the United States the National Meteorological Center is located in Washington D.C. From here the information is further disseminated to numerous regional and

Figure 11-1

Curve follower used for drawing isobars on weather maps. (Courtesy of NOAA.)

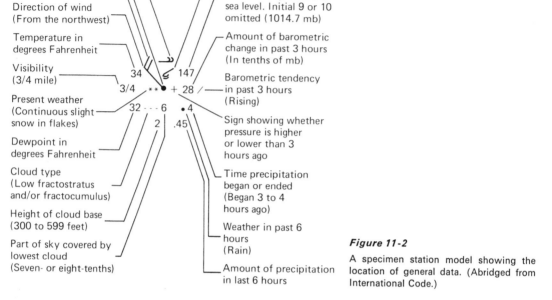

Cloud type
(High cirrus)

Total amount of clouds
(Sky completely covered)

Wind speed
(18 to 22 knots)

Direction of wind
(From the northwest)

Temperature in
degrees Fahrenheit

Visibility
(3/4 mile)

Present weather
(Continuous slight
snow in flakes)

Dewpoint in
degrees Fahrenheit

Cloud type
(Low fractostratus
and/or fractocumulus)

Height of cloud base
(300 to 599 feet)

Part of sky covered by
lowest cloud
(Seven- or eight-tenths)

Cloud type
(Middle altocumulus)

Barometric pressure at
sea level. Initial 9 or 10
omitted (1014.7 mb)

Amount of barometric
change in past 3 hours
(In tenths of mb)

Barometric tendency
in past 3 hours
(Rising)

Sign showing whether
pressure is higher
or lower than 3
hours ago

Time precipitation
began or ended
(Began 3 to 4
hours ago)

Weather in past 6
hours
(Rain)

Amount of precipitation
in last 6 hours

34   147
3/4 ·** ● + 28 /
32 - - - 6   ● 4
    2 | .45

Figure 11-2

A specimen station model showing the
location of general data. (Abridged from
International Code.)

local meteorological centers where it is used to provide more detailed forecasts for those respective areas.

The actual production of surface weather charts is accomplished by first
plotting the data from selected observing stations. By international agreement
the data are plotted by using the symbols illustrated in Figure 11-2. Normally
the data that are plotted include: temperature, dew point, pressure and its
tendency, cloud cover (height, type, and amount), wind speed and direction,
and both current and past weather. They are always plotted in the same
position around the station symbol for easy reading. For example, from
Figure 11-2 we see that the temperature is plotted in the upper-left corner
of the sample model and hence it will always appear in that location. The
only exception to this arrangement is the wind arrow that is oriented with
the direction of airflow. A more complete weather station model and a key
for decoding weather symbols are found in Appendix B. Before you read the
next section, which uses weather maps to trace the movement of a wave
cyclone, it will be advantageous for you to become familiar with the station
model if you have not already done so.

Once the data have been plotted, isobars and fronts are added to the
weather charts (Figure 11-3). On surface maps isobars are usually plotted at
intervals of 4 millibars in order to keep the map from becoming too cluttered.
The positions of the isobars are estimated as accurately as possible from the

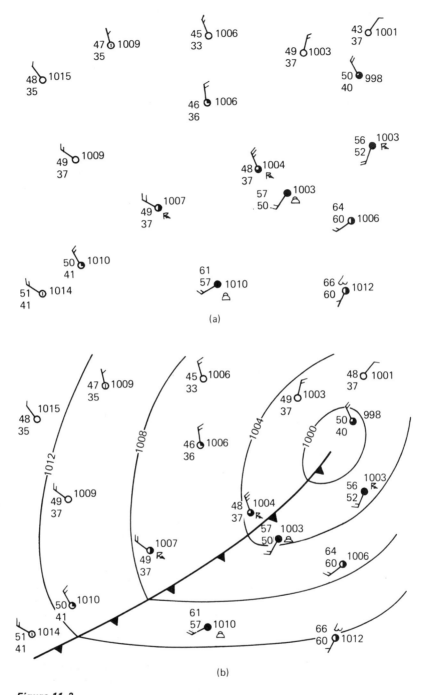

Figure 11-3

Simplified weather chart including (*a*) stations giving data for temperature, dew point, wind direction, sky cover, and barometric pressure and (*b*) isobars and cold front.

pressure readings available. For example, note in Figure 11-3 that the 1012-millibar isobar, which is located near the lower-left corner of the chart, is found about half-way between the stations reporting 1010 millibars and 1014 millibars, as you would expect. Frequently, observational errors and other complications require that the analyst smooth the isobars so that they comform to the overall picture. Many irregularities in the pressure field are caused by local influences and have little bearing on the larger circulation that the charts are attempting to depict. Once the construction of the isobars is complete, the centers of high and low pressure are designated.

Fronts, being boundaries separating contrasting air masses, can often be identified on weather charts by locating places of rather abrupt changes in conditions. Since several elements change across a front, all are examined so that the frontal position might be located most accurately. Some of the most easily recognized changes that can aid in identifying a front on a surface chart are listed below.

1. marked temperature contrast over a short distance
2. wind arrows veering (turning in a clockwise direction) by as much as 90 degrees
3. humidity variations commonly occurring across a front that can be detected by examining dew-point temperatures
4. clouds and precipitation patterns giving clues to the positions of fronts

Notice in Figure 11-3 that all of the conditions listed above are easily detected across the frontal zone, but not all fronts are as easily detected as those on our sample map, for, in some cases, surface contrasts may be rather subdued. When this happens, charts of the upper air, where flow is less complex, become an invaluable tool for detecting fronts. Other useful aids for determining the position of frontal zones include the most recently available weather charts. Generally, by examining these charts it is possible to estimate the current position of a preexisting front from the wind field.

In order to describe the atmosphere as completely as possible, surface charts and charts for several levels of the atmosphere are produced. Upper-air charts are drawn on a regular basis for 850-, 700-, 500-, 300-, 200-, and 100-millibar surfaces (Figure 11-4). Recall that on these charts height contours instead of isobars are used to depict the pressure field. Generally, a contour interval of 60 meters is used. The charts also contain isotherms as dashed lines at 5-degree intervals. This series of upper-air charts provides a three-dimensional view of the atmosphere from which the forecaster can better evaluate current conditions.

In addition to the aforementioned charts, the National Weather Service provides numerous other special-information charts. One of the most widely distributed of these is the Daily Weather Map that is constructed for 7:00 a.m. EST (1200 GMT). This map consists of a surface weather chart, a 500-millibar height contour chart, a highest and lowest temperatures chart, and a

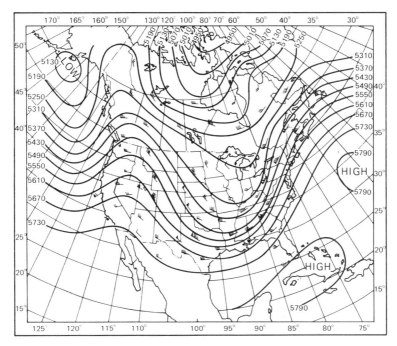

Figure 11-4
A 500-millibar height contour chart.

precipitation chart.* We can learn a great deal about the weather from careful examination of these weather charts. When understood, they provide a pictorial view of weather systems as they migrate across the United States. The following section uses the daily maps as an aid in examining the weather over the United States during a 3-day period. This discussion is intended to not only illustrate the information provided by weather charts but also to provide additional reinforcement of the wave cyclone model discussed earlier.

weather of a wave cyclone

March of 1975 followed a rather uneventful February, but the circulation of the westerlies changed with the onset of spring. We have selected a cyclonic storm that was generated during this particular month to illustrate the nature of a real wave cyclone. Our "sample" storm was one of three cyclones to migrate across the United States during the latter half of March. This cyclone reached our west coast on March 21 at a position over the Pacific a few hundred kilometers northwest of Seattle, Washington. Like many Pacific storms, this one rejuvenated over the western United States and moved east-

---

*The Daily Weather Maps are for sale by the Public Documents Department, U.S. Government Printing Office, Washington D.C., 20402, at a reasonable annual subscription rate. These maps are mailed weekly and contain a series of charts for the preceding week.

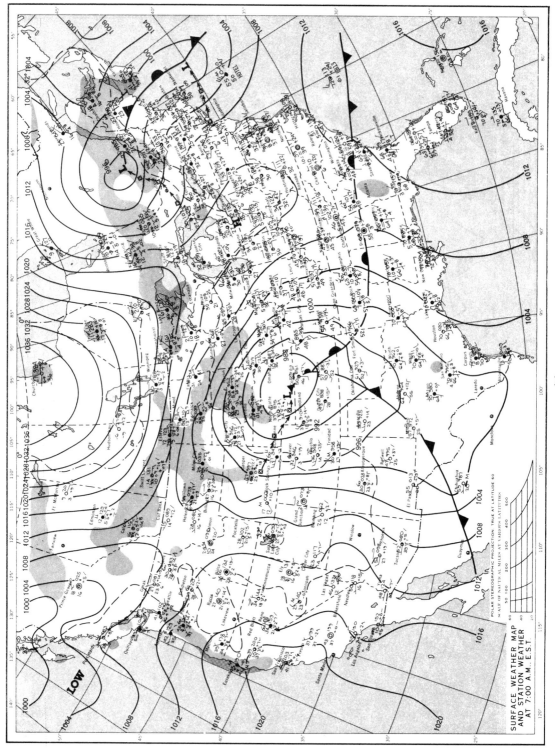

SURFACE WEATHER MAP
AND STATION WEATHER
AT 7:00 A.M. E.S.T

(a)

(b)

Figure 11-5

(a) Surface weather map for March 23, 1975. (b) Satellite photograph showing the cloud patterns on that day. (Courtesy of NOAA.)

ward into the Plains states. By the morning of March 23 it was centered over the Kansas–Nebraska border, as shown in Figure 11-5. At this time the central pressure had reached 985 millibars and its well-developed cyclonic circulation exhibited a warm front and a cold front. During the next 24 hours the forward motion of the storm's center became sluggish and it curved slowly northward to a position in northern Iowa, and the pressure continued to deepen to 982 millbars (Figure 11-6). Although the storm center advanced slowly, the associated fronts moved vigorously toward the east and somewhat northward. The northern sector of the cold front overtook the warm front and generated an occluded front, which by the morning of March 24 was oriented in nearly an east–west direction, as can be seen in Figure 11-6. This period in the storm's history marked one of the worst blizzards to ever hit

SURFACE WEATHER MAP
AND STATION WEATHER
AT 7:00 A.M., E.S.T.

POLAR STEREOGRAPHIC PROJECTION TRUE AT LATITUDE 60

(a)

(b)

Figure 11-6

(*a*) Surface weather map for March 24, 1975. (*b*) Satellite photograph showing the cloud patterns on that day. (Courtesy of NOAA.)

the north central states. While the winter storm was brewing in the north, the cold front marched from northwestern Texas to the Atlantic Ocean. During its 2-day trek to the ocean this violent cold front, with an associated squall line, generated numerous severe thunderstorms and spawned 19 tornadoes. By March 25 the low pressure had diminished in intensity (1000 millibars) and had split into two centers (Figure 11-7). Although the remnant low from this system, which was situated over the Great Lakes, generated some weather for the remainder of March 25, by the following day it had completely dissipated.

After a brief view, let us now revisit the passage of this cyclone in more detail by using the weather charts for March 23 through March 25 (Figures 11-5, 11-6, 11-7) as our guide. The weather map for March 23 clearly depicts

251

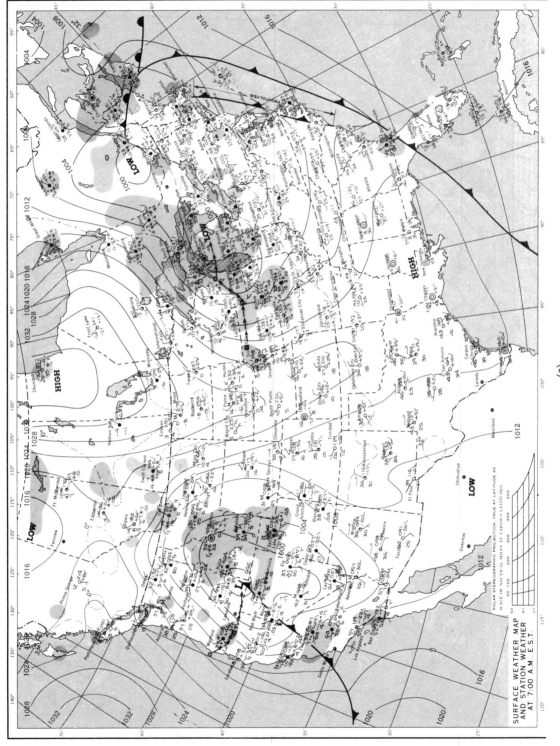

SURFACE WEATHER MAP
AND STATION WEATHER
AT 7:00 A.M. E.S.T.

(a)

POLAR STEREOGRAPHIC PROJECTION TRUE AT LATITUDE 60
SCALE OF NAUTICAL MILES AT VARIOUS LATITUDES

(b)

Figure 11-7

(a) Surface weather map for March 25, 1975. (b) Satellite photograph showing the cloud patterns on that day. (Courtesy of NOAA.)

a developing cyclone. The warm sector of this system, as exemplified by Fort Worth, Texas, is under the influence of a warm, humid air mass having a temperature of 70°F and a relative humidity of approximately 80 percent. (NOTE: The relative humidity is determined from the air temperature and dew point by using relative humidity and dewpoint tables). Notice in the warm sector that winds are from the south and are overrunning cooler air situated north of the warm front. By contrast, the air behind the cold front is 20°F to 40°F cooler than the air of the warm sector and is flowing in a southeasterly direction as depicted by the data for Roswell, New Mexico. This cold surge consists of air with below freezing temperatures in the area just north of the Canadian border.

Prior to March 23 this system had generated some precipitation in the

Northwest and as far south as California. However, on the morning of March 23, when the first map was constructed, very little activity was occurring along the fronts, but as the day progressed, the storm intensified and changed dramatically. The map for March 24 illustrates the typical mature cyclone that developed during the next 24 hours. Careful examination of a few of the past weather symbols will provide some insight into the fury of this storm. In particular, look at stations located ahead of the cold front and north of the occluded front. Further, notice on the map of March 24 the closely spaced isobars that indicate the strength of this system as it affected the circulation of the entire eastern two-thirds of the United States. A quick glance at the wind arrows reveals a strong counterclockwise flow converging on the low. The activity in the cold sector of the storm just north of the occluded front produced one of the worst March blizzards since the Great Blizzard of March, 1951. In the Duluth–Superior area winds were measured up to 81 miles per hour. Unofficial estimates of wind speeds in excess of 100 miles per hour were made on the aerial bridge connecting these cities. Twelve inches of snow were blown into 10-to 15-foot drifts and closed some roads for 3 days. One large supper club in Superior was destroyed by fire because drifts prevented fire-fighting equipment from reaching the site of the blaze.

At the other extreme was the weather produced by the cold front as it closed in on the warm, humid air flowing into the warm sector. By the late afternoon of March 23 the cold front had generated a hailstorm in parts of eastern Texas. As the cold front and squall line moved eastward, they affected all of the southeastern United States except southern Florida. Throughout this region numerous thunderstorms were spawned. Although high winds, hail, and lightning caused extensive damage, the 19 tornadoes generated by the storm caused even greater death and destruction. The path of the front can be easily traced from the reports of storm damage. By the evening of March 23 hail and wind damage was reported as far east as Mississippi and Tennessee. Early on the morning of March 24 golf-ball-size hail was reported in downtown Selma, Alabama. About 6:30 A.M. that day the "Governor's Tornado" struck Atlanta, Georgia. Here the storm displayed its worst temper. Damages were estimated at over $50 million, three lives were lost, and 152 persons were injured. The 12-mile path of the "Governor's Tornado" cut through an affluent residential area of town which included the Governor's Mansion (hence the name). It is noted in the official report on this tornado that no mobile homes lay in the tornado's path. Why do you suppose that fact was worth noting in a report on tornado destruction? The last damage along the cold front was reported at 4:00 A.M. on March 25 in northeastern Florida. Here, hail and a small tornado caused minor damage. Thus, a day and a half and some 1200 kilometers after the cold front became active in Texas, it left the United States and entered the Atlantic.

By the morning of March 24 we can see that cold polar air had penetrated deep into the United States behind the cold front (Figure 11-6). Fort Worth, Texas, which just the day before was situated in the warm sector, was now experiencing cool northwesterly winds. Below freezing temperatures had moved as far south as northern Oklahoma. However, notice that by March 25 Fort Worth was again experiencing a southerly flow. We can conclude that this is a result of the decaying cyclone that no longer dominated the circulation in the region. We can also safely assume that a warming trend was experienced in Fort Worth over the next day or so. Also notice on the map for March 25 that a high was situated over southwestern Mississippi. The clear skies and calm conditions we associate with the center of a subsiding air mass are well illustrated here.

The observant student might have already noticed another cyclone moving in from the Pacific on March 25 as the other was exiting the country. This storm developed in a manner similar to the one just described, but it was centered somewhat farther north. As you should have already guessed, another blizzard struck the northern Plains states and a few tornadoes slipped through Texas, Arkansas, and Kentucky, while precipitation dominated the weather pattern in the central and eastern United States.

After having examined the general weather associated with the passage of this cyclone from March 23 through March 25, let us now look at the weather experienced at a single location during this same time period. For this purpose we have selected Peoria, a city in central Illinois located just north of the Springfield station shown on the weather charts. Before you read the description of Peoria's weather, try to answer the following questions by using Table 11-1 which provides weather observations at 3-hour intervals during this time period. Also use the three weather charts provided and recall the general wind and temperature changes expected with the passage of fronts. If you need to refresh your memory concerning the idealized weather associated with a wave cyclone, refer to Chapter 9.

1. What type of clouds were probably present in Peoria during the early morning hours of March 23?

2. At approximately what time did the warm front pass through Peoria?

3. List two lines of evidence that indicate that a warm front did pass through Peoria.

4. How did the wind and temperature change during the early morning hours of March 23 indicate the approach of a warm front?

5. Explain the slight temperature increases experienced between 6:00 p.m. and 9:00 p.m. on March 23.

6. By what time had the cold front passed through Peoria?

7. List some changes that indicate the passage of the cold front.

8. Since the cold front had already gone through Peoria by noon on March 24, how do you account for the snow shower that occurred during the next 24 hours?

9. Did the thunderstorm in Peoria occur with the passage of the warm front, the cold front, or the occluded front?

10. Basing your answer on the apparent clearing skies late on March 25, would you expect the low temperature on March 26 to be lower or higher than on March 25? Explain.

Table 11-1

Weather Data for Peoria on March 23 to 25

		Temperature (°F)	Wind Direction	Cloud Coverage (tenths)	Visibility (miles)	Weather and Precipitation
March 23	00:00	43	ENE	5	15	
	3:00 A.M.	43	ENE	8	15	
	6:00 A.M.	42	E	5	12	
	9:00 A.M.	50	ESE	10	10	
	12:00 P.M.	61	SE	10	12	
	3:00 P.M.	64	SE	10	10	Thunderstorm with rain showers
	6:00 P.M.	64	SE	10	6	Haze
	9:00 P.M.	65	S	10	10	Thunderstorm with rain
March 24	00:00	57	WSW	10	15	
	3:00 A.M.	47	SW	2	15	
	6:00 A.M.	42	SW	6	15	
	9:00 A.M.	39	SW	8	15	
	12:00 P.M.	37	SSW	10	15	Snow showers
	3:00 P.M.	33	SW	10	10	Snow showers
	6:00 P.M.	30	WSW	10	10	Snow showers
	9:00 P.M.	26	WSW	10	6	Snow showers
March 25	00:00	26	WSW	10	15	
	3:00 A.M.	25	WSW	10	8	Snow shower
	6:00 A.M.	24	WSW	10	1	Snow
	9:00 A.M.	26	W	10	3	Snow
	12:00 P.M.	25	W	10	12	Snow
	3:00 P.M.	27	WNW	10	12	Snow
	6:00 P.M.	25	NW	9	12	
	9:00 P.M.	24	NW	4	15	

Read the following discussion to obtain a more complete description of the weather you just reviewed.

We begin our weather observations in Peoria just after midnight on March 23. The sky contains cirrus and cirrocumulus clouds, and cool winds from the ENE dominate. As the early morning hours pass, we observe a slight wind shift toward the Southeast and a very small drop in temperature. (Recall that veering winds are a sign of an approaching front.) Three hours after sunrise altocumulus clouds are replaced by stratus and nimbostratus clouds that darken the sky. However, the warm front passes without incident. The 20°F increase in temperature and the wind shift from an easterly to a southerly flow, which occurred between 6:00 A.M. and noon on March 23, marked the passage of the warm front as Peoria entered the warm sector. The pleasant 60°F temperatures were welcomed in Peoria which experienced a day 14°F above normal. But the mild weather was short-lived because the part of the cyclone that passed Peoria was near the apex of the storm where the cold front is generally close behind the warm front. By that afternoon the cold front had generated numerous thunderstorms from cumulonimbus clouds imbedded in the warm sector just ahead of the cold front. Strong winds, half-inch hail, and one tornado caused some local damage in the Peoria area. The temperature remained unseasonably warm during the thunderstorm activity. Later, during the evening of March 23, the passage of the cold front was marked by a wind shift, rapidly clearing skies, and a temperature drop of nearly 20°F, all a period of less than 4 hours. Throughout March 24 southwest winds brought cold air around the back side of the intensifying storm, as shown in Figure 11-6. Although the surface fronts had passed, this intense cyclone, with its occluded front aloft, was generating snowfall over a wide area. It is not unusual for an occluding cyclone to slow its movement as this one did. For about 24 hours snow flurries dominated Peoria's weather picture. By noon on March 25 the storm had lost its punch and the pressure began to climb. The skies began to clear and the winds became northwesterly as the once tightly wound storm weakened. The mean temperature on March 25 was 16°F below normal and the ground was covered with snow.

The effect of a spring cyclone on the weather of a midlatitude location should be very apparent from this example. Within a period of 3 days Peoria's temperatures changed from unseasonably warm to unseasonably cold. Thunderstorms with hail were followed by snow showers. It should also be evident that the north–south temperature gradient, which is most pronounced in the spring, generates these intense storms. Recall that it is the role of these storms to transfer heat from the tropics poleward. However, because of the earth's rotation, this latitudinal heat exchange is made very complex, for the Coriolis force gives the winds a zonal (west-to-east) orientation. If the earth rotated more slowly, or not at all, a more leisurely north–south flow would exist and would possibly reduce the latitudinal temperature gradient. Thus, the tropics would be cooler and the poles warmer, and the midlatitudes would not experience such intense storms.

weather forecasting

In most instances weather prediction is the ultimate goal of atmospheric research, the exceptions being the rather rare attempts at controlling the weather. It is also the most advanced area in meteorology. Because of the complex and highly quantitative nature of modern weather forecasting, we can only highlight the approaches used here, which include but are not limited to the traditional synoptic approach, statistical methods, and numerical weather prediction. The object of each of these methods is to not only project the location and possible intensification of existing pressure systems but also to determine the formation of new storm centers.

Synoptic weather forecasting was the primary method used in making weather predictions until the late 1950s. As the name implies, synoptic weather charts are the basis of these forecasts. From the careful study of weather charts over many years a set of empirical rules was established to aid

257

the forecaster in estimating the rate and direction of weather system movements. For example, when the forecaster knows the type of weather being generated along a front and is able to predict its motion, a rather accurate forecast for the affected area can be made, but since cyclonic systems change so quickly, these forecasts are generally accurate only on a short-range basis of a few hours, or perhaps a day.

Early attempts to predict cyclone development from synoptic charts relied heavily upon the analysis of surface fronts. However, since the discovery of the relationship between the flow aloft and surface weather, these efforts have been supplanted by the use of upper-air data. Recall, for example, that wave cyclones can develop without the prior existence of surface fronts. More recently it has been shown that other methods can more accurately predict the future state of the atmosphere than can be accomplished by synoptic analysis. This is particularly true for forecasts made for periods longer than 1 or 2 days. Nevertheless, empirical rules applied to synoptic charts are still used by local forecasters in their attempt to pinpoint the occurrence of specific events such as the arrival time of a storm.

Modern weather forecasting relies heavily upon *numerical weather prediction* (NWP). The word "numerical" is misleading since all types of weather forecasting are based on some quantitative data and therefore could fit under this heading. The basis for numerical weather prediction rests on the fact that the gases of the atmosphere obey a number of known physical principles. Ideally, these physical laws can be used to predict the future state of the atmosphere given the current conditions. This is analogous to predicting future positions of the moon based upon physical laws and the knowledge of its current position. However, the large number of variables that must be accounted for when considering the dynamic atmosphere makes this task extremely difficult. In order to simplify the problem, numerical models were developed which omit some of the variables by assuming that certain aspects of the atmosphere do not change with time. Although these models do not fully represent the "real" atmosphere, their usefulness in prediction has been well established. Most of these modern approaches strive to predict the flow pattern aloft. From this information meteorologists project favorable sites for cyclogenesis. However, even the most simplified models require such a vast number of calculations to be performed that they could only be used after the advent of high-speed computers (Figure 11-8). As faster computers are developed and improvements in atmospheric models are made, we can expect even greater advances in the area of weather prediction. Only a decade ago it was not uncommon to hear people say, "It should be nice tomorrow; the weather report calls for rain." This attitude no longer prevails, thanks partly to advances in NWP.

Although the accuracy of numerical weather predictions had been shown to greatly exceed results obtained from more traditional methods, the prognostic charts obtained by these techniques are rather general. Hence, the detailed aspects of the weather must still be determined by applying traditional methods to these charts. Furthermore, numerical forecasts are limited

Figure 11-8
High-speed computers and communication facilities used to formulate prognostic charts. (Courtesy of NOAA.)

by deficiencies in observational data. Stated another way, a NWP can be no more complete nor more accurate than the data that go into making it.

Statistical methods are often used in conjunction with and to supplement the NWP. Statistical procedures involve the study of past weather data in order to uncover those aspects of the weather that are good predictors of future events. Once these relationships are established, current data can then be used to project future conditions. This procedure can be used to predict the overall weather, but it is most often used to determine one aspect of the weather at a time. For example, it is frequently used to project the maximum temperature for the day at a given location. This is done by first compiling statistical data relating temperature to wind speed and direction, cloud cover, humidity, and to the season of the year. These data are displayed on charts that provide a reasonable estimate of the maximum temperature for the day from these aspects of the current conditions.

Long-range forecasting is another area in which statistical studies have proven valuable. Currently, the National Weather Service prepares weekly and monthly weather outlooks. These are not weather forecasts in the usual sense; they are estimates of the rainfall and temperatures that should be expected during these periods. These projections only indicate whether or not the region will experience near normal conditions. Detailed forecasts for more than a few days are currently beyond the capability of the National Weather Service.

The general monthly extended forecasts are produced by first constructing a mean 700-millibar contour chart for the coming month. This requires taking into account the statistical records for that season of the year and altering them based upon the known effects of such things as ocean temperatures and snow cover. Once this chart is compiled, the relationships between the flow aloft and the development and movement of surface weather patterns are considered in making a prediction for each segment of the United States.

Another statistical approach to weather forecasting is called the *analog method*. The idea here is to locate in the weather records conditions that are as nearly analogous to the current conditions as possible. Once these are found, the sequence of weather events should parallel those of the past situation. Although this seems to be a very straightforward method of prediction, it is not without its drawbacks. No two periods of weather are identical in all respects, and there are just too many variables to match. Even when two periods seem to match very well, the sequence of weather that follows may be very different in each case. The main problem with this method may well be the lack of complete enough information, which as you recall greatly limited the usefulness of numerical weather predictions.

In summary, weather forecasts are produced by using several methods. The National Weather Service primarily uses numerical weather prediction methods to generate large-scale prognostic charts. These charts are then disseminated to regional and local forecast centers that use traditional synoptic and statistical methods to generate more specific forecasts.

weather forecasting and upper-level flow

In Chapter 9 we demonstrated that a strong correlation exists between cyclonic disturbances and the wavy flow in the westerlies aloft. By comparing surface charts and charts showing flow aloft we can also see that upper-air patterns are somewhat simpler than the surface circulation. Because of these and other facts, modern weather forecasting depends heavily upon predicting the weather from changes in the upper-level flow. Once the future state of the flow pattern aloft is determined, this information is used to make predictions concerning changes in the surface pressure systems.

Although a great deal is still unknown about the wavy flow of the westerlies, some of its most basic features are understood with some degree of certainty. Among the most obvious features of the flow aloft are the changes that occur seasonally. The change in wind speed is reflected on upper-air charts by more closely spaced contour lines in the cool season. The seasonal fluctuation of wind speeds is a consequence of the seasonal variation of the temperature gradient. The steep temperature gradient across the middle latitudes in the winter months corresponds to stronger flow aloft. In addition, the position of the polar jet stream fluctuates seasonally. Its mean position migrates southward with the approach of winter and northward as summer nears. By midwinter the jet core may penetrate as far south as central Florida.

Since the paths of cyclonic systems are guided by the flow aloft, we can expect the southern tier of states to encounter most of their severe storms in the winter. During the hot summer months the storm track is situated across the northern states and some cyclones never leave Canada. The northerly storm track associated with summer also applies to Pacific storms, which move toward Alaska during the warm months, thus producing a rather long dry season for much of our west coast. In addition to the seasonal changes in storm tracks, the number of cyclones generated is also seasonal, with the largest number occurring in the cooler months when the temperature gradients are greatest. This fact is in agreement with the role of cyclonic storms in the distribution of heat across the midlatitudes.

Even in the cool season the westerly flow goes through an irregular cyclic change. There may be periods of a week or more when the flow is nearly west to east, as shown in Figure 11-9(a). Under these conditions, relatively mild temperatures occur, and few disturbances are experienced in the region south of the jet stream. Then, without warning, the upper flow begins to meander wildly and produces large-amplitude waves and a general north-to-south flow [Figure 11-9(b)]. This change allows for an influx of cold air southward that intensifies the temperature gradient and the flow aloft. During these periods cyclonic activity dominates the weather picture. For a week or more the cyclonic storms redistribute large quantities of heat across the midlatitudes by moving cold air southward and warm air northward. This redistribution

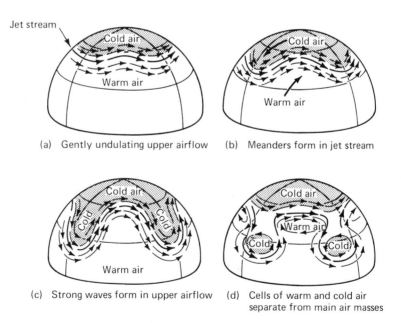

(a) Gently undulating upper airflow   (b) Meanders form in jet stream

(c) Strong waves form in upper airflow   (d) Cells of warm and cold air separate from main air masses

Figure 11-9

Cyclic changes that occur in the upper-level airflow of the westerlies. The flow, which has the jet stream as its axis, starts out nearly straight and then develops meanders that are eventually cut off. (After J. Namais, NOAA.)

261

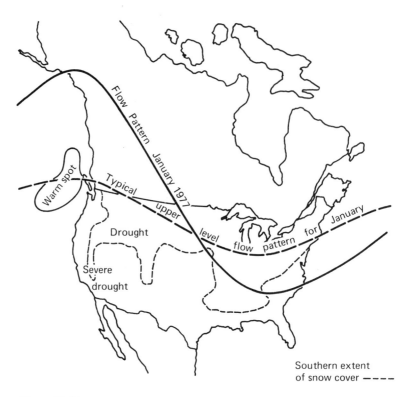

Southern extent
of snow cover ----

Figure 11-10

The unusually high amplitude experienced in the flow pattern of the prevailing westerlies during the winter of 1977 brought warmth to Alaska, drought to the west, and frigid temperatures to the central and eastern United States.

eventually results in a weakened temperature gradient and a return to a flatter flow aloft and less intense weather at the surface. These cycles, consisting of alternating periods of calm and stormy weather, can last from 1 to 6 weeks.

Because of the rather irregular behavior of the circulation patterns of the flow aloft, long-range weather prediction still remains essentially beyond the forecaster's reach. Nonetheless, numerous attempts are being made to more accurately predict changes in the upper-level flow on a long-term basis. It is hoped from this research that such eluding questions as the following can be answered: "Will next winter be colder than normal?" "Will California experience a drought next year?"

Let us now consider an example of the influence of the flow aloft on the weather for an extended period of time. As our example, we will examine an atypical winter. In a normal January an upper-air ridge is situated over the Rocky Mountains and a trough extends across the eastern two-thirds of the United States, as is shown in Figure 11-10. This "typical" flow pattern is believed to be caused by the mountains. During January, 1977, the normal flow pattern was greatly accentuated, as illustrated in Figure 11-10. The greater amplitude of the upper-level flow caused an almost continuous influx of cold air into the deep south, producing record low temperatures throughout much of the eastern and central United States. Because of dwindling natural gas supplies, many industries experienced layoffs. Much of Ohio was hit so hard that 4-day weeks and massive shutdowns were ordered. Most of the East was in the deep freeze, but the westernmost states were under the influence of a strong ridge of high pressure. Generally mild temperatures and clear skies dominated their weather picture. But that was no blessing because this ridge of high pressure blocked the movement of Pacific storms that usually provides much needed winter precipitation. The shortage of moisture was especially serious in California where January is the middle of its 3-month rainy season. Throughout most of the western states the winter rain and snow that supply water for summer irrigation was far below normal. This dilemma was compounded by the fact that the previous year's precipitation had also been far below normal and many reservoirs were almost empty. Although much of the country was concerned about economic disaster caused either by a lack of moisture or frigid temperatures, the highly accentuated flow pattern channeled unseasonably warm air into Alaska. Even Fairbanks, which generally experiences temperatures as low as 40°F below zero, had a rather mild January with numerous days having above freezing temperatures.*

Although the effects of the upper-level flow on the weather are well documented, as in the example above, for the most part the causes of these fluctuations still elude meteorologists. Numerous attempts have been made to relate temperature variations to such diverse phenomena as sunspot cycles and volcanic activity. One attempt to relate the flow pattern experienced in 1977 to ocean temperatures has been given considerable attention. Jerome Namias of the Scripps Institution suggested prior to the winter of 1977 that above average ocean temperatures in the eastern Pacific may cause a greater than average amplitude in the wavy pattern of the westerlies. It was also suggested that once snow is distributed farther south, the increases in albedo will further support the southward migration of the jet stream axis in this region. In other words, cold temperatures tend to perpetuate themselves by determining the position of the flow aloft.

forecast accuracy

At one time or another most people have asked the question, "Why are the weather forecasts that are given on radio and television so often in error?" As contradictory as it might sound, the answer lies to some degree in our desire for more accurate predictions. When weather forecasting was in its infancy, the public was pleased when even an occasional forecast came true. Today, we expect more accurate predictions, and hence we concentrate more on the relatively few incorrect forecasts and take for granted the more accurate predictions. Also, as weather forecasting became more accurate, attempts

---

*For an excellent review of the winter weather of 1976–1977 see: Thomas Y. Canby, "The Year the Weather Went Wild," *National Geographic*, CLII, No. 6 (1977), 798–829.

were made to make the forecasts even more specific. Instead of just predicting the likelihood of precipitation, a modern forecast often gives the expected times of occurrence and the probable amount. Just the other day I overheard a colleague complain that the forecast called for 3 to 6 inches of snow, and we got only an inch. At one time this forecast would have been considered correct, but today it is viewed, at least by some, as inaccurate.

In general, we are safe in saying that modern weather forecasts are relatively accurate for the immediate future (from 6 to 12 hours), and they are generally good for a day or two. However, after a forecast becomes a few days old, its accuracy decreases rapidly. Even today, specific forecasts are only made for periods of 3 to 5 days in advance and they are constantly being revised.

When making very short-range predictions of a few hours, it is difficult to improve on *persistence forecasts*. These forecasts assume that the weather occurring upstream will persist and move on and will affect the areas in its path in much the same way. Persistence forecasts are used by local forecasters in determining such events as the time of the arrival of a thunderstorm that is moving toward their region. Persistence forecasts do not account for changes that might occur in the intensity or in the path of a weather system, and they do not predict the formation or dissipation of cyclones. Because of these limitations, and the rapidity with which weather systems change, persistence forecasts break down after 12 hours, or a day at most.

For periods of up to 5 days forecasts made by numerical weather prediction methods and supplemented by traditional techniques are difficult to beat. They take into consideration both the formation and the movement of pressure systems. However, beyond 5 days specific forecasts made by the more sophisticated methods prove to be no more accurate than projections made from past climatic data. The reasons for the limited range of modern forecasting techniques are many. As stated earlier, the network of observing stations is rather incomplete. Not only are large areas of the earth's land–sea surface inadequately monitored, but data gathering in the middle and upper troposphere is meager at best, except in a few regions. Further, the laws governing the atmosphere are not completely understood and the current models of the atmosphere are not as complete as possible. Nevertheless, NWP has greatly improved the forecaster's ability to project changes in the upper-level flow. When the flow aloft can be tied more fully to surface conditions, weather forecasting should improve greatly.

You may have asked, "Just how accurately is the weather currently being forecast?" This question is more difficult to answer than you might imagine. One of the major problems is establishing when a forecast is correct. For example, if a forecast predicts a minimum temperature of 10°C and the temperature falls to 9°C, only 1 degree off, is that forecast incorrect? When a forecast calls for snow in Wisconsin and only the northern two-thirds of the state receives snow, is that forecast incorrect or is it two-thirds correct? The problems of assessing forecast accuracy are many, as you can see from these examples.

The only aspect of the weather that is predicted as a percent probability is rainfall. Here statistical data are used to indicate the number of times precipitation occurred under similar conditions. Although the occurrence of precipitation can be predicted with about 80 percent accuracy, predictions on the amount and time of occurrence of precipitation are still fairly unreliable. Probably temperatures and wind directions can be most accurately predicted.

**satellites in
weather
forecasting**

The field of meteorology entered the space age on April 1, 1960, when the first weather satellite, TIROS 1, was launched. (TIROS stands for Television and Infra-Red Observation Satellite.) The early versions of this generation of weather satellites rotated for stability and, consequently, for nearly 75 percent of the time their cameras were pointed away from the earth. Nevertheless, in its short life span of only 79 days, TIROS 1 radioed back thousands of pictures to the earth. A year before the ninth and last TIROS was launched in 1965 the first of the second-generation Nimbus (Latin for cloud) satellites was orbiting the earth (Figure 11-11). The later Nimbus satellites were equipped with infrared "cameras" capable of detecting cloud coverage at night. Another series of satellites were placed into polar orbits so that they circled the earth in a north-to-south direction. The *polar satellites* orbit the earth at rather low altitudes (a few hundred kilometers) and require only 100 minutes per orbit. By properly orienting the orbits, these satellites drift about 15 degrees westward over the earth's surface during each orbit. Thus, they are able to obtain photo coverage of the entire earth twice each day and coverage of a

Figure 11-11
Nimbus satellite (Courtesy of NOAA.)

Figure 11-12

Lined up in perfect order between Hawaii and Central America are three Pacific storms. On the left is Hurricane Kate; in the center is a tropical depression that never reached hurricane force; and on the right is Hurricane Liza which within hours smashed into LaPaz and caused millions of dollars in damage. (Courtesy of NOAA.)

rather large region in only a few hours. By 1966 *geostationary satellites* were placed over the equator. These satellites, as their name implies, remain over a fixed point because their rate of travel corresponds to the earth's rate of rotation. However, in order to keep a satellite positioned over a given site, the satellite must orbit at a great distance from the earth's surface (about 35,000 kilometers). At that altitude the speed required to keep a satellite in orbit will also keep it moving with the rotating earth. But at this distance some of the detail is lost on the photographs.

Up to this time weather satellites have not become the modern weather forecasters some had hoped. Nevertheless, they have greatly added to our knowledge of weather patterns. Most importantly, they have helped to fill gaps in observational data, especially over some parts of the oceans. For example, examine the clarity with which the clouds outline the fronts of the wave cyclone shown in Figure 9-11. These wishbone-shaped whirls can be easily traced by satellites as they migrate across even the most remote portions of our planet. Of similar significance are the frequent discoveries of developing hurricanes that are made before they enter our surface observational network (Figure 11-12). In addition, infrared photography from satellites has proven to be very useful in determining regions of possible precipitation within a cyclone. For example, compare the infrared photograph in Figure 11-14 with the photograph taken in the visible wavelengths in Figure 11-13. Notice that in the photograph taken in the visible wavelengths it is not possible to distinguish one cloud type from another, that is, it is not possible to determine which clouds may be producing rain or snow and which

266

clouds may be yielding only drizzle or no precipitation at all. All of the clouds in the visible wavelength photograph appear white. However, the infrared photograph does give us a strong indication of which clouds are the most probable precipitation producers. The reason is that in the infrared picture warm objects appear dark and cold objects appear white. Since high cloud tops are colder than low cloud tops the infrared photograph allows us to distinguish between the high thick (cold) clouds that may be producing precipitation and the thinner lower (warm) clouds that can, at best, only produce light drizzle.

Although weather satellites are still more weather observers than weather forecasters, a number of ingenious developments have made them more than just TV cameras pointed at the earth from space. It is expected that satellites will soon be able to directly or indirectly determine wind speeds, humidity, and temperatures at various altitudes. Presently, geostationary satellites are used to estimate wind speeds from cloud movements. Further, satellites are

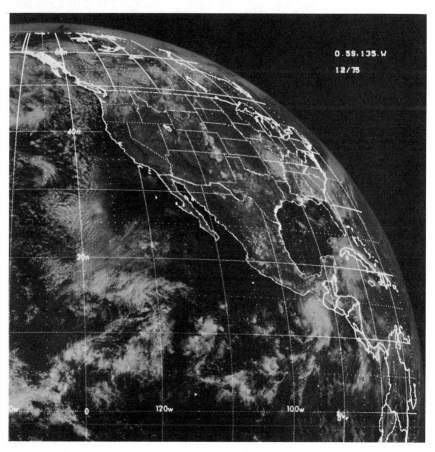

Figure 11-13

Satellite view of the cloud distribution on June 4, 1976. Compare this with the infrared photograph in Figure 11-14. (Courtesy of NOAA.)

Figure 11-14

Infrared photograph of the same cloud pattern as shown in Figure 11-13. On this image some of the clouds between the equator and 10 degrees north latitude appear much whiter than others. These are the thicker, vertically developed clouds that have cold tops. Another band of rain clouds can be easily seen running from Wyoming to New Mexico. (Courtesy of NOAA.)

being equipped with instruments designed to measure temperatures at various elevations. The principle used to perform this task involves the fact that different wavelengths of energy are absorbed and radiated from different layers of the atmosphere. For example, on a clear night the atmosphere does not absorb radiation from the earth's surface with a wavelength of 10 micrometers (recall that this is the atmospheric window). Thus, if we measure the radiation of 10 micrometers, we will have a method of determining the earth's surface temperature. Since other wavelengths indicate the altitudes within the atmosphere from which they originate, they too can be used to measure temperatures. This method of obtaining temperature readings is simpler in principle than it is in practice. Nevertheless, with refinements it is hoped that this method may supplement conventional radiosonde data. Satellites are also useful in tracking and transmitting information gathered from instruments attached to balloons. A project currently underway endeavors to send numerous long-life, constant-elevation balloons into the middle and upper troposphere. If successful, satellites could monitor the temperature and humidity instruments attached to these balloons as they migrate around the globe.

Plans are also underway to develop an extensive network of weather satellites consisting of several geostationary and polar orbiting types. The geostationary satellites would provide a continuous view of the entire earth and the polar satellites would monitor temperatures and other data. This system would also be connected by computers to the entire network of surface stations. This total picture should greatly help fill some of the current gaps in observational data that are needed to obtain accurate forecasts.

REVIEW 1. Compare the tasks of the weather analyst with those of a weather forecaster.

2. List the three types of weather forecasting procedures used by the National Weather Service and discuss the basis of each procedure.

3. Describe the statistical approach called the analog method. What are its drawbacks?

4. What features of the wavy flow of the westerlies exhibit seasonal variations?

5. What is a persistence forecast?

6. Why is the name "numerical weather forecasting" misleading?

7. The map in Figure 11-15 has the positions of several weather stations plotted on it. Using the weather data for March 5, 1964, which are given in Table 11-2, complete the following:

Figure 11-15

Table 11-2

Weather Data for March 5, 1964

Location	Temperature (°F)	Pressure (mb)	Wind Direction	Sky Cover (tenths)
Wilmington, NC	57	1006	SW	7
Philadelphia, PA	59	1001	S	10
Hartford, CT	47	1007	SE	Sky obscured
International Falls, MN	−12	1008	NE	0
Pittsburgh, PA	52	995	WSW	10
Duluth, MN	−1	1006	N	0
Sioux City, IA	11	1010	NW	0
Springfield, MO	35	1011	WNW	2
Chicago, IL	34	985	NW	10
Madison, WI	23	995	NW	10
Nashville, TN	40	1008	SW	10
Louisville, KY	40	1002	SW	5
Indianapolis, IN	35	994	W	10
Atlanta, GA	49	1010	SW	7
Huntington, WV	52	998	SW	6
Toronto, Canada	44	985	W	4
Seven Islands, Canada	24	1008	E	10
Albany, NY	50	998	SE	7
Savanna, GA	63	1012	SW	10
Jacksonville, FL	66	1013	WSW	10
Norfolk, VA	67	1005	S	10
Cleveland, OH	49	988	SW	4
Little Rock, AR	37	1014	WSW	0
Cincinnati, OH	41	997	WSW	10
Detroit, MI	44	984	SW	10
Montreal, Canada	42	993	E	10
Quebec, Canada	34	999	NE	Sky obscured

a. On a copy of Figure 11-15 plot the temperature, wind direction, pressure, and sky coverage by using the international symbols given in Appendix B.

b. Using Figure 11-3 as a guide, complete this weather map by adding isobars at 4-millibar intervals, the cold front and the warm front, and the symbol for low pressure.

c. From your knowledge of the weather associated with a middle-latitude cyclone in the spring of the year, describe the weather conditions experienced on this date at each of the following locations:

1. Philadelphia, Pennsylvania
2. Seven Islands, Canada
3. Sault Ste. Marie, Michigan
4. Sioux City, Iowa

VOCABULARY REVIEW

weather forecasting

weather analysis

synoptic weather charts

World Meteorological Organization

synoptic weather forecasting

numerical weather prediction (NWP)

Ch. 11
_____

weather analysis

statistical methods (in forecasting)

long-range forecasting

analog method

persistence forecasts

TIROS 1

polar satellites

geostationary satellites

Chapter Twelve

THE CHANGING CLIMATE

the climate of cities

The most apparent human impact on climate is the modification of the atmospheric environment by the building of cities. The construction of every factory, road, office building, and house destroys existing microclimates and creates new ones of great complexity. As far back as the early nineteenth century, Luke Howard, the Englishman who is most remembered for his cloud classification scheme, recognized that the weather in London was different from that of the surrounding rural countryside, at least in terms of reduced visibility and increased temperature. Indeed, with the coming of the Industrial Revolution in the 1800s, the trend toward urbanization accelerated, leading to significant changes in the climate in and near most cities. At the beginning of the last century, only about 2 percent of the world's population lived in cities of more than 100,000 people. Today, not only is the total world population dramatically larger, but a far greater percentage of people reside in cities. On a worldwide basis, perhaps a quarter of the population is urban and in many regions (including the United States, Western Europe, and Japan) the percentage is more than two or three times that figure.

As Table 12-1 illustrates, the climatic changes produced by urbanization involve all major surface conditions. Some of these changes are very obvious and relatively easy to measure. Others are more subtle and sometimes difficult to measure. The amount of change in any of these elements, at any time, depends upon several variables including: the extent of the urban complex, the nature of industry, site factors such as topography and proximity to water bodies, time of day, season of the year, and existing weather conditions.

272

Table 12-1

Average Climatic Changes Produced by Cities

Element	Comparison with Rural Environment
Particulate matter	10 times more
Temperature	
Annual mean	0.5–1.5°C higher
Winter	1–2°C higher
Heating degree-days	10% fewer
Solar radiation	15–30% less
Ultraviolet, winter	30% less
Ultraviolet, summer	5% less
Precipitation	5–15% more
Thunderstorm frequency	16% more
Winter	5% more
Summer	29% more
Relative humidity	6% lower
Winter	2% lower
Summer	8% lower
Cloudiness (frequency)	5–10% more
Fog (frequency)	60% more
Winter	100% more
Summer	30% more
Wind speed	25% lower
Calms	5–20% more

SOURCE: Helmut E. Landsburg, "City Air—Better or Worse," *Symposium: Air Over Cities*, U.S. Public Health Service, Taft Sanitary Eng. Center, Cincinnati, Ohio, Tech. Rept. A62-5; Helmut E. Landsburg, "Man-Made Climate Changes," *Science*, CLXX, 3964; Stanley A. Changnon, Jr., "Atmospheric Alterations from Man-Made Biospheric Changes," *Modifying the Weather*, Western Geophysical Series, Vol. 9 (Toronto: University of Victoria, 1973).

the urban heat island

An examination of Tables 12-2 and 12-3 reveals the most studied and well-documented urban climatic effect—the *urban heat island*. The term simply refers to the fact that temperatures within cities are generally higher than in rural areas. This is shown in Table 12-2 by using mean temperatures and in Table 12-3 by using a less common statistic, the number of days on which the temperature either equals or exceeds 32°C or is lower than or equal to 0°C. Although used less frequently, the data in Table 12-3 are nonetheless revealing. They show us that, in addition to the fact that mean temperatures are higher in the city, these locales also have a greater number of hot days and fewer cold days than outlying areas. The magnitude of the temperature differences as shown by these tables is probably even greater than the figures indicate, because studies have shown that temperatures observed at suburban airports are usually higher than those in truly rural environments.

As is typical, the data for Philadelphia show that the heat island is most pronounced when minimum temperatures are examined. While mean maximum temperatures are only 0.3°C to 0.4°C higher in the city, minimums are from 1.2°C to 1.7°C higher. Figure 12-1(a) shows the distribution of average

Table 12-2

Average Temperatures (in °C) for Philadelphia Airport
and Downtown Philadelphia (10-Year Averages)

	Downtown	Airport
Annual mean	13.6°	12.8°
Mean June Maximum	28.2°	27.8°
Mean December Maximum	6.7°	6.4°
Mean June Minimum	17.7°	16.5°
Mean December Minimum	−0.4°	−2.1°

SOURCE: Hans Neuberger and John Cahir, *Principles of Climatology* (New York: Holt, Rinehart and Winston, Inc., 1969), p. 128.

Table 12-3

Average Annual Number of Hot Days (≥ 32°C) and Cold
Days (≤ 0°C) at Downtown (D) and Airport (A) Stations
for Five American Cities

City	≤ 0°C	≥ 32°C
Philadelphia, PA		
D	73	32
A	89	25
Washington, D.C.		
D	68	39
A	72	33
Indianapolis, IN		
D	106	36
A	124	23
Baltimore, MD		
D	62	35
A	96	33
Pittsburgh, PA		
D	96	19
A	124	9

SOURCE: Hans Neuberger and John Cahir, *Principles of Climatology* (New York: Holt, Rinehart, and Winston, Inc., 1969), p. 128.

minimum temperatures in the Washington, D.C. metropolitan area for the 3-month winter period over a 5-year span and it also illustrates a well-developed heat island. The warmest winter temperatures occurred in the heart of the city, while the suburbs and surrounding countryside experienced

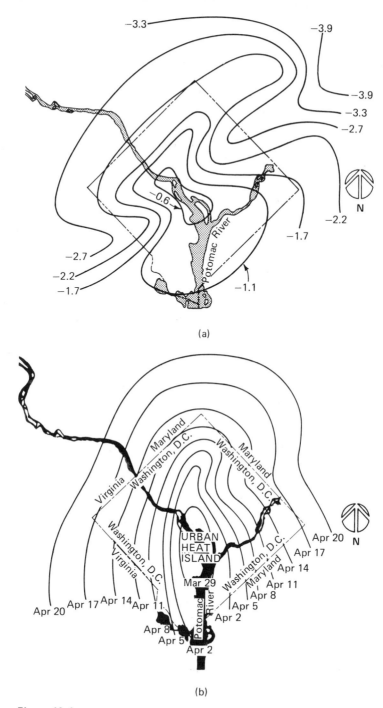

(a)

(b)

Figure 12-1

(a) The heat island of Washington, D.C. as shown by average minimum temperatures (°C) during the winter season (December through February). The city center had an average minimum that was nearly 4°C higher than some outlying areas. (b) Average dates of the last freezing temperatures in the spring for the Washington, D.C. area. [After Clarence A. Woolum, "Notes from a Study of the Microclimatology of the Washington, D.C. Area for the Winter and Spring Seasons," *Weatherwise*, XVII, no. 6 (1964), 264, 267.]

average minimum temperatures that were as much as 3.3°C lower. It should be remembered that these temperatures are averages, for on many clear, calm nights the temperature difference between the city center and the countryside was considerably greater, often 11°C or more. Conversely, on many overcast or windy nights the temperature differential approached 0°C. The map in Figure 12-1(*b*) resulted from the same study of metropolitan Washington, D.C. and reveals another interesting aspect of the heat-island effect. It shows that over the 5-year study period the last winter freeze occurred 18 days earlier in the central part of the city than in outlying areas.

Although the heat island is most obvious on clear, calm nights, the effect is still evident in long-term mean values. Isotherms for the Paris region, for example, show a pronounced heat island of about 1.6°C (Figure 12-2). A comparison of annual mean temperatures in and around London reveals similar results. While the center of London has an annual mean of 11°C, the suburbs average 10.3°C and the adjacent rural countryside has an annual mean of 9.6°C.

The radical change in the surface that results when rural areas are transformed into cities is a significant cause of the urban heat island. First, the tall buildings and the concrete and asphalt of the city absorb and store greater quantities of solar radiation than do the vegetation and soil typical of rural areas. In addition, because the city surface is impermeable, the runoff of water following a rain is rapid, resulting in a severe reduction in the evaporation rate. Hence, heat that once would have been used to convert liquid water to a gas now goes to further increase the surface temperature.

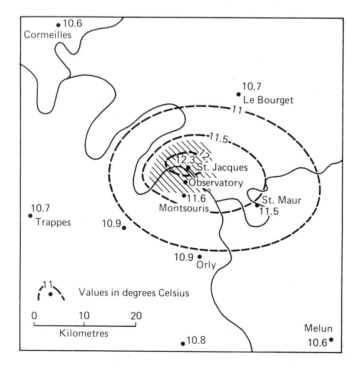

Figure 12-2

The urban heat island of Paris as shown by mean annual isotherms (°C). [After Helmut E. Landsberg, "Man-Made Climatic Changes," *Science,* 170 (1970), 1270. © 1970 by the American Association for the Advancement of Science.]

At night, while both the city and countryside cool by radiative losses, the stonelike surface of the city gradually releases the additional heat accumulated during the day, keeping the urban air warmer than that of the outlying areas.

A portion of the urban temperature rise must also be attributed to waste heat from sources such as home heating and air conditioning, power generation, industry, and transportation. Many studies have shown that the magnitude of man-made energy in metropolitan areas can be a significant fraction of the energy received from the sun at the surface. For example, investigations in Sheffield, England, and Berlin showed that the annual heat production in those cities was equal to approximately one-third of that received from solar radiation. Another study of densely built-up Manhattan in New York City revealed that during the winter the quantity of heat produced from combustion alone was two and one-half times greater than the amount of solar energy reaching the ground. In summer the figure dropped to one-sixth. It is interesting to note that during the summer there is a mutual reinforcement between the higher nighttime temperatures of the city and the man-made heat that helped create them. That is, the higher temperatures result in the increased use of air conditioners, which, in turn, use energy and further increase the amount of urban heat. During the winter the nighttime warmth of urban areas, produced in large part by heavy energy consumption, is beneficial because less energy is needed to heat buildings.

In addition to the primary causes for the heat island that were just described, other factors are also influential and worthy of mention. For example, the "blanket" of pollutants over a city, including particulate matter, water vapor, and carbon dioxide, contributes to the heat island by absorbing a portion of the upward-directed long-wave radiation emitted at the surface and reemitting some of it back to the ground. A somewhat similar effect results from the complex three-dimensional structure of the city. The vertical walls of office buildings, stores, and apartments do not allow radiation to escape as readily as in outlying rural areas where surfaces are relatively flat. As the sides of these structures emit their stored heat, a portion is reradiated between buildings instead of upward and is therefore slowly dissipated.

In addition to retarding the loss of heat from the city, tall buildings also alter the flow of air. Because of the greater surface roughness, wind speeds within an urban area are reduced. Estimates from available records suggest a decrease on the order of about 25 percent from rural values. The lower wind speeds decrease the city's ventilation by inhibiting the movement of cooler outside air which, if allowed to penetrate, would reduce the higher temperatures of the city center. Indeed, when regional winds are strong enough, the dampening effect of the city's vertical structures is not sufficient to stop the ventilating effect. Studies have shown that when the regional wind exceeds a critical value, the heat island can no longer be detected. This critical value appears to be closely related to the size of the urban area; the larger the city, the greater the wind speed necessary to eliminate the heat island (Table 12-4).

277

Table 12-4

Critical Wind Speeds for the Elimination of the Heat Island
at Various Cities

City	Population	Year of Survey	Critical Wind Speed (km/hr)
London, England	8,500,000	1959–1961	43
Montreal, Canada	2,000,000	1967–1968	40
Bremen, Germany	400,000	1933	29
Hamilton, Canada	300,000	1965–1966	22–29
Reading, England	120,000	1951–1952	14–25
Kumagaya, Japan	50,000	1956–1957	18
Palo Alto, California	33,000	1951–1952	11–18

SOURCE: James T. Peterson, *The Climate of Cities: A Survey of Recent Literature,* National
Air Pollution Control Administration Publication No. AP-59, Oct., 1969, p. 15.

In summary, the existence of the urban heat island can be linked to the
following causative factors:

1. the rocklike materials from which the city is made which have large
thermal capacities (ability to store heat) and create impervious surfaces that
lead to the rapid removal of precipitation

2. heat generated by artificial sources such as industry, motor vehicles,
and domestic heating

3. increased atmospheric pollution that inhibits the loss of upward-
directed radiation from the surface

4. the tall buildings of cities that create a three-dimensional structure
that alters the flow of air and creates a complex geometry for heat exchange

urban-induced precipitation

Most climatologists agree that cities influence the occurrence and amount
of precipitation in their vicinities. Several reasons have been proposed to
explain why an urban complex might be expected to increase precipitation.

1. The urban heat island creates thermally induced upward motions that
act to diminish the atmosphere's stability.

2. Clouds may be modified by the addition of condensation nuclei and
freezing nuclei from industrial discharges.

3. Because of the obstructions to airflow caused by the increased roughness of the urban landscape, there is an increase in low-level turbulence.

Several studies comparing urban and rural precipitation have concluded
that the amount of precipitation over a city is about 10 percent greater than
over the nearby countryside. However, more recent investigations have
shown that although cities may indeed increase their own rainfall totals, the
greatest effects may occur downwind of the city center.

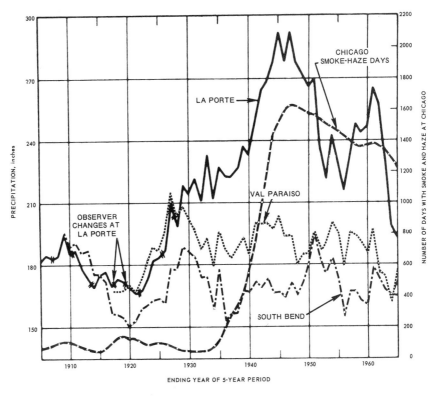

Figure 12-3

Precipitation values at selected Indiana stations and smoke-haze days at Chicago both plotted as 5-year moving totals. [From Stanley A. Changnon, Jr., "The La Porte Weather Anomaly—Fact or Fiction?" *Bulletin of the American Meteorological Society*, XLIX, no. 1 (1968), 5.]

A striking and controversial example of such a downwind effect was examined by Stanley Changnon in the late 1960s.* Records indicated that since 1925 La Porte, Indiana, located 48 kilometers downwind of the large complex of industries at Chicago, experienced a notable increase in total precipitation, number of rainy days, number of thunderstorm days, and number of days with hail (Table 12-5). The magnitude of the changes and the absence of such change in the surrounding area led to widespread public attention. Many questioned whether the anomaly was real or simply the result of such factors as observer error and changes in the exposure of instruments. However, after completing his study, Changnon concluded that the observed differences were real and were likely produced by the large industrial complex west of La Porte. Among the reasons for this conclusion was the fact that the number of days with smoke and haze (a measure of atmospheric pollution) at Chicago after 1930 corresponded very well with the La Porte precipitation curve. An examination of Figure 12-3, for example, shows a marked

---

*Stanley A. Changnon, Jr., "The La Porte Weather Anomaly—Fact or Fiction?," *Bulletin of the American Meteorological Society*, XLIX, No. 1 (1968), 4–11. (*Note:* the term anomaly refers to something that deviates in excess of normal variation.)

increase in smoke and haze days after 1940 when the La Porte curve began its sharp rise. Further, the decrease in smoke and haze days after a peak in 1947 also generally matches a drop in the La Porte curve. A second urban-related factor, steel production, was also found to correlate with the La Porte precipitation curve. Records showed that seven peaks in steel production between 1923 and 1962 were all associated with highs in the La Porte curve.

Table 12-5

Difference Between Average La Porte Weather Values for
1951–1955 and the Means at Surrounding Stations Expressed
as a Percent of the Means

Annual precipitation	31
Warm season precipitation	28
Annual number of days with precipitation	
$\geq$ 0.25 inch	34
Annual number of thunderstorm days	38
Annual number of hail days	246

SOURCE: Stanley A. Changnon, Jr., "The La Porte Weather Anomaly— Fact or Fiction?" *Bulletin of the American Meteorological Society,* XLIX, No. 1 (1968), 11.

The conclusions were accepted very cautiously by many and were criticized by others who believed that, because of the high natural rainfall variability of the area, there was neither sufficient nor accurate enough data to make a sound determination. Nevertheless, this study was a pioneering effort that illustrated the possible effect of human activity on local climates.

Since the La Porte investigation subsequent studies of the complex problem of urban effects on precipitation have generally confirmed the view that cities increase rainfall in downwind areas. The results of one study involving eight American cities revealed increases ranging from 9 to 27 percent over outlying rural areas at six of the cities. In addition, the incidence of thunderstorms and hailstorms also showed an increase (Table 12-6).

A second, more comprehensive urban rain investigation is centered in St. Louis, Missouri. Here, an elaborate field experiment, involving a wide variety of equipment and several research groups, is attempting to determine how, when, and where an urban complex influences atmospheric behavior, especially precipitation. Preliminary studies have already indicated that the St. Louis urban-industrial complex definitely affects the distribution of precipitation in and downwind of the city's center. Data showed that during a 27-year period (1941–1968) there was increased average precipitation of about 10 percent in the downwind area, with values for individual stations ranging from 6 to 15 percent. Although factors such as topography were examined as possible causes for the significant downwind increase in rainfall, the evidence seems to point to the city as the primary cause. Discussing the higher summer rainfall at a downwind station relative to the urban area,

Table 12-6

Areas of Maximum Increases (Urban–Rural Difference)
in Summer Rainfall and Severe Weather Events

	Rainfall		Thunderstorms		Hailstorms	
Name	Percent-age	Loca-tion*	Percent-age	Loca-tion*	Percent-age	Loca-tion*
St. Louis	+15	b	+25	b	+276	c
Chicago	+17	c	+38	a, b, c	+246	c
Cleveland	+27	c	+42	a, b	+ 90	c
Indianapolis	0	—	0	—	0	—
Washington, D.C.	+ 9	c	+36	a	+ 67	b
Houston	+ 9	a	+10	a, b	+430	b
New Orleans	+10	a	+27	a	+350	a, b
Tulsa	0	—	0	—	0	—

SOURCE: Stanley A. Changnon, Jr., "Atmospheric Alterations From Man-Made Biospheric Changes," *Modifying the Weather,* Western Geographical Series, vol. 9 (Toronto: University of Victoria, 1973).
*a = within city perimeter; b = 5 to 15 miles downwind, c = 15 to 40 miles downwind.

Huff and Changnon state, "The trend was most pronounced in the last 15 years of the sampling period, 1954–1968. If urban-induced, this observed trend would be expected with gradual expansion of the urban-industrial complex; if topographically related, the trend should not occur."*

In addition to the fact that the magnitude of the downwind precipitation increase climbed as industrial development expanded, a second line of evidence also supports the hypothesis that cities are responsible for these increases. If cities are really modifying precipitation, the effects should be greater on weekdays when urban activities are most intense than on weekends when much of the activity ceases. This is indeed the case. In the St. Louis study analysis of precipitation on weekdays versus weekends revealed a significantly greater frequency of rain per weekday than per weekend day in the affected downwind area. The graph in Figure 12-4 shows similar findings that have been reported for Paris. Here a rise in precipitation can be noted from Monday through Friday, while the amounts for Saturday and Sunday are considerably smaller. Referring to this relationship and to city-induced precipitation in general, Changnon has stated that, "The reality of these urban rain changes has been questioned as has the importance of the various mechanisms involved, but urban differences in rain on weekends and weekdays have been important in proving the reality of such urban effects and in indicating that the urban aerosol explanation for them is the most satisfactory."†

---

*F. A. Huff and S. A. Changnon, Jr., "Climatological Assessment of Urban Effects on Precipitation at St. Louis," *Journal of Applied Meteorology*, XI, No. 3 (1972), 829.

†Stanley A. Changnon, Jr., "Atmospheric Alterations From Man-Made Biospheric Changes," *Modifying the Weather*, Western Geographical Series, Vol. 9 (Toronto: University of Victoria, 1973), p. 143.

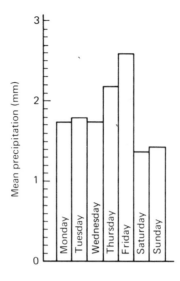

Figure 12-4

Average precipitation by day of the week in Paris for an 8-year period (1960–1967). Notice the gradual increase from Monday to Friday and the sharp drop in precipitation totals on Saturday and Sunday. The average for weekdays was 1.93 millimeters compared to only 1.47 on weekends, a difference of 24 percent. (After I. Dettwiller in Helmut E. Landsberg, "Inadvertent Atmospheric Modification," in *Weather and Climate Modification,* edited by Wilmot N. Hess, John Wiley & Sons, Inc., 1974, p. 755.)

other urban effects

In earlier sections we saw that the greater air pollution in cities contributes to the heat island by inhibiting the loss of long-wave radiation at night and has a "cloud seeding" effect that is believed to increase precipitation in and downwind of cities. These influences, however, are not the only ways in which pollutants influence urban climates. For example, the blanket of particulates over most large cities significantly reduces the amount of solar radiation reaching the surface. As Table 12-1 indicates, the overall reduction in the receipt of solar energy is 15 percent, while short wavelength ultraviolet is decreased by up to 30 percent. This weakening of solar radiation is, of course, variable. During air pollution episodes the decrease will be much greater than for periods when ventilation is good. Furthermore, particles are most effective in reducing solar radiation near the surface when the sun angle is low because the length of the path through the dust increases as the sun angle drops. Hence, for a given quantity of particulate matter, solar energy will be reduced by the largest percentage at higher-latitude cities and during the winter.

Even though relative humidities are generally lower in cities, because temperatures are higher and evaporation is reduced, the frequency of fogs and the amount of cloudiness are nevertheless greater (see Table 12-1). It is likely that the great quantities of condensation nuclei produced by human activities in urban areas is a major contributor to these phenomena. When hygroscopic (water-seeking) nuclei are plentiful, water vapor readily condenses on them, even when the air is not yet saturated. Although cities have been shown to increase precipitation, some meteorologists are concerned that because of a large oversupply of condensation nuclei too many particles will compete for available moisture. As a consequence, cloud droplets will be smaller and the coalescence process will be inhibited. However, as air

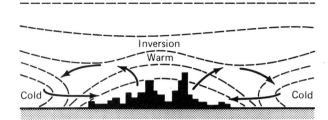

Figure 12-5
Idealized scheme of nighttime circulation above a city on a clear, calm night. The dashed lines are isotherms and the arrows represent the wind (country breeze). [After Helmut E. Landsberg, "Man-Made Climatic Changes," *Science,* 170 (1970), 1271. © 1970 by the American Association for the Advancement of Science.]

pollution abatement efforts increase, this possibility should diminish as should the city's effect on fogs and clouds.

Finally, mention should be made of another urban-induced phenomenon termed the *country breeze.* As the name implies, this circulation pattern is characterized by a light wind blowing into the city from the surrounding countryside. It is best developed on relatively clear, calm nights, that is, on nights when the heat island is most pronounced. Heating in the city creates upward air motion, which, in turn, initiates the country-to-city flow (Figure 12-5). One investigation in Toronto showed that the heat island created a rural–city pressure difference that was sufficient to cause an inward and counterclockwise circulation centered on the downtown area. When this circulation pattern exists, pollutants emitted near the urban perimeter tend to drift in and concentrate near the city's center.

is our climate changing?

Not too many years ago the concept of *climatic change* was perceived to be a subject that had little but academic importance, for the problems most often investigated related primarily to the remote past. "What caused the Ice Age?" seemed to be the most debated question.* Indeed, in the nineteenth and early twentieth centuries it was assumed that climatic changes were a thing of the past, or at least that such changes occurred only over vast periods of geologic time. All of the observed departures from mean values were thought to be nothing more than what we now call statistical noise. However, since that time, and especially in the past decade or two, scientists have come to recognize that climate is inherently variable on virtually all time scales. It is no longer described as static but rather as dynamic. Furthermore, it is not just the scientific community that has shown increasing interest in climatic change. In recent years governments as well as the general public have also become aware of and shown an interest and concern in the possible variability of our planet's climate.

As to why climatic change has become a much discussed topic and a matter of considerable concern for the future, we can point to the following reasons:

---

*It is still a much discussed problem that has not yet found a definitive answer.

283

1. Recent detailed reconstructions of past climates show that climate has varied on all time scales, which suggests that climate in the future will more likely be different from the present than stay the same.

2. Increased interest and research into human activities and their effect on the environment have created suspicions that people have or will inadvertently change the climate.

3. There is observational evidence that global temperatures have been dropping for the past three decades and, in addition, that, at least in some respects, world climate has become more variable.

The last statement is perhaps the most important reason for the public's awareness and government's increasing concern for the issue. For several decades the climate had been generally favorable for crop production. For the most part, stockpiles of food grew faster than the population, and energy consumption (at least in developed countries) was not a problem—costs were relatively low and supplies seemed limitless. Then in the 1970s several events focused attention on global climate and its possible change. Devastating droughts in Africa, India, China, and the Soviet Union, for example, caused a serious depletion of grain reserves and led to sharply higher food prices. At the same time, other world regions were recording excessive rainfalls and floods. In January, 1977, many cities in the United States experienced the coldest winter on record, and natural gas (at any price) became critically scarce; many schools and factories were forced to close in order to conserve fuel. Concurrently, several western states were experiencing severe drought conditions; water rationing became the rule in several locales.

At this point it should be noted and emphasized that the fact that a given winter was warmer than the last or that a given summer was among the driest on record does not prove that climate is changing. The list of weather peculiarities over the past several years is lengthy, yet it provides no clear-cut indication of our planet's immediate or long-term climatic future. Many years of data are required to indicate a trend in the climate, but even then there is often considerable disagreement among atmospheric scientists about its meaning and cause.

As was mentioned previously, observational data indicate that the mean global temperature has dropped by approximately 0.5°C since the 1940s. It is the feeling of many scientists who have studied climatic history that this cooling trend will continue into the twenty-first century. Some even strongly believe that the downward trend in global temperatures signals the beginning of another ice age. Other scholars, however, have just the opposite point of view. They argue that the cooling has or will soon "bottom out." They then foresee a gradual rise in world temperatures. Thus, instead of an impending ice age, some of the advocates of global warming believe that existing glaciers may shrink and sea level may rise. Still another group of climatologists is reserving judgment about the nature and intensity of any future climate change on earth. Although this group realizes that climatic change is an historic fact and indeed inevitable, it believes that the current evidence of change is inadequate. Predicting the future trend of global climate now, this group feels, would be premature, for the interrelationships within the earth's climatic system between the atmosphere, the oceans, and the continents are far too complex to be characterized by existing data.

How can so many scientists disagree about the facts? Who is right? Will our planet become *Hothouse Earth* or will it experience *The Cooling*?* The fact of the matter is no one can really be sure. Richard J. Kopec has summed up the situation this way:

> The facts, per se, are irrefutable. It is their interpretation and the predictions resulting from these interpretations which have caused divergent views among climatologists. As yet, there is no deterministic, predictive model of our planet's climate, and until one is developed, predictions are as valid as the logic producing them. In addition, the periods of time involved in climatic predictions cover centuries and the validity of climate forecasting is not easily tested. As a result, there exist several opposing, and equally convincing, viewpoints concerning this very important topic.†

Tens of thousands of pages have been written about climatic change and tens of thousands more can be expected. In the limited space available here it is not possible to discuss the large number of past climate changes that science has documented and it is not possible to deal fully with the many thoughts about the future. Such an encyclopedic recitation of facts would serve little purpose here. Instead, we will briefly examine climates of the past and look at some of the techniques scientists have devised to decipher these fluctuations. Then we will review several of the major theories that have been proposed as possible answers to the question of why climates change.

deciphering past climates

Since instrumental records of climate elements go back only a couple of hundred years (at best), how do scientists find out about climates and climatic changes prior to that time? The obvious answer is that they must reconstruct past climates from indirect evidence, that is, they must examine and analyze phenomena that respond to and reflect changing atmospheric conditions. In the following discussion we will briefly look at some of these techniques. Keep in mind that these reconstructions may capture no more than the most general features of climate.

Today the hippopotamus is confined to the tropics of East Africa, but 100,000 years ago herds of these giant mammals lived far to the north in present-day England. This is known because paleontologists have unearthed

_____

*The author of *Hothouse Earth*, H. A. Wilcox (New York: Praeger Publishing, 1975), focuses on theories that predict a global warming, while L. Ponte, author of *The Cooling* (Englewood Cliffs, N.J.: Prentice-Hall, Inc., 1976), presents a case for a possible new ice age.

†Richard J. Kopec, editor, *Atmospheric Quality and Climatic Change*, Studies in Geography No. 9 (Chapel Hill: University of North Carolina, 1976), p. 2.

abundant fossil bones and teeth of these warmth-loving animals. Pollen of the tundra plant *Armeria sibirica* has been collected from a layer of sediment in southeastern Massachusetts that has been dated by radiocarbon techniques at 12,000 years old. Today this plant is found only in the Arctic tundra of northern Canada. Obviously, the climate in England 100,000 years ago and the conditions in Massachusetts 12,000 years ago do not coincide with the present-day environments of these places. England must have been more tropical, and Massachusetts must have resembled Arctic Canada.

The fossilized remains of animal and plant life such as those just cited provide important clues from which inferences are made about the duration and geographic extent of climatic conditions in the geologic past. However, since fossil evidence is meager or nonexistent in the geologic record prior to the beginning of the Paleozoic era (about 550 million years ago), the very early history of our planet, from its formation 4.6 billion years ago until the start of the Paleozoic era, is not sufficiently well-known to determine these most ancient of climate changes. In addition to fossils, other types of "traditional" geologic evidence give indications of past climatic conditions. Coal beds in Antarctica, glacial features in Africa, solidified sand dunes outcropping in humid regions, and countless other examples have been made availabe to climatology from the science of geology. It should be noted before going further that much of this geologic evidence (both fossils and other types) provides us with only a broad and generalized picture of past climatic changes on a scale of thousands to millions of years, as Figure 12-6 illustrates. Such data give us no clues to short-term variations in climate.

Among the most interesting and important techniques for analyzing the earth's climatic history on a scale of hundreds to thousands of years are the study of ocean floor sediments and oxygen isotope analysis. Both methods are relatively recent developments used to reconstruct past temperatures and each, in part, is related to the other.

Although sea-floor sediments are of many types, most contain the remains of organisms that once lived near the sea surface (the ocean–atmosphere interface). When such near-surface organisms die, their shells slowly settle to the floor of the ocean where they become part of the sedimentary record. One reason why sea-floor sediments are useful recorders of worldwide climatic

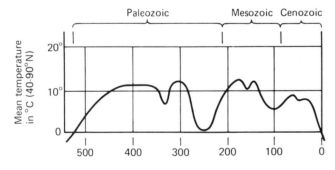

Figure 12-6

Variations of temperature during geologic time as estimated from fossil evidence. (After C. E. P. Brooks, "Geologic and Historical Aspects of Climatic Change," in *Compendium of Meteorology*, Boston: American Meteorological Society, 1951.)

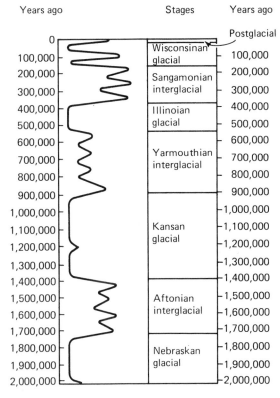

Colder   Warmer

Years ago   Stages   Years ago

Postglacial

Figure 12-7

Temperature variations during the ice age were revealed by the study of small fossils in deep-sea sediments. [After D. B. Ericson and G. Wollin, "Pleistocene Climates and Chronology in Deep-Sea Sediments," *Science*, CLVII, no. 3859 (1968), 1233. © 1968 by the American Association for the Advancement of Science.]

change is that the numbers and types of organisms living near the sea surface change as the climate changes. This principle is explained by Richard Foster Flint as follows:

> ... we would expect that in any area of the ocean/atmosphere interface the average annual temperature of the surface water of the ocean would approximate that of the contiguous atmosphere. The temperature equilibrium established between surface seawater and the air above it should mean that ... changes in climate should be reflected in changes in organisms living near the surface of the deep sea .... When we recall that the sea-floor sediments in vast areas of the ocean consist mainly of shells of pelagic foraminifers, and that these animals are sensitive to variations in water temperature, the connection between such sediments and climatic change becomes obvious.*

Thus, in seeking to understand climatic change as well as other environmental transformations, scientists have become increasingly interested in the huge reservoir of data in sea-floor sediments. One reconstruction of Ice Age climatic fluctuations based on deep-sea sediments is shown in Figure 12-7.

Glacial and Quaternary Geology (New York: John Wiley & Sons, Inc., 1971), p. 718.

Since 1970 the specially constructed research vessel, *Glomar Challenger*, has greatly aided this quest, for it is capable of drilling and collecting deep-sea sediments that could not be reached earlier (Figure 12-8).

The second technique, *oxygen isotope analysis*, is based on precise measurement of the ratio between two isotopes of oxygen—^{16}O, which is the most common, and the heavier ^{18}O. Since the lighter isotope, ^{16}O, evaporates more readily from the oceans, precipitation (and hence the glacial ice that it may form) is enriched in ^{16}O. Of course, this leaves a greater concentration of the heavier isotope, ^{18}O, in the ocean water. Thus, during periods when glaciers are extensive the concentration of ^{18}O in seawater increases, and, conversely, during warmer interglacial periods when the amount of glacial ice drops dramatically the amount of ^{18}O relative to ^{16}O in ocean water also drops. As certain microorganisms secrete their shells of calcium carbonate ($CaCO_3$), the prevailing $^{18}O/^{16}O$ ratio is reflected in the composition of these hard parts. Consequently, periods of glacial activity can be determined from variations in the oxygen isotope ratio found in shells of certain microorganisms buried in deep-sea sediments.

A second use of the $^{18}O/^{16}O$ ratio technique is applied to the study of cores taken from ice sheets, such as the one that covers Greenland. In this application another cause for variation in the oxygen isotope ratio is used, namely, that the ratio is influenced by temperature. More ^{18}O is evaporated from the oceans when temperatures are high and less is evaporated when temperatures are low. Thus, the heavy isotope is more abundant in the precipitation of warm eras and less abundant during colder periods. Using this principle, scientists studying the layers of ice and snow in Greenland have been able to produce a record of past temperature changes. A portion of this record is shown in Figure 12-9.

There are many other methods that have been used to gain insight into past climates. For example, since climate has a major effect upon soil develop-

Figure 12-8

The *Glomar Challenger,* the research vessel for the National Science Foundation's Deep Sea Drilling Project. This ship is capable of conducting drilling operations in the open ocean. (Courtesy of Deep Sea Drilling Project, Scripps Institution of Oceanography.)

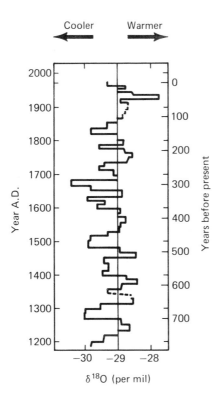

Cooler ← → Warmer

Figure 12-9

Temperature variations as revealed by differences in the $^{18}O/^{16}O$ ratio in Greenland ice cores. A decrease of 1 per mil in the ^{18}O content corresponds to a 1.5°C drop in air temperature. [After Wallace S. Broecker, "Climatic Change: Are We on the Brink of a Pronounced Global Warming?" *Science*, CLXXXIX, no. 3859 (1975), 462. © 1975 by the American Association for the Advancement of Science.]

ment and the growth and nature of vegetation, the study of buried soils (*paleosols*) and the analysis of the yearly growth rings of trees have been used to examine local climate history.

Still another method is the use of historical documents. Although one might at first have the notion that such records should readily lend themselves to climatic analysis, this is not the case. The reason is that most manuscripts were written for purposes other than climatic description. Furthermore, writers often had a tendency to neglect periods of relatively stable atmospheric conditions and to mention only events such as droughts, severe storms, and other extremes. Nevertheless, records of such things as crops, floods, and the migration of people have furnished useful evidence of the possible influence of a changing climate.

Before ending this discussion about how past changes in climate elements are deciphered, mention should be made of the time period represented by records of instrumental observations. Although one might expect that such records would cause no speculation about the nature of recent climatic trends and fluctuations, this is not the case. The reasons for this unexpected problem are summarized by Howard J. Critchfield:

> ... climatic records are not readily subjected to objective study. In the first place, weather observers are subject to human failings, and even small errors affect calculations that may involve equally small trends. The exposure and height above ground of instruments also materially affect results.

Removal of a weather station to a new location practically destroys the value of its records for purposes of studying climatic change. But even if a station remains in the same location for a century, the changes in vegetation, drainage, surrounding buildings, and atmospheric pollution are likely to produce a greater effect on the climatic record than any true climatic changes. Thus, very careful checking and comparison of climatic records are necessary to detect climatic fluctuations.*

theories of climatic change

The theories that have been proposed to explain climatic change are many and varied, to say the least. Several have gained relatively wide support, only to subsequently lose it, and then, in some cases, regain it again. It is safe to say that most, if not all, theories are controversial. This is to be expected when one considers them in light of a point made earlier, namely, that there is presently no deterministic predictive model of the earth's climate; hence each theory is as valid as the logic behind it. In this section we will examine several current theories that to varying degrees have gained support from a portion of the scientific community. Four of the theories deal with "natural" causes of climatic change, that is, causes unrelated to human activities. These include changes brought about by continental drift and volcanic activity as well as variations caused by fluctuations in solar output and changes in the earth's orbit. Two other theories relate primarily to possible man-made climatic changes. These will examine the possible effects of rising carbon dioxide levels caused primarily by combustion processes and the influence of increased dust loading of the atmosphere caused by human activities.

As you read this section, you will find that there is often more than one logical way to explain the same climatic change. Further, it should be pointed out that no single theory explains climatic change on all time scales. A theory that explains variations over millions of years, for example, is generally not satisfactory when dealing with fluctuations over a span of just hundreds of years. When (or perhaps, if) the time comes when our atmosphere and its changes through time are fully understood, we will likely see that many of the factors discussed here, as well as others, will be found to influence climatic change.

drifting continents and climatic change

Over the past 15 to 20 years a revolutionary theory has emerged from the science of geology—the *plate tectonic theory*. This theory, which has gained very wide acceptance among scientists, states that the outer portion of the earth is made up of several individual pieces, called plates, which move in relation to one another upon a partially molten zone below. With the exception of the plate that encompasses the Pacific Ocean Basin, the other large plates are made up of both continental and oceanic crust (Figure 12-10).

General Climatology, 3rd ed. (Englewood Cliffs, N.J.: Prentice-Hall, Inc., 1974), p. 376.

WORLD PLATE BOUNDARIES

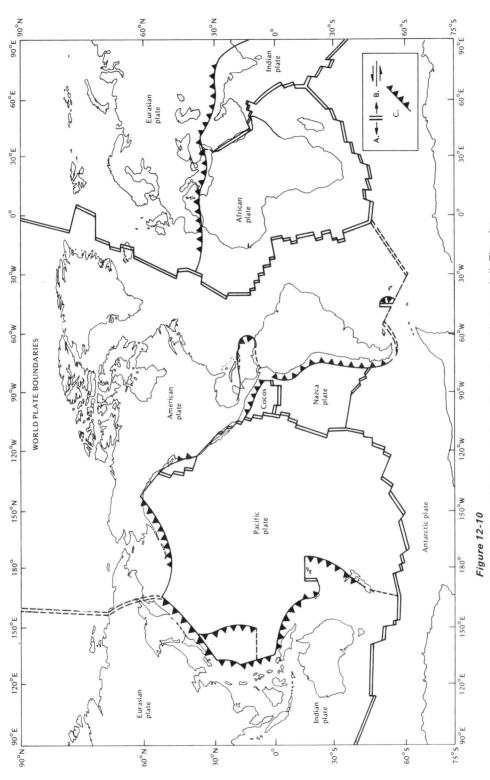

Figure 12-10

The mosaic of rigid plates that constitutes the earth's outer shell. The plate tectonic theory states that the outer portion of the earth, called the lithosphere, is made up of several individual plates. These plates move in relation to one another upon a partially molten zone below. (After F. J. Sawkins, Clement G. Chase, David G. Darby, George Rapp, Jr., *The Evolving Earth,* © 1974 by The Macmillan Company, reprinted by permission of the publisher.)

Thus, as plates move, the continents also change positions. Not only does this theory provide the geologist with explanations about many previously misunderstood processes and features, it also provides the climatologist with a probable explanation for some hitherto unexplainable climatic changes. For example, glacial features in present-day Africa, Australia, South America, and India indicate that these regions experienced an ice age near the end of the Paleozoic era, about 230 million years ago. For many years this puzzled scientists. Was the climate in these relatively tropical latitudes once like it is today in Greenland and Antarctica? Until the plate tectonic theory was formulated and proven, there was no reasonable explanation. Today scientists realize that the areas containing these ancient glacial features were joined together as a single "supercontinent" that was located at high latitudes far to the south of their present positions. Later, this land mass broke apart and its pieces, each moving on a different plate, drifted toward their present locations. Hence, large fragments of glaciated terrain ended up in widely scattered subtropical locations (Figure 12-11).

It is now believed that during the geologic past continental drift accounted for many other equally dramatic climate changes as land masses shifted in relation to one another and moved to different latitudinal positions. Changes in oceanic circulation also must have occurred, altering the transport of heat and moisture, and consequently the climate as well. Since the rate of plate movement is very slow, on the order of a few centimeters per year, appreciable changes in the positions of the continents occur only over great spans of geologic time. Thus, climatic changes brought about by continental drift are extremely gradual and happen on a scale of millions of years. As a result, the theory of plate tectonics is not useful for explaining climatic variations that occur on shorter time scales such as tens, hundreds, or thousands of years. Other explanations must be sought to explain these changes.

volcanic dust theory

Like plate tectonics, the *volcanic dust theory* is a "geological" theory of climatic change. First proposed many years ago, it is still regarded as a plausible explanation for some aspects of climatic variability.

Explosive volcanic eruptions emit great quantitites of fine-grained debris into the atmosphere. Some of the biggest are sufficiently powerful to inject dust and ash high into the stratosphere where it is spread around the globe and where it may remain for several years. The basic premise of the volcanic dust theory is that this suspended volcanic material will filter out a portion of the incoming solar radiation which, in turn, will lead to lower air temperatures. Thus, many have speculated that periods of high volcanic activity should be accompanied by cool climatic periods, perhaps even ice ages, while times of reduced volcanic activity should coincide with warmer eras. This idea is, in part, supported by temperature trends in the twentieth century. From 1915 until the explosive eruption of Mt. Agung on the island of Bali

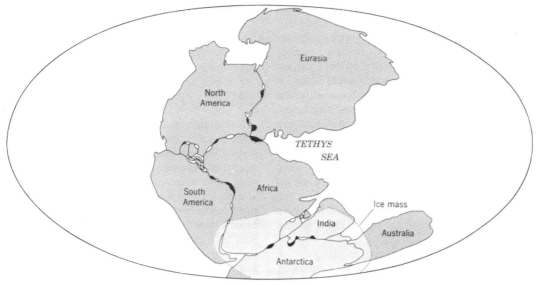

A

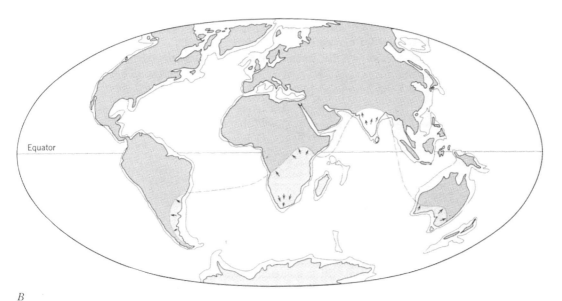

B

Figure 12-11

(a) The supercontinent Pangaea showing the area covered by glacial ice 300 million years ago. (b) The continents as they are today. The shading outlines areas where evidence of the old ice sheets exists. The dashed line joining the glaciated regions indicates how large the ice sheet would have had to be if the continents had been in their present positions at the time of glaciation. (After R. F. Flint and B. J. Skinner, *Physical Geology*, 2nd edition, © 1977 by John Wiley & Sons, Inc., reprinted by permission of the publisher.)

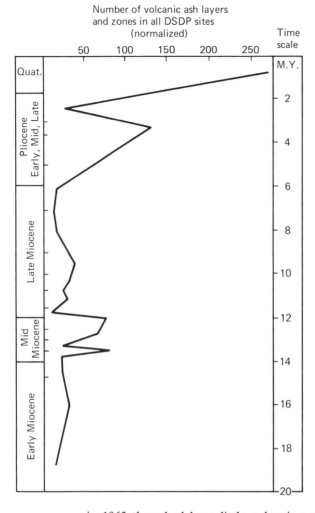

Number of volcanic ash layers
and zones in all DSDP sites
(normalized)

Time scale

50   100   150   200   250

M.Y.

Quat.

Pliocene
Early, Mid, Late

Late Miocene

Mid Miocene

Early Miocene

2

4

6

8

10

12

14

16

18

20

Figure 12-12

Variations in volcanic activity over the past 20 million years are revealed by the number of volcanic ash layers. Notice the dramatic increase over the past two million years. (After James P. Kennett and Robert C. Thunell, "Global Increase in Quaternary Explosive Volcanism," *Science,* CLXXXIX, no. 4176 (1975), 499. Copyright 1975 by the American Association for the Advancement of Science.)

in 1963 there had been little volcanic activity. It is a well-known fact that global temperatures increased during much of this period. As a result, many scientists have postulated that this era of rising temperatures was connected to the lack of fine volcanic debris in the atmosphere. Some studies have tended to support this conclusion, but they have also shown that the cooling that began in the mid-1940s could not be explained by volcanic activity. That is, the drop in global temperatures began before the resumption of increased explosive volcanism.

The volcanic dust theory is most often mentioned as a possible cause for the Ice Age. However, although a veil of volcanic dust may indeed cause world temperatures to drop, for many years no one could substantiate the theory with facts. In 1965 Arthur Holmes, a distinguished British geologist, stated that no correlation had yet been demonstrated between glacial periods and times of prolonged or exceptionally intense volcanic activity. As a result of recent investigations, however, the possibility that volcanic eruptions have triggered glacial activity has gained new life. Analysis of deep-sea sediment cores recovered by the *Glomar Challenger* as a part of the National Science Foundation's Deep Sea Drilling Project have shown that there was a much higher rate of explosive volcanism during the past 2 million years than during the previous 18 million years (Figure 12-12). Consequently, the study reveals a reasonably good correlation between a period of greatly increased volcanic activity and an interval in earth history that was characterized by rapidly changing climatic conditions, namely, the most recent ice age.

The Ice Age was not a single period of continuous glaciation. Instead, it was characterized by alternating periods of glacial advance and retreat. The warm periods between advances are termed interglacial periods. Thus, if volcanic activity was the primary forcing mechanism for the Ice Age, there must have been alternating periods of explosive volcanism and relatively quiet conditions. At present, the data are not sufficiently clear to deal with these glacial–interglacial episodes. Hence, the full importance of volcanic activity as a possible cause for the Ice Age is still something of a mystery and a matter for continued speculation. Stephen H. Schneider, an active researcher and writer in the field of climatic change, summarizes the current uncertainty of many scientists regarding the volcanic dust theory as follows:

> . . . although the role of volcanoes in forcing climatic change may prove critical—in which case I should be trying to learn to predict volcanic eruptions and not spend most of my time building atmospheric models—as yet we cannot be sure the extent to which the potential climatic effects of volcanic blasts are detectable in climatic records.*

astronomical theory

The *astronomical theory* is based on the premise that variations in incoming solar radiation are a principal factor in controlling the climate of the earth. This hypothesis was first developed and strongly advocated by the Yugoslavian astronomer Milutin Milankovitch. He formulated a comprehensive mathematical model based on the following elements:

1. variations in the shape (*eccentricity*) of the earth's orbit about the sun
2. changes in *obliquity*, that is, changes in the angle that the axis makes with the plane of the ecliptic (plane of the earth's orbit)
3. *precession*, that is, the wobbling of the earth's axis

Although variations in the distance between the earth and sun are of minor significance in understanding current seasonal temperature fluctuations, they may play a very important role in producing global climatic changes on a time scale of tens of thousands of years. A difference of only 3 percent exists between aphelion, which occurs on about July 4 in the middle

The Genesis Strategy (New York: Plenum Press, 1976), p. 134.

of the Northern Hemisphere summer, and perihelion, which takes place in the midst of the Northern Hemisphere winter on about January 3. This small difference in distance means that the earth receives about 6 percent more solar energy in January than in July. However, this is not always the case. The shape of the earth's orbit changes during a cycle that astronomers say takes between 90,000 and 100,000 years: It stretches into a longer ellipse and then returns to a more circular shape. When the orbit is very eccentric, the amount of radiation received at closest approach (perihelion) would be on the order of 20 to 30 percent greater than at aphelion. This would most certainly result in a substantially different climate from what we now have.

In Chapter 2 the inclination of the earth's axis to the plane of the ecliptic was shown to be the most significant cause for seasonal temperature changes. At present the angle that the earth's axis makes with the plane of its orbit is 23.5 degrees. However, this angle changes. During a cycle that averages about 41,000 years the tilt of the axis varies between 22.1 degrees and 24.5 degrees. Because this angle varies, the severity of the seasons must also change. The smaller our tilt, the smaller the temperature difference between winter and summer. It is believed that such a reduced seasonal contrast could promote the growth of ice sheets. Since winters could be warmer, more snow would fall because the capacity of air to hold moisture increases with temperature. Conversely, summer temperatures would be cooler, meaning that less snow would melt. The result could be the growth of ice sheets.

Like a partly run-down top, the earth is wobbling as it spins on its axis. At the present time the axis points toward the star Polaris (often called the North Star). However, about the year A.D. 14,000 the axis will point toward the bright star Vega, which will then be the North Star (Figure 12-13). Since the period of precession is about 26,000 years, Polaris will once again be the North Star by the year 27,000. As a result of this cyclical wobble of the axis, a climatically very significant change must take place. When the axis is tilted

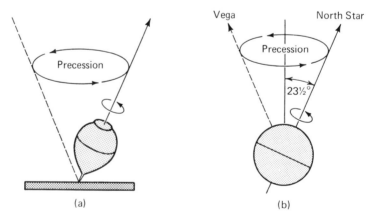

(a)   (b)

Figure 12-13

(a) Precession as illustrated by a spinning top. (b) Precession of the earth will eventually result in the Northern Hemisphere experiencing winter near aphelion.

toward Vega in about 12,000 years, the orbital positions at which the winter and summer solstices occur will be reversed. Consequently, the Northern Hemisphere will experience winter near aphelion (when the earth is farthest from the sun) and summer will occur near perihelion (when our planet is closest to the sun). Thus, seasonal contrasts will be greater because winters will be colder and summers will be warmer than at present.

Using these factors, Milankovitch calculated variations in insolation and the corresponding surface temperature of the earth back into time in an attempt to correlate these changes with the climatic fluctuations of the Ice Age. In explaining climatic changes that result from these three variables, it should be pointed out that they cause little or no variation in the total annual amount of solar energy reaching the ground. Instead, their impact is felt because they change the degree of contrast between the seasons.

Since Milankovitch's pioneering work the time scales for orbital and insolational changes have been recalculated several times. Past errors have been corrected and measurements have been made with greater precision. Over the years the astronomical theory has been widely accepted, then largely rejected, and now, in light of recent investigations, is very popular again.

Among recent studies that have added credibility and support to the astronomical theory is one in which deep-sea sediments were examined.* Through oxygen isotope analysis and other statistical analyses of certain climatically sensitive microorganisms, a chronology of temperature changes going back 450,000 years was established. This time scale of climatic change was then compared to astronomical calculations of eccentricity, obliquity, and precession in order to determine if a correlation did indeed exist. It should be noted that the study was not aimed at identifying or evaluating the mechanisms by which the climate is modified by the three orbital variables. The goal was to see if past climatic and astronomical changes corresponded.

Although the study was very involved and mathematically complex, the conclusions were straightforward. The authors found that major variations in climate over the past several hundred thousand years were closely associated with changes in the geometry of the earth's orbit, that is, cycles of climatic change were shown to correspond closely with the periods of obliquity, precession, and orbital eccentricity. More specifically, they stated: "It is concluded that changes in the earth's orbital geometry are the fundamental cause of the succession of Quaternary ice ages."† Further, the study went on to predict the future trend of climate, but with two qualifications: (1) that the prediction apply only to the natural component of climatic change and ignore any human influence and (2) that it be a forecast of long-term trends because it must be linked to factors that have periods of 20,000 years

---

*J. D. Hays, John Imbrie, and N. J. Shackleton, "Variations in the Earth's Orbit: Pacemaker of the Ice Ages," *Science*, CXCIV, No. 4270 (1976), 1121–1132.

†J. D. Hays et al., p. 1131. The term Quaternary refers to the period on the geologic calendar that encompasses the last few million years.

and longer. Thus, even if the prediction is correct, it contributes little to our understanding of climate changes over periods of tens to hundreds of years because the cycles in the astronomical theory are too long for this purpose. With these qualifications in mind, the study predicts that the long-term trend (over the next 20,000 years) will be toward a cooler climate and extensive glaciation in the Northern Hemisphere.

If the astronomical theory does indeed explain alternating glacial–interglacial periods, a question immediately arises as to why glaciers have been absent throughout most of earth history. Prior to the plate tectonic theory there was no widely accepted answer. In fact, this question was a major obstacle for the supporters of Milankovitch's hypothesis. Today, however, there is a plausible answer. Since glaciers can form only on the continents, land masses must exist somewhere in the higher latitudes before an ice age can commence. Long-term temperature fluctuations are not great enough to create widespread glacial conditions in the tropics. Thus, many now believe that ice ages have only occurred when the earth's shifting crustal plates have carried the continents from tropical latitudes to more poleward positions.

variable sun theory

So far two of the three theories discussed have dealt with variations in the receipt of solar radiation. In one case (the volcanic theory), changes in the transparency of the atmosphere were believed to alter the amount of energy received at the surface. The astronomical theory was based on the fact that changes in the amount of solar energy received by the earth are caused by external factors, that is, factors unrelated to the atmosphere. In this section yet another theory relying on variations in the receipt of solar energy is presented. Like the astronomical theory, the *variable sun theory* relies upon fluctuations in solar energy that are caused by factors external to the atmosphere.

One of the most persistent theories of climatic change is based on the idea that the sun is a variable star and that its output of energy varies through time. The effect of such changes would seem to be direct and easily understood— increases in solar output would cause the atmosphere to warm and reductions would result in cooling. This hypothesis is very appealing because it can be used to explain climatic changes of any length or intensity. However, there is at least one major drawback. No major variations in the total intensity of solar radiation have yet been measured outside the atmosphere.

Such measurements were an impossible task until satellite technology became available. However, we have not yet used satellites to monitor solar output. Even when this occurs, records for many years will be necessary before we begin to get a feeling for how variable (or invariable) energy from the sun really is. Thus, for the time being science must still rely on ground-based estimates of the sun's energy output. The greatest difficulty with this method is determining the influence of the atmosphere on the amount of

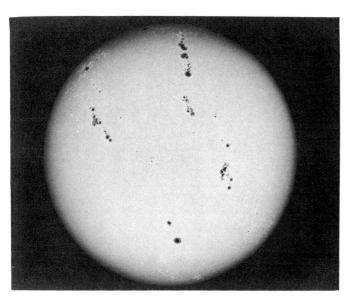

Figure 12-14

Sunspots. (Courtesy of Hale Observatories.)

radiation measured. To do this, observations are made at different times of the year. As one might expect, as the sun angle gets lower, the amount of radiation that is absorbed or scattered by the atmosphere increases and the quantity received at the earth's surface declines. A graph is then constructed of the intensity of solar energy as a function of the amount of air that must be traversed (which is governed by the sun angle). This curve is then extrapolated to the point of no air at all. In this manner, the *solar constant* has been estimated to be about 2 calories per square centimeter per minute.* The difficulty with this method is that scientists cannot attain the necessary degree of precision to establish that meaningful fluctuations (which may be slight) actually occur. Further, since each ground-based observation must be corrected for large atmospheric effects, the source of any variance is extremely difficult to determine.

Theories of climatic change based on a variable sun merge with those concerning sunspot cycles. The most conspicuous and best-known features on the surface of the sun are the dark blemishes called *sunspots* (Figure 12-14). Although their origin is uncertain, sunspots have been established to be huge magnetic storms that extend from the sun's surface deep into the interior. Further, these spots are associated with the sun's ejection of huge masses of particles which upon reaching the upper atmosphere interact with the gases there to produce auroral displays.

Along with other solar activity, the number of sunspots increases and decreases on a regular basis, creating a cycle of about 11 years. A curve of

*The solar constant is a standard energy measurement used in meteorology and refers to the amount of solar radiation striking a surface perpendicular to the sun's rays at the outer edge of the atmosphere when the earth is at an average distance from the sun.

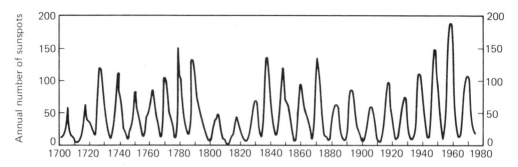

Figure 12-15

Mean annual sunspot numbers, 1700 to the present.

the annual number of sunspots, beginning in the early 1700s and continuing to the present appears to be very regular (Figure 12-15). In fact, until recently few scientists doubted that sunspots and the 11-year cycle were enduring features of the sun. However, articles by the solar astronomer, John A. Eddy, give evidence that solar activity may vary considerably and that it may indeed be responsible for some climatic changes.*

Papers published by two astronomers in the late nineteenth century had indicated that the interval from about 1645 to 1715 was a period of prolonged sunspot absence. Eddy set out to see if this were indeed the case, for if it were true, it would have some interesting ramifications. For instance, it would mean that the sun was not the constant star that so many believed it to be. In addition, it had implications for climatic change because the proposed sunspot minimum in the late seventeenth and early eighteenth centuries also corresponded very closely with the coldest portion of a period known in climatic history as the "Little Ice Age." It is a well-documented cold period that saw alpine glaciers in Europe advance farther than at any time since the last major glaciation. By examining historical records of observations made during this period, as well as making use of indirect evidence of solar activity, Eddy accomplished his task. His indirect evidence included, among other things, records of auroral displays and the abundance of the radioactive isotope carbon-14 (^{14}C) in tree rings. Since the number of nights when auroral displays occur has a strong positive correlation with sunspot activity, records of auroras served as an independent check of solar activity. The same thing was true for the ^{14}C record locked in trees.

The ^{14}C history is useful in its own right as a measure of past solar activity, as has been demonstrated by a number of investigators. The isotope is continually formed in the atmosphere through the action of cosmic rays, which,

*Much of the remaining discussion in this section is based on two articles by John Eddy: "The Maunder Minimum," *Science*, CXCII, No. 4245 (1976), 1189–1203; "The Case of the Missing Sunspots," *Scientific American*, CCXXXVI, No. 5 (1977), 80–92.

in turn, are modulated by solar activity. When the sun is active, some of the incoming galactic cosmic rays are prevented from reaching the earth. At these times, corresponding to maxima in the sunspot cycle, less than the normal amount of ^{14}C is produced in the atmosphere and less is found in tree rings formed then. When the sun is quiet, terrestrial bombardment by galactic cosmic rays increases and the ^{14}C proportion in the atmosphere rises.*

By comparing graphs, as shown in Figure 12-16, Eddy showed that every decline in solar activity corresponded to a period of glacial advance in Europe and that every increase in solar activity was matched by a time of retreating glaciers. Furthermore, periods like the medieval maximum were found to correlate well with warm periods. Referring to these excellent matches, Eddy stated: "These early results in comparing solar history with climate make it appear that changes on the sun are the dominant agent of climatic changes lasting between 50 and several hundred years."†

It should be remembered that Eddy's study was not concerned with short-term changes in the sunspot cycle. Thus, the investigation revealed little about how the 11-year cycle might influence short-term weather. However, others have examined sunspot cycles of various lengths in an attempt to find positive correlations and relationships. Some have been successful. For instance, one study showed a possible connection between periods of drought on the Great Plains and the minimum of the double (22-year) sunspot cycle. Nevertheless, it must be noted that no completely acceptable theory has yet been formulated to explain how solar variations are translated into climatic change.

carbon dioxide and climatic change

Up to this point we have examined four potential causes of climatic change. Although each differed considerably from the others, all four had at least one thing in common—they were unaffected by human activities. They relied upon "natural" variables to produce climatic fluctuations. The two remaining theories, however, differ from the others because they examine man's possible impact on global climate.

In Chapter 1 we learned that although carbon dioxide (CO_2) represents only 0.03 percent of the gases that make up clean, dry air, it is nevertheless a meteorologically significant component. The importance of carbon dioxide lies in the fact that it is transparent to incoming short wavelength solar radiation, but it is not transparent to longer wavelength outgoing terrestrial radiation. A portion of the energy leaving the ground is absorbed by carbon

_____
*Eddy, "The Maunder Minimum," p. 1195.
†Eddy, "The Case of the Missing Sunspots," pp. 88, 92.

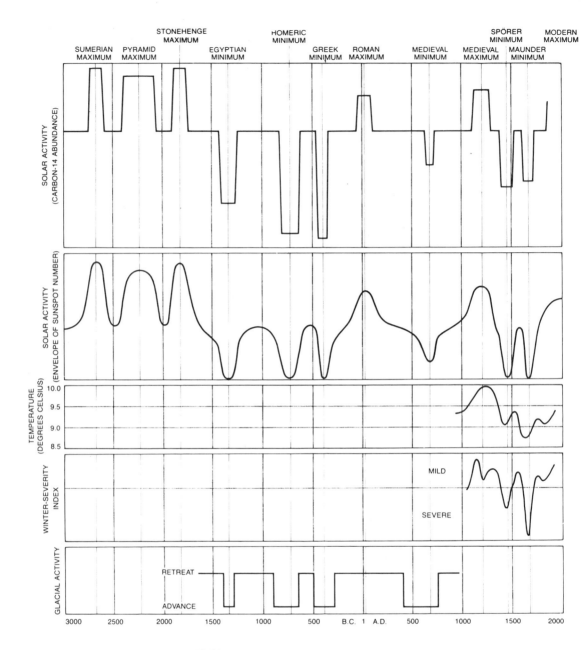

Figure 12-16

A comparison of the carbon-14 derived history of the sun and the history of global climate based on historical records and the advance and retreat of alpine glaciers. Notice that every decrease in solar activity (minimum) matches a time of glacial advance in Europe and that each rise in solar activity (maximum) corresponds to a time of glacier retreat. The figure indicates that for almost 5000 years all of the climatological curves seem to rise and fall in response to the long-term level of solar activity. (From John A. Eddy, "The Case of the Missing Sunspots." © 1977 by Scientific American, Inc. All rights reserved.)

302

dioxide and subsequently reemitted, part of it toward the surface, keeping the air near the ground warmer than it would be without carbon dioxide. Thus, along with water vapor, carbon dioxide is largely responsible for the so-called greenhouse effect of the atmosphere. Since carbon dioxide is an important heat absorber, it logically follows that any change in the air's carbon dioxide content should alter temperatures in the lower atmosphere. This is the basis of the carbon dioxide theory of climatic change.

Paralleling the rapid growth of industrialization, which began in the nineteenth century, has been the consumption of fossil fuels (coal, natural gas, and petroleum). The combustion of these fuels has added great quantities of carbon dioxide to the atmosphere. Although some of this excess is taken up by plants or is dissolved in the ocean, about 50 percent remains in the atmosphere. Consequently, from 1860 to 1970 there was an increase of about 10 percent in the carbon dioxide content of the air (Figure 12-17). Naturally, one would expect that as a result global temperatures should have climbed steadily during this period, but temperatures rose only until the mid-1940s and since then they have slowly dropped. Is this a refutation of the carbon dioxide theory? Most climatologists believe that it is not. Rather, many suggest that the cooling trend since the 1940s would have been even greater had it not been for the increased greenhouse effect provided by the rising carbon dioxide content. Others simply say that the effects of the 10-percent carbon dioxide increase over more than 100 years were not large enough to show up in any clear way in the climatic record.

If you examine the curve in Figure 12-17, you will see that the amount of

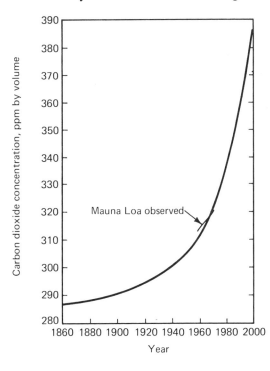

Figure 12-17

Changes in atmospheric carbon dioxide resulting from the combustion of fossil fuels, 1860–2000. (From L. Machta and K. Telegadas, "Inadvertent Large-Scale Weather Modification," in *Weather and Climate Modification,* Wilmot N. Hess, editor, © 1974 by John Wiley & Sons, Inc., reprinted by permission of the publisher.)

carbon dioxide in the air has increased at an increasing rate. Furthermore, the graph shows the curve continuing in this manner into the twenty-first century. Current estimates now indicate that the atmosphere's current carbon dioxide content of almost 330 parts per million (ppm) will approach 400 parts per million by the year 2000 and will approximately double by the year 2040. With such an increase in carbon dioxide, the enhancement of the greenhouse effect would be much more dramatic and measurable than in the past. Thus, if the estimated increases come to pass, calculations show that we can probably expect global temperatures to increase by nearly 1°C by the year 2000 and by about 2°C by 2040.

What if the cooling trend that began in the 1940s continues? Will it mask the effect of the proposed carbon dioxide increases? The answer, of course, depends upon the magnitude of the temperature decline. According to some studies, this question will not have to be raised because the cooling trend will terminate soon. Studying the oxygen isotope record in the Greenland ice core, Wallace Broecker of the Lamont-Doherty Geological Observatory suggests that the present cooling is only one of a long series of similar climatic fluctuations over the past millenium (Figure 12-9). He says, "By analogy with similar events in the past, the present cooling will . . . bottom out during the next decade or so. Once this happens, the CO_2 effect will tend to become a significant factor and by the first decade of the next century we may experience global temperatures warmer than any in the last 1000 years."[*]

So far in this discussion we have assumed that global temperatures would rise if the carbon dioxide content were to increase. However, the possibility exists that increasing the amount of carbon dioxide in the atmosphere could ultimately result in a temperature decline. This could occur if, because of higher temperatures, cloud cover were to increase because of increased evaporation. The increased cloud cover would, in turn, reduce the amount of solar radiation reaching the ground, which would result in a temperature decrease. Calculations show that an increase in the world's cloud cover of just 0.6 percent could decrease the low-level temperature by 0.5°C.

Because the atmosphere is a very complex interactive physical system, scientists must consider many possible outcomes, like the one just mentioned, when one of the system's elements is altered. These various possibilities, termed *climatic feedback mechanisms*, not only add more complexities to climatic modeling, but they also add greater uncertainty to climatic predictions. In the case of a global warming, as might result from increased carbon dioxide, other possible feedback mechanisms also exist. For example, warming the lower atmosphere may also warm the surface of the ocean. Since warm water cannot hold as much carbon dioxide in solution as cold water, more carbon dioxide could be released to the air. This might result in a positive feedback because more atmospheric carbon dioxide would speed

---

[*]Wallace S. Broecker, "Climatic Change: Are We on the Brink of a Pronounced Global Warming?" *Science*, CLXXXIX No. 4201, (1975), 460–461.

up the warming. Yet another possibility exists if a warming should occur. Although higher temperatures mean increased evaporation, more clouds may not result. If this is so, the greater amount of water vapor in the air could increase the absorption of long wavelength terrestrial radiation, thus compounding the temperature increase because of an even more effective atmospheric greenhouse. Hence, unlike the first example, these latter two possibilities would reinforce the warming trend instead of offset it.

man-generated dust and climatic change

Human activities such as transportation, solid waste disposal, industrial processes, and slash-and-burn agriculture are all capable of generating great quantities of particulate matter (referred to by some as aerosols or turbidity and by others simply as dust). Do these particles influence climate? If so, will they cause global temperatures to rise or drop? A candid answer to the first question is, "Perhaps," and to the second, "We are not sure!" Indeed, some scientists strongly believe that the addition of man-generated dust will cause world temperatures to drop, while others present a strong case for just the opposite effect. In the following discussion we will briefly examine both points of view.

A well-known meteorologist, Reid A. Bryson, has for many years been a strong proponent of the idea that increased dustiness of the atmosphere caused by man is the primary cause of the global cooling that began in the 1940s. In part, the logic that advocates of this hypothesis use is as follows:

1. Since the late nineteenth century and continuing until recently sunspot activity has been increasing. If it is assumed that increased numbers of sunspots reflect an increasing solar energy output, temperatures should have continued to rise beyond the 1940s. Since this is not the case, other factors must be operative.

2. A rise in atmospheric carbon dioxide should cause temperatures to increase. Although the carbon dioxide content of the atmosphere has continued to climb since the 1940s, as it had for 60 to 70 years before, temperatures have fallen. Hence, some other factor(s) must be overshadowing the carbon dioxide effect.

3. A rise in atmospheric turbidity should cause global temperatures to fall by scattering away increased amounts of solar radiation. As illustrated by Figure 12-18, dust concentrations appear to have grown enormously since the 1930s, following the same trend as industrialization, mechanization, urbanization, and population. In addition, volcanic activity has increased somewhat over previous levels since the 1950s, adding to atmospheric particulate concentrations.

Thus, to explain the cooling trend in light of increased sunspot numbers and carbon dioxide, which are assumed to cause temperatures to rise, Bryson

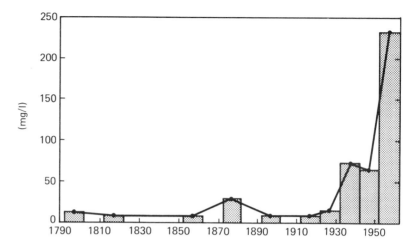

Figure 12-18

Dust trapped in the glaciers of the Caucasus Mountains. Note the sharp rise since the 1930s. (After Reid A. Bryson and John E. Kutzbach, *Air Pollution,* Commission on College Geography Resource Paper No. 2, p. 8. Copyright by the Association of American Geographers.)

concludes that " . . . increasing global air pollution, through its effect on the reflectivity of the earth, is currently dominant and is responsible for the temperature decline of the past decade or two."*

Although the proponents of the foregoing ideas presume that the net effect of increased aerosols is to scatter away a portion of the incoming solar energy, others are not as confident about this conclusion. Although it seems likely that the reflectivity of the atmosphere is altered by the addition of aerosols, the nature of the change in the world's overall albedo is difficult to determine. For example, in regions where particles are bright in relation to the surface below (which is believed to be the case over the oceans) the amount of light reflected is likely to increase and cause the air to cool. However, where the particles are relatively dark in comparison to the surface (a situation that is most probable over land masses), the absorption of solar radiation will increase, albedo will decrease, and temperatures will rise. Thus, determining whether the world's mean temperature will rise or fall depends upon the ratio of absorption to scattering on a global scale.

Unlike volcanic dust, which may be thrown high into the stratosphere where it can remain for several years, the mean residence time for man-made dust particles, which are confined largely to the troposphere, is only a few days. Within that span of time they are usually washed out of the air. Keeping this fact in mind, and realizing that the main source of these particles is the continents, some meteorologists reason that the primary effect of the dust must be upon the air temperatures over land masses. If this is the case, the

---

*Reid A. Bryson, " 'All Other Factors Being Constant . . .'; A Reconciliation of Several Theories of Climatic Change," *Weatherwise*, XXI, No. 2 (1968), 61.

net influence of man-made aerosols is probably a small warming (remember that earlier it was stated that over the continents aerosols are relatively dark in comparison to the surface). If this conclusion is correct, some other explanation will have to be found for the cooling trend that began in the 1940s.

In addition to the reasoning mentioned in the preceding paragraph, a recent study that analyzed several atmospheric dust-loading events in southern Arizona also supports the hypothesis that an increase in particulate matter in the troposphere will result in a warming of the lower atmosphere.* The researchers found that when dust was injected into the atmosphere, the amount of solar radiation received at the earth's surface was only slightly reduced. On the other hand, their data indicated that the dust acted, "to significantly 'close' the atmospheric 'window' that otherwise allows a good deal of the terrestrial thermal radiation of the earth to escape to space."† In other words, absorption by the dust particles made the atmospheric greenhouse more effective. Commenting on the climatological consequences of man's industrial pollution of the atmosphere, the authors state that, "If, and when, this source of tropospheric particulates ever becomes climatologically significant, our findings imply that the resultant temperature trend will definitely be one of warming not cooling. Thus, whereas many groups assigned to assess the problem have looked on this aspect of intensified industrialization as acting as a 'brake' on the warming influence of increased carbon dioxide production, just the opposite is actually the case—the two phenomena complement each other."

Of the many influences of pollution on the atmospheric environment, its effects on global climate are certainly the most nebulous. Will increases in atmospheric carbon dioxide cause a global warming that, in turn, may be reinforced by positive feedback mechanisms or will increased cloud cover negate the impact of carbon dioxide? Is man-made dust causing a global cooling or is its net effect a warming? Is it possible that the climate would fluctuate without changes in any of these factors? In the preceding discussions of man's possible impact on climate we have seen that science has thus far been unsuccessful in providing adequate answers to these questions. Although the effects of pollutants on the world's climate are not yet well understood, they may nevertheless be very significant, for even slight changes in temperature could be responsible for important shifts in weather patterns. Such changes could have a dramatic influence upon human activities such as agriculture and perhaps could even threaten survival in some regions.

How important is man's role in bringing about climatic change? Although many differences of opinion and much uncertainty exists in the scientific community concerning this question, there seems to be general agreement on at least one point. W. W. Kellogg and S. H. Schneider stated it this way:

*Sherwood B. Idso and Anthony J. Brazel, "Planetary Radiation Balance as a Function of Atmospheric Dust: Climatological Consequences," *Science*, CXCVIII, No. 4318 (1977), 731–733.

†A discussion of the atmospheric window may be found in the section on the heat budget in Chapter 2.

"Human influence on climate, which may already be appreciable, can only be properly assessed when the natural forces at play are understood."* In a report resulting from a conference on man's impact on climate, the following point was made: " . . . the uncertainty lies in the fact that we are competing with powerful natural sources that are also causing the climate to change, as it has changed in the past. We must understand these factors well enough to be able to put man's activities into proper perspective relative to nature's."†

climatic prediction

For several decades scientists have been attempting to forecast short-term climate changes, often, it would appear, with some degree of success. How is it done? Can we be confident that such predictions will correctly forecast future climatic variations?

Since no satisfactory mathematical models of the earth's atmosphere exist, climate forecasters must base their predictions of future climate on extrapolation, that is, estimating the probable continuation of a curve that shows past events. For example, earlier in this chapter mention was made that the current cooling trend may soon end. The basis of this forecast was the extrapolation of a 700-year long record of temperature trends as revealed by oxygen isotope analysis of Greenland ice cores (Figure 12-9). However, caution must be exercised when using extrapolation. For instance, if in the mid-1940s we had extrapolated the temperature curve of the past 60 years, we would likely have extended it upward. Measurements since that time would have proven us wrong. Herein lies one of the dangers in using this technique. For a prediction to have some degree of reliability, it must be based on long curves that show change in a regular manner. If the troughs and peaks are spaced in some sort of repeating pattern, extrapolation may be justified.

In 1951 H. C. Willett, using cyclic trends tentatively associated with sunspot activity, forecast a drop in temperatures over much of the globe. He indicated that decreases should be especially evident over the Spitzbergen–Greenland–Iceland region and to a lesser degree over northern Europe and the eastern United States. He also suggested that the downward trend would not occur in the eastern Mediterranean and the northwestern United States. As it turned out, most of these temperature predictions were correct. However, the forecast of the sunspot cycle (the proposed cause) proved to be almost entirely in error. In 1961 G. Siren forecast future temperature trends by using an analysis of the growth of pine trees in the northern forests of Finland between 1181 and 1960. His forecast (extrapolation) has been verified for at least the first decade of the prediction. Other predictions have made use of such things as variations in particle radiation from solar flares

---

*"Climate Stabilization: For Better or for Worse?" *Science*, CLXXXVI, No. 4170 (1974), 1163.

†*Inadvertent Climate Modification; Report of the Study of Man's Impact on Climate* (Cambridge: Massachusetts Institute of Technology, 1971), p. 27.

and the extrapolation of circulation patterns based on patterns derived from historical accounts.

Can we be confident about these forecasts? Most atmospheric scientists do not believe we can. As was evident in each of the preceding examples, and, in fact, as a general rule, climate forecasts have not been based on meteorological reasoning. For this reason, and despite the general success of many of the predictions, climate variability cannot yet be predicted with confidence. Unfortunately, our ignorance of the true causes of climatic change greatly limits the usefulness of such predictions. The viewpoint expressed by J. Murray Mitchell, Jr., a scientist with the National Oceanic and Atmospheric Administration, probably reflects the current feelings of many about climate prediction and theories of climatic change.

> We have tended to pick out one, or a few, simplistic causative mechanisms to account for the facts of past climatic change, and to rely on these as a guide to future climatic developments. In truth, there are undoubtedly many causative mechanisms—some internal and some external to the system—that govern climatic variability and change. No climate mechanism can be realistically dealt with in isolation of the others. All must be considered together in a suitably general physical framework before we can claim to understand much of anything about climatic changes. Since we do not yet have a good grasp of all causative mechanisms, and we are not yet certain what constitutes a suitably general physical framework to encompass those mechanisms, it is rather absurd for us to speak of having "explained" the changes of global-average temperature in the past century (for example) as the result of increasing atmospheric pollution, or of changing solar activity. Where interesting statistical relationships appear to exist, as between climatic variations and such plausible forcing phenomena as volcanic eruptions, carbon dioxide trends, or sunspot numbers, these should by all means be reported as possible clues to climatic behavior. Hypotheses to account for climatic change, based on such relationships, should be advanced. But let them be recognized for what they are: statistical relationships and hypotheses. And let them be recognized also for what they clearly are not: self-sufficient explanations of climate behavior and reliable bases for climate predictions.*

REVIEW

1. Which temperature statistic (maximums, minimums, or means) reveals the greatest rural–urban temperature difference?

2. Compare a "typical" rural surface with a "typical" city surface. List two ways in which this difference adds to the urban heat island.

3. How do each of the following factors contribute to the urban heat island: heat production, "blanket" of pollutants, three-dimensional city structure?

---

*J. Murray Mitchell, Jr., "What Can We Say About Future Trends in Our Climate?" *Atmospheric Quality and Climatic Change*, Studies in Geography No. 9, edited by Richard J. Kopec [Chapel Hill: University of North Carolina, Department of Geography, (1976)], pp. 164–65.

4. During what season is heat production most important as a factor affecting the heat island?

5. List three factors that are the likely causes of greater precipitation in and downwind of cities.

6. a. What is the La Porte anomaly?
   b. What evidence led Changnon to conclude that the La Porte situation was related to the Chicago urban-industrial complex?

7. A study of 27 years of precipitation data in the St. Louis region led researchers to conclude that the urban-industrial complex caused precipitation totals to increase up to 15 percent. List two lines of evidence that supported this conclusion.

8. Describe the reasons for the following urban climatic characteristics:
   a. Reduced solar radiation
   b. Reduced relative humidity
   c. Increased fog and cloud frequency
   d. Reduced wind speeds and more days when calms prevail
   e. Fewer heating degree-days

9. What is a "country breeze"? Describe its cause.

10. List some of the reasons that have made climatic change a topical subject.

11. List and describe some of the methods scientists use to gain insight into the climates of the past.

12. How does the plate tectonic theory (continental drift) help explain the previously "unexplainable" glacial features in present-day Africa, South America, and Australia?

13. Can continental drift explain short-term climatic changes? Explain.

14. What is the basic premise of the volcanic dust theory?

15. Is it possible that increased volcanic activity contributed to ice age cooling? What evidence supports your answer?

16. Can the volcanic dust theory presently be used to explain the alternating glacial and interglacial periods of the ice age?

17. List and describe each of the three variables in the astronomical theory of climatic change.

18. Do recent studies of sea-floor sediments tend to confirm or refute the astronomical theory? What do these studies predict for the future?

19. How is the solar constant determined from ground-based measurements?

20. Why are scientists not confident about possible variations in the solar constant based on measurements made at the earth's surface?

21. What indirect evidence was used in John Eddy's investigations to doublecheck observed variations in solar activity (sunspots)?

22. Was a positive correlation found between sunspots and climatic change? If so, briefly describe these findings.

23. What is the basic principle of the carbon dioxide theory of climatic change?

24. Is the fact that the atmosphere has been cooling for the past three decades necessarily a refutation of the carbon dioxide theory? Explain.

25. What are climatic feedback mechanisms? Give some examples.

310

26. Will man-generated dust cause global temperatures to rise or fall? Make a case for each point of view.

27. How is extrapolation used in predicting climates of the future? Can we be confident of these predictions? Why or why not?

VOCABULARY
REVIEW

urban heat island

country breeze

climatic change

oxygen isotope analysis

paleosol

plate tectonic theory

volcanic dust theory

astronomical theory

eccentricity

obliquity

precession

variable sun theory

solar constant

sunspots

climatic feedback mechanisms

Chapter Thirteen

WORLD CLIMATES

Although previous chapters have examined the spatial and seasonal variations of the major elements of weather and climate, we have not yet investigated the combined effects of these variations in different parts of the world. The varied nature of the earth's surface and the many interactions that occur between atmospheric processes give every location on our planet a distinctive, even unique, climate. Our intention, however, is not to describe the unique climatic character of countless different locales; such a task would require many volumes. Instead, the purpose of this chapter is to introduce the reader to the major climatic regions of the world. The discussion will examine large areas and use particular places only to illustrate the characteristics of these major climatic regions. In addition, for those regions that are most likely to be unfamiliar to the majority of students (in particular the tropical, desert, and polar realms), a brief description of the natural landscape is also provided. Even so, the chapter should be considered a summary that highlights many of the most striking features in a generalized way.

In the first chapter of this book it was pointed out that a common misconception is to think of climate as simply "the average state of the atmosphere." Although averages such as mean annual temperature and precipitation are certainly important to climatic descriptions, variations and extremes must also be included to accurately portray the character of an area. To demonstrate this important point, the list below, which was compiled largely from the *Climatic Atlas of the United States**, shows some of the important data used in the study of climate.

Mean monthly and annual temperatures

Mean annual ranges of temperature

Normal daily maximum, minimum, average, range, and extremes, monthly

Mean annual number of days maximum temperature 32°C and above

Mean annual number of days minimum temperature 0°C and below

Mean length of frost-free period (days)

Normal total heating degree-days, monthly and annual

Normal total precipitation, monthly and annual

Greatest precipitation in 24 hours

Greatest and least monthly and annual precipitation totals

Mean total snowfall

Mean monthly total snowfall

Mean number of days with 0.25 millimeter or more of precipitation, monthly and annual

Mean dew-point temperature, monthly and annual

Mean relative humidity, monthly and annual

Pan and lake evaporation

Mean percentage of possible sunshine, monthly and annual

Mean total hours of sunshine, monthly and annual

Mean sky cover, monthly and annual

Prevailing direction, mean speed, and fastest mile of wind, monthly and annual

Mean frequency of winds from different directions

Normal sea-level pressures, monthly and annual

Temperature and precipitation are the most important elements in a climatic description because they have the greatest influence on people and their activities and they also have an important impact on the broad-scale distribution of such things as vegetation and soils. Nevertheless, as the list illustrates, other factors are also important for a complete climatic description. When possible, some of these factors will be introduced into our discussion of world climates.

climatic classification

The distribution of the major atmospheric elements is, to say the least, complex. Because of the many differences from place to place as well as from time to time at a particular locale, it is unlikely that any two sites on the

---

*U.S. Department of Commerce, Environmental Data Service, NOAA, 1968.

earth's surface experience exactly the same weather conditions. Since the number of places on earth is infinite, it becomes readily apparent that the number of different climates must be extremely large. Of course, having a great diversity of information to investigate is not unique to the study of the atmosphere, it is a problem that is basic to all of science. In order to cope with such variety, it is not only desirable but also essential to devise some means of classifying the vast array of data to be studied. By establishing groups consisting of items that have certain important characteristics in common, order and simplicity are introduced. Bringing order to large quantities of information not only aids comprehension and understanding, but it also facilitates analysis and explanation.

Probably the first attempt at climatic classification was made by the ancient Greeks who divided each hemisphere into three zones: torrid, temperate, and frigid. The basis of this simple scheme was earth–sun relationships and the boundaries were the four astronomically important parallels, the Tropic of Cancer and the Tropic of Capricorn and the Arctic and Antarctic circles. Hence, the globe was divided into winterless climates and summerless climates and an intermediate type that had features of the two extremes.

Few other attempts were made until the beginning of the twentieth century. Since that time, however, many climate classification schemes have been devised. It should be remembered that the classification of climates (or of anything else) is not a natural phenomenon, but the product of human ingenuity. The value of any particular classification is determined largely by its intended use. A system designed for one purpose is not necessarily applicable to another.

In this chapter we will use a classification devised by the Russian-born German climatologist Wladimir Koeppen (1846–1940). As a tool for presenting the general world pattern of climates, it has been the best-known and most used system for more than 50 years. The reasons for its wide acceptance are many. For one, it uses only easily obtained data: mean monthly and annual values of temperature and precipitation. Further, the criteria are unambiguous and relatively simple to apply, and they divide the world into climatic regions in a realistic way.

Koeppen believed that the distribution of natural vegetation was the best expression of the totality of climate. Consequently, the boundaries he chose were largely based on the limits of certain plant associations. Five principal groups were recognized; each group was designated by a capital letter as follows:

A Humid tropical. Winterless climates; all months having a mean temperature above 18°C
B Dry. Climates where evaporation exceeds precipitation; there is a constant water deficiency
C Humid middle-latitude. Mild winters; the average temperature of the coldest month is below 18°C but above −3°C

D Humid middle-latitude. Severe winters; the average temperature of the coldest month is below $-3°C$ and the warmest monthly mean exceeds $10°C$

E Polar. Summerless climates; the average temperature of the warmest month is below $10°C$.

Notice that four of the major groups (A, C, D, E) are defined on the basis of temperature characteristics and the fifth, the B group, has precipitation as its primary criterion. Each of the five groups is further subdivided by using the criteria and symbols presented in Table 13-1.* In addition, the world

Table 13-1

Koeppen System of Climatic Classification

Letter Symbol 1st	2nd	3rd	
A			Average temperature of the coldest month is 18°C or higher
	f		Every month has 6 cm of precipitation or more
	m		Short dry season; precipitation in driest month less than 6 cm but equal to or greater than $10 - R/25$ (R is annual rainfall in cm)
	w		Well-defined winter dry season; precipitation in driest month less than $10 - R/25$
	s		Well-defined summer dry season (rare)
B			Potential evaporation exceeds precipitation. The dry/humid boundary is defined by the following formulas as follows
			(NOTE: R is average annual precipitation in cm and T is average annual temperature in °C)
			$R < 2T + 28$ when 70% or more of rain falls in warmer 6 months
			$R < 2T$ when 70% or more of rain falls in cooler 6 months
			$R < 2T + 14$ when neither half year has 70% or more of rain
	S		Steppe ⎤
	W		Desert ⎦ —The BS/BW boundary is 1/2 the dry/humid boundary
		h	Average annual temperature is 18°C or greater
		k	Average annual temperature is less than 18°C

*When classifying climatic data using Table 13-1, you should first determine whether or not the data meet the criteria for the E climates. If the station is not a polar climate, proceed to the criteria for B climates. If your data do not fit into either the E or B groups, check the data against the criteria for A, C, and D climates, in that order.

Table 13-1 continued

Letter Symbol			
1st	2nd	3rd	
C			Average temperature of the coldest month is under 18°C and above −3°C
	w		At least ten times as much precipitation in a summer month as in the driest winter month
	s		At least three times as much precipitation in a winter month as in the driest summer month; precipitation in driest summer month less than 4 cm
	f		Criteria for w and s cannot be met
		a	Warmest month is over 22°C; at least 4 months over 10°C
		b	No month above 22°C; at least 4 months over 10°C
		c	One to 3 months above 10°C
D			Average temperature of coldest month is −3°C or below; average temperature of warmest month is greater than 10°C
	s		Same as under C
	w		Same as under C
	f		Same as under C
		a	Same as under C
		b	Same as under C
		c	Same as under C
		d	Average temperature of the coldest month is −38°C or below
E			Average temperature of the warmest month is below 10°C
	T		Average temperature of the warmest month is greater than 0°C and less than 10°C
	F		Average temperature of the warmest month is 0°C or below

distribution of climates according to the Koeppen classification is shown in Figure 13-1. You will be referred to this figure several times as the climates of the earth are discussed in the following pages.

Although a strength of the Koeppen system is the relative ease with which boundaries are determined, these boundaries should not be viewed as fixed. To the contrary, all climatic boundaries, no matter what classification is used, shift their positions from one year to the next (Figure 13-2). The boundaries shown on maps such as the world map in Figure 13-1 are simply average locations based on data collected over many years. Thus, a climatic boundary should be regarded as a broad transition zone and not a sharp line.

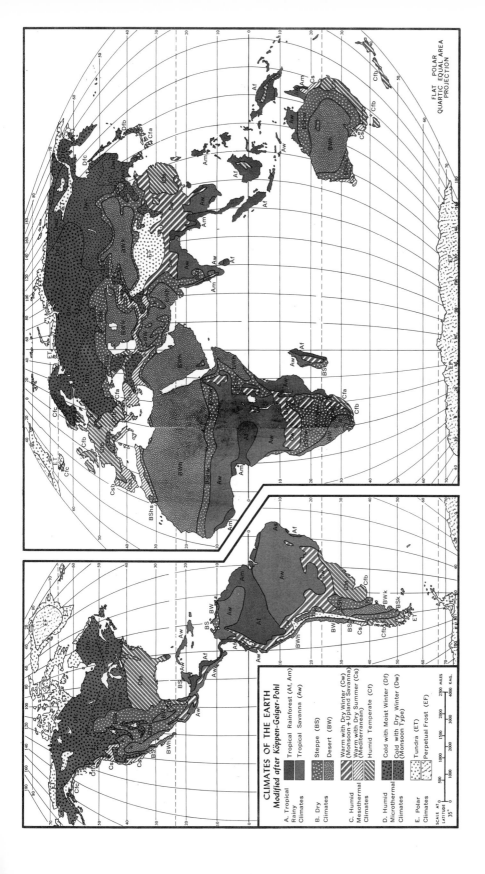

Figure 13-1

Climates of the earth. (From Glenn T. Trewartha, *An Introduction to Climate,* 4th ed., © 1968 by McGraw-Hill Book Company, reprinted by permission of the publisher.)

CLIMATES OF THE EARTH
Modified after Köppen-Geiger-Pohl

A. Tropical Rainy Climates
Tropical Rainforest (Af, Am)
Tropical Savanna (Aw)

B. Dry Climates
Steppe (BS)
Desert (BW)

C. Humid Mesothermal Climates
Warm with Dry Winter (Cw) (Monsoon + Upland Savanna)
Warm with Dry Summer (Cs) (Mediterranean)
Humid Temperate (Cf)

D. Humid Microthermal Climates
Cold with Moist Winter (Df)
Cold with Dry Winter (Dw) (Monsoon Type)

E. Polar Climates
Tundra (ET)
Perpetual Frost (EF)

SCALE AT 35° LATITUDE
MILES 0 500 1000 1500 2000 2500
KMS. 0 1000 2000 3000 4000

FLAT POLAR QUARTIC EQUAL AREA PROJECTION

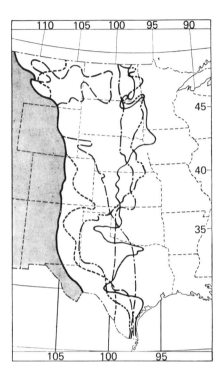

Figure 13-2

Yearly fluctuations in the dry/humid boundary during a 5-year period. (From C. E. Koeppe and G. C. DeLong, *Weather and Climate,* © 1958 by McGraw-Hill Book Company, reprinted by permission of the publisher.)

climatic controls: a summary

If the earth's surface were completely homogeneous, the map of world climates would be very simple, looking much like the ancient Greeks must have pictured it—a series of latitudinal bands girdling the globe in a symmetrical pattern on each side of the equator. This, of course, is not the case. The earth is not a homogeneous sphere and hence many factors exist that act to disrupt the symmetry that was just described.

At first glance, the world climate map (Figure 13-1) reveals what appears to be a scrambled, or even haphazard pattern, with similar climates located in widely separated parts of the earth. A closer examination will show that although they are far apart, similar climates generally have similar latitudinal and continental positions. This fact suggests that there is order in the distribution of climatic elements and that the pattern of climates is not a matter of chance. This is indeed the situation and reflects a regular and dependable operation of the major climatic controls. Thus, before we begin a discussion of the earth's major climates, it will be worthwhile to review each of the major controls.

latitude

The single greatest cause for temperature differences are fluctuations in the amount of solar radiation received at the earth's surface. Although nonperiodic variations in such factors as cloud coverage and the amount of dust

318

in the air may be locally influential, seasonal changes in sun angle and length of daylight are the most important factors controlling the global distribution of temperature. Since all places situated along the same parallel have identical sun angles and lengths of day, variations in the receipt of solar energy are largely a function of latitude. Further, since the vertical rays of the sun migrate between the Tropic of Cancer and the Tropic of Capricorn there is a latitudinal shifting of temperatures. Temperatures in the tropical realm are consistently high because the vertical rays of the sun are never far away. As one moves farther poleward, however, greater seasonal fluctuations in the receipt of solar energy are reflected in larger annual temperature ranges.

land and water

The distribution of land and water must be considered second in importance only to latitude as a control of temperature. Land heats more rapidly and to higher temperatures than water and it cools more rapidly and to lower temperatures than water. Consequently, variations in air temperatures are much greater over land than over water. This differential heating of the earth's surface has led to climates being divided into two broad classes based upon their position in relation to water and land. *Marine climates* are considered mild for their latitude because the moderating effect of water produces summers that are warm but not hot and winters that are cool but not cold. By contrast, *continental climates* tend to be much more extreme. Although a marine station and a continental station along the same parallel in the middle latitudes may have similar annual mean temperatures, the annual temperature range will be far greater at the continental station.

The differential heating and cooling of land and water can also have a significant effect on pressure and wind systems and hence on seasonal precipitation distribution as well. High summer temperatures over the continents can produce low-pressure areas that allow the inflow of moisture-laden maritime air. Conversely, the high pressure that forms over the chilled continental interiors in winter causes a reverse flow of dry air toward the oceans.

geographic position and prevailing winds

To fully understand the influence of land and water on the climate of an area, the position of that area on the continent and its relationship to the prevailing winds are most important considerations. The moderating influence of water is much more pronounced along the windward side of a continent, for here the prevailing winds may carry the maritime air masses far inland. On the other hand, places located on the lee side of a continent, where the prevailing winds blow from the land toward the ocean, are likely to have a more continental temperature regime.

319

mountains and highlands

The location of mountains and highlands plays an important part in the distribution of climates. This impact may be illustrated by examining the situation in western North America. Here the north–south trending mountain chains serve as major barriers. They do not allow the moderating influence of maritime air masses to reach far inland. Consequently, although stations may lie within a few hundred kilometers of the Pacific, their temperature regime is essentially continental. Further, these topographic barriers trigger orographic rainfall on their windward slopes, often leaving a dry rain shadow on the leeward side. Similar effects may be seen in South America and Asia where the towering Andes and the massive Himalayan system serve as major barriers. By way of comparison, western Europe does not have a mountain barrier to obstruct the free movement of air from the North Atlantic. As a result, moderate temperatures and sufficient precipitation mark the entire region.

Where highlands are extensive, they create their own climatic regions. Because of the drop in temperature with increasing altitude, areas such as the Tibetan Plateau, the Altiplano of Bolivia, and the uplands of East Africa are cooler and drier than their latitudinal locations alone would indicate.

ocean currents

The effect of ocean currents on the temperatures of adjacent land areas can be significant. Poleward moving currents such as the Gulf Stream and the Kuroshio currents in the Northern Hemisphere and the Brazilian and East Australian currents in the Southern Hemisphere cause air temperatures to be warmer than would otherwise be expected. This influence is especially pronounced in the winter. Conversely, currents such as the Canaries and California currents in the Northern Hemisphere and the Humboldt and Benguela currents south of the equator reduce the temperatures of bordering coastal zones. In addition, the chilling effect of these cold currents acts to stablize the air masses moving across them. The result is marked aridity and often considerable advection fog.

pressure and wind systems

The world distribution of precipitation shows a very close relationship to the distribution of the earth's major pressure and wind systems. Although the latitudinal distribution of these systems does not generally take the form of simple "belts," it is still possible to identify a zonal arrangement of precipitation from the equator to the poles. In the realm of the equatorial low the convergence of warm, moist, and unstable air makes this zone one of heavy rainfall. In the regions dominated by the subtropical highs general aridity prevails. Farther poleward, in the middle-latitude zone dominated by the

320

irregular subpolar low, the influence of the many traveling cyclonic disturbances again causes precipitation totals to rise. Finally, in polar regions,
where temperatures are low and the air can hold only small quantities of
moisture, precipitation totals decline.

The seasonal shifting of the pressure and wind belts, which follows the
movement of the sun's vertical rays, has a significant effect on areas situated
in intermediate positions. Such regions are alternately influenced by two
different pressure and wind systems. For example, a station located poleward
of the equatorial low and equatorward of the subtropical high will experience
a summer rainy period as the low migrates poleward and a wintertime drought
as the high moves equatorward. This latitudinal shifting of pressure belts is
largely responsible for the seasonality of precipitation in many regions.

**the wet
tropics
(Af and Am)**

The constantly high temperatures and year-round rainfall in the wet tropics
combine to produce the most luxuriant vegetation found in any climatic
realm—the *tropical rain forest* or *selva* (Figure 13-3). Unlike the forests that
most of us living in North America are accustomed to, the tropical rain forest
is made up of broadleaf trees that are green throughout the year. In addition,
instead of being dominated by just a few species, the selva is characterized
by many. It is not uncommon for hundreds of different species to inhabit a
single square kilometer of the forest. As a consequence, the individuals of a
single species are often widely spaced. Standing on the shaded floor of the
forest looking upward, one would see tall, smooth-barked, vine-entangled
trees, the trunks unbranching in their lower two-thirds, forming an almost
continuous canopy of foliage above. A closer look would reveal a three-
layered structure. Nearest the ground, perhaps 5 to 15 meters above, the narrow crowns of some rather slender trees are visible. Rising above these
relatively short components of the forest, a more continuous canopy of foliage
is seen in the 20- to 30-meter height range. Finally, visible through an occa-

Figure 13-3

Unexcelled in luxuriance and characterized by hundreds of different species
per square kilometer, the tropical rain forest, or selva, is a broadleaf evergreen
forest that dominates the wet tropics.

sional opening in the second layer, a third layer may be seen at the very top of the forest. Here the crowns of the trees tower 40 meters or more above the forest floor. Much sun streams through the highest level to the lower layers below, but little light penetrates to the ground. As a result, plant foliage is relatively sparse on the dimly lit forest floor. Where considerable light does make its way to the ground, as along river banks or in man-made clearings, an almost impenetrable growth of tangled vines, shrubs, and short trees is found. The familiar term *jungle* is used to describe such sites.

The environment of the wet tropics that was just described characterizes almost 10 percent of the earth's land area. An examination of Figure 13-1 shows that Af and Am climates form a discontinuous belt astride the equator that typically extends 5 to 10 degrees into each hemisphere. The poleward margins are most often marked by diminishing rainfall, but occasionally decreasing temperatures mark the boundary. Because of the general decrease in temperature with the height in the troposphere, this climatic region is restricted to elevations below 1000 meters. Consequently, the major interruptions near the equator are principally cooler highland areas. It should also be noted that the rainy tropics tend to have a greater latitudinal extent along the eastern side of continents (especially South America) and along some tropical coasts. The greater width on the eastern side of a continent is due primarily to its windward position on the weak western side of the subtropical high, a zone dominated by neutral or unstable air. In other cases, as along the eastern side of Central America, the coast, backed by interior highlands, intercepts the flow of the trade winds. Orographic uplift thus greatly enhances the rainfall total.

Data for some representative stations in the wet tropics are shown in Table 13-2. A brief examination of the numbers reveals the most obvious features that characterize the climate in these areas:

1. Temperatures usually average 25°C or more each month. Consequently, not only is the annual mean high, but the annual range is also very small.

Table 13-2

Data for Wet Tropical Stations

	J	F	M	A	M	J	J	A	S	O	N	D	Yr.
Singapore 1°21′N; 10 m												
Temp. (°C)	26.1	26.7	27.2	27.6	27.8	28.0	27.4	27.3	27.3	27.2	26.7	26.3	27.1
Precip. (mm)	285	164	154	160	131	177	163	200	122	184	236	306	2282
Belém, Brazil 1°18′S; 10 m												
Temp. (°C)	25.2	25.0	25.1	25.5	25.7	25.7	25.7	25.9	25.7	26.1	26.3	25.9	25.7
Precip. (mm)	340	406	437	343	287	175	145	127	119	91	86	175	2731
Douala, Cameroon 4°N; 13 m												
Temp. (°C)	27.1	27.4	27.4	27.3	26.9	26.1	24.8	24.7	25.4	25.9	26.5	27.0	26.4
Precip. (mm)	61	88	226	240	353	472	710	726	628	399	146	60	4109

2. The total precipitation for the year is high, often exceeding 200 centimeters.

3. Although rainfall is not evenly distributed throughout the year, tropical rain forest stations are generally wet in all months. If a dry season exists, it is a very short one.

temperature characteristics

Since places with an Af or Am designation lie near the equator, the reason for the uniform temperature rhythm experienced in such locales is clear: The intensity of insolation is consistently high. The vertical rays of the sun are always relatively close and changes in the length of daylight throughout the year are slight; hence, seasonal temperature variation is minimal. The small differences that exist between the warmest and coldest months often reflect changes in the amount of sky coverage rather than the position of the sun. In the case of Belém, for example, it can be seen that the highest temperatures occur during the months when rainfall (and hence cloud cover) is least.

Daily temperature variations greatly exceed seasonal differences. The graph showing average daily maximum and minimum temperatures in Figure 13-4 illustrates this nicely. While annual temperature ranges in the wet tropics rarely exceed 3°C, daily temperature ranges are from two to five times greater. Hence, there is a greater variation between day and night than there is seasonally. Although rapid nighttime cooling does not generally occur, the drop in temperature is nevertheless often sufficient to produce early morning fogs

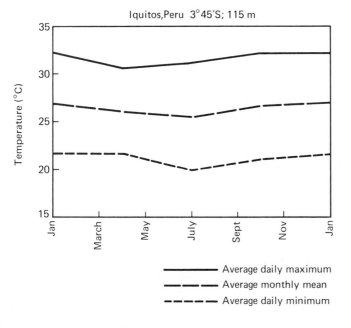

Iquitos, Peru 3°45′S; 115 m

——————— Average daily maximum

– – – – – Average monthly mean

– – – – – Average daily minimum

Figure 13-4

As is typical of the wet tropics, the daily temperature range at Iquitos, Peru (3 degrees, 45 minutes south latitude) is much greater than the annual range.

and heavy dew. Since the daytime air is often nearly saturated, only a small drop in temperature is necessary to bring about condensation.

It should be remembered that the unique feature of the temperatures experienced in the rainy tropics is not their values, for monthly and daily means there are no greater than those in many American cities during the summer. For example, the highest temperature recorded at Jakarta, Indonesia over a 78-year period and at Belém, Brazil over a 20-year period has been only 36.6°C, contrasted with extremes of 40.5°C and 41.1°C at Chicago and New York City, respectively. Rather, the unique feature is the day in and day out, month in and month out regularity of the temperature regime. Although the thermometer may not indicate abnormal or extreme conditions, the warm temperatures combined with the high moisture content of the air and meager winds make sensible temperatures particularly high. The reputation of the wet tropics as being oppressive and monotonous is for the most part well deserved.

precipitation characteristics

The regions dominated by Af and Am climate generally receive from 175 to 250 centimeters of rain each year. However, a glance at the data in Table 13-2 reveals more variability in rainfall than in temperature, both seasonally and from place to place. The rainy nature of the equatorial realm is in part related to the extensive heating of the region and the consequent thermal convection. In addition, this is the zone of the converging trade winds, often referred to as the *zone of intertropical convergence*, or simply the *ITC*. Therefore, the thermally induced convection coupled with convergence leads to widespread ascent of the warm, humid, unstable air. Conditions near the equator are thus ideal for the formation of precipitation.

Rain typically falls on more than 50 percent of the days each year. In fact, at some stations three-quarters of the days experience some rain. There is a marked diurnal regularity to the rainfall at many places. Cumulus clouds begin forming in late morning or early afternoon. The buildup continues until about 3 or 4 p.m., the time when temperatures are highest and thermal convection is at a maximum, at which time the cumulonimbus towers yield showers. Figure 13-5, showing the diurnal distribution of the rainfall at Kuala Lumpur, exemplifies this pattern. At many marine stations the cycle is different, with the rainfall maximum coming at night. The reason is that the lapse rate is steepest and hence instability is greatest during the dark hours instead of in the afternoon. The lapse rate steepens at night because the radiation heat loss from the air at heights of 600 to 1500 meters is greater than near the surface where the air continues to be warmed by conduction and low-level turbulence from the water.

Portions of the rainy tropics are wet throughout the year; that is, according to the Koeppen scheme, there is at least 6 centimeters of rain each month. However, there are extensive areas (those having the Am designation) that

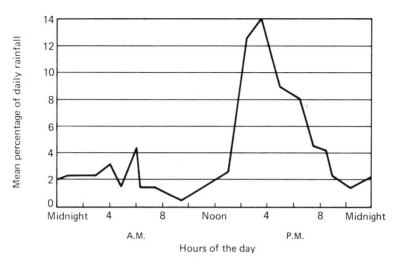

Figure 13-5

The diurnal distribution of rainfall at Kuala Lumpur, Malaya, with its midafternoon maximum, illustrates the typical pattern at many wet tropical stations. [After Ooi Jin-bee, "Rural Development in Tropical Areas," *Journal of Tropical Geography*, XII, no. 1 (1959).]

are characterized by a brief dry season of 1 or 2 months. In spite of the short dry season, the annual precipitation total in Am regions closely corresponds to the total in areas that are wet year-round (Af). Since the dry period is too brief to deplete the supply of soil moisture, the rain forest is maintained. Although an explanation of the seasonal pattern of precipitation in wet tropical climates is very complex and not yet fully understood, month-to-month variations are, at least in part, caused by the seasonal migration of the ITC that follows the latitudinal shift of the vertical rays of the sun.

tropical wet and dry (Aw)

In the latitude zone poleward of the rainy tropics and equatorward of the tropical deserts (from about 5 degrees or 10 degrees to 15 degrees or 20 degrees) lies a transitional climatic region referred to as *tropical wet and dry*. Along the margins nearest the equator the dry season is short and the boundary between Aw and the rainy tropics is difficult to define. Along the poleward side, however, the dry season is prolonged and conditions merge into those of the semiarid realm.

Here, in the tropical wet and dry climate the rain forest gives way to the *savanna*, a tropical grassland with scattered deciduous trees (Figure 13-6). In fact, Aw is often referred to as the savanna climate. This name, however, may not be appropriate, for many ecologists doubt that these grasslands are climatically induced. Instead, it is believed that woodlands once dominated and that subsequently the savanna grasslands were created in response to the seasonal burnings carried out by native human populations.

325

Figure 13-6

The tropical savanna, with its stunted, drought-resistant trees scattered amid a grassland, probably resulted from seasonal burnings carried out by native human populations.

temperature characteristics

An examination of the temperature data in Table 13-3 reveals only modest differences between the wet tropics and the tropical wet and dry regions. Because of the somewhat higher average latitude of most Aw stations, annual mean temperatures are slightly lower. In addition, the annual temperature range, although still small, is higher, varying from 3°C to perhaps 10°C. The daily temperature range, however, still exceeds the annual variation. Since seasonal fluctuations in humidity and sky coverage are more pronounced in Aw areas, diurnal temperature ranges vary noticeably during the year. They are generally small during the rainy season when humidity and cloud cover are at a maximum and large during periods of drought when the skies are clear and the air is dry. Furthermore, because of the more persistent summertime cloudiness, many Aw stations experience their warmest temperatures at the end of the dry season, just prior to the summer solstice. Hence, in the Northern Hemisphere March, April, and May are often warmer than June and July.

Table 13-3

Data for Tropical Wet and Dry Stations

	J	F	M	A	M	J	J	A	S	O	N	D	Yr.
					Calcutta 22°32′N; 6 m								
Temp. (°C)	20.2	23.0	27.9	30.1	31.1	30.4	29.1	29.1	29.9	27.9	24.0	20.6	26.94
Precip. (mm)	13	24	27	43	121	259	301	306	290	160	35	3	1582
					Cuiaba, Brazil 15°30′S; 165 m								
Temp. (°C)	27.2	27.2	27.2	26.6	25.5	23.8	24.4	25.5	27.7	27.7	27.7	27.2	26.5
Precip. (mm)	216	198	232	116	52	13	9	12	37	130	165	195	1375

precipitation characteristics

Since temperature regimes among the A climates are very similar, the primary factor that distinguishes the Aw climate from Af and Am is precipitation. Aw stations typically receive from 100 to 150 centimeters of rainfall each year. This overall amount is often appreciably less than in the wet tropics. However, the most distinctive characteristic of this climate is not the yearly precipitation total but the markedly seasonal character of the rainfall—wet summers followed by dry winters.

The reason for the alternating wet and dry periods is the intermediate position of the Aw climatic region between the zone of intertropical convergence, with its sultry weather and convective thundershowers, and the stable, subsiding air of the subtropical highs. Following the spring equinox the ITC, along with the other wind and pressure belts, shifts poleward following the migration of the vertical rays of the sun (Figure 13-7). With the advance of the ITC into a region, the summer rainy season commences and features weather patterns typical of the wet tropics. Later, with the retreat of the ITC, the subtropical high advances into the region and brings with it intense drought conditions. During the dry season the landscape takes on a parched appearance and nature seems to become dormant as the trees lose their leaves and the abundant tall grasses turn brown and wither. The length of the dry season depends primarily upon distance from the ITC. Typically, the farther an Aw station is from the equator, the shorter the period of ITC control and the longer the locale is under the influence of the stable subtropical high. Consequently, with an increase in latitude, the dry season gets longer while the length of the wet period diminishes.

The importance of the movement of the ITC to an understanding of the rainfall distribution in the tropics cannot be overemphasized. Table 13-4, which shows precipitation data for six African stations, reinforces this point. Notice that at Maiduguri and Francistown there are single rainfall maxima that occur when the ITC reaches its most poleward positions. Between these stations double maximums represent the passage of the ITC on its way to and from these extreme locations. It is important to remember that these

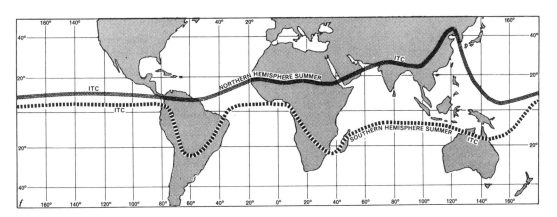

Figure 13-7

The seasonal migration of the ITC strongly influences precipitation distribution in the tropics. (From Joseph E. Van Riper, *Man's Physical World*, 2nd ed., © 1971 by McGraw-Hill Book Company, reprinted by permission of the publisher.)

327

statistics represent long-term averages and that on a year-to-year basis the movement of the ITC is far from regular. There is, nevertheless, little doubt that in the tropics rainfall follows the sun.

Table 13-4
Rainfall Regimes and the Movement of the ITC in Africa

	J	*F*	*M*	*A*	*M*	*J*	*J*	*A*	*S*	*O*	*N*	*D*
				Maiduguri, Nigeria 11°51′N								
Precip. (mm)	0	0	0	7.6	40.6	68.6	180.3	**220.9**	106.6	17.7	0	0
				Yaounde, Cameroon 3°53′N								
Precip. (mm)	22.8	66.0	147.3	170.1	**195.6**	152.4	73.7	78.7	213.4	**294.6**	116.8	22.9
				Kisangani, Zaire 0°26′N								
Precip. (mm)	53.3	83.8	**177.8**	157.5	137.2	114.3	132.0	165.1	182.9	**218.4**	198.1	83.8
				Luluabourg, Zaire 5°54′S								
Precip. (mm)	137.2	142.2	**195.6**	193.0	83.8	20.3	12.7	58.4	116.8	165.1	**231.1**	226.0
				Zomba, Malawi 15°23′S								
Precip. (mm)	274.3	**289.6**	198.1	76.2	27.9	12.7	5.1	7.6	17.8	17.8	134.6	**279.4**
				Francistown, Botswana 21°13′S								
Precip. (mm)	**106.7**	78.7	71.1	17.8	5.1	2.5	0	0	0	22.9	58.4	86.4

the monsoon

In much of India, Southeast Asia, and portions of Australia the alternating periods of rainfall and drought characteristic of the Aw precipitation regime are associated with a phenomenon called the *monsoon*. The term is derived from an Arabic word, "mausim," which means "season" and typically refers to wind systems that have a pronounced and fairly sudden seasonal reversal of direction. During the summer conditions are conducive to rainfall as humid, unstable air moves from the oceans toward the land. In winter a dry wind, having its origins over the continent, blows toward the sea.

In part, the monsoonal circulation system develops in response to the differences in annual temperature variations between continents and oceans. In principle, the processes associated with the monsoon are similar to those described in connection with the land and sea breeze (Chapter 7) except that the scales, in both time and space, are much larger. During spring in the Northern Hemisphere an irregular area of thermally induced low pressure gradually develops over the interior of South Asia. It is further strengthened by the poleward advance of the ITC. Hence, the summertime circulation is from the higher pressure over the ocean toward the lower pressure over the continent. As winter approaches, the winds reverse direction as the ITC migrates southward and a deep anticyclone develops over the chilled continent. By midwinter (Northern Hemisphere) the winds are from the north and converge on Australia and South Africa.

the Cw variant

An examination of Figure 13-1 will reveal areas adjacent to the wet and dry tropics in southern Africa, South America, northeastern India, and China that are designated as Cw. Although C has been substituted for A, indicating that these regions are subtropical instead of tropical, the Cw climate is nevertheless just a variant of Aw, for the only major difference is somewhat lower temperatures. In Africa and South America Cw climates are highland extensions of Aw. Because they occupy elevated sites, the warmer temperatures of the adjacent wet and dry tropics are reduced. In India and China Cw areas are middle-latitude extensions of the tropical monsoon realm. In some cases, especially in India, the Cw areas are barely poleward enough to have winter temperatures below those of the A climates.

the dry (B) climates

The dry regions of the world cover some 42 million square kilometers, or about 30 percent, of the earth's land surface. No other climatic group covers so large a land area. The major deserts listed in Table 13-5 alone account for 1 square kilometer out of every 7 square kilometers on the continents.

Table 13-5

Major Deserts of the World

Desert	Approximate Area	
	km² (thousands)	mi² (thousands)
Sahara	9100	3500
Australian	3400	1300
Arabian	2600	1000
Turkistan	2000	750
North American	1300	500
Patagonian	680	260
Indian	600	230
Kalahari-Namib	570	220
Gobi-Takla Makan	520	200
Iranian	390	150
Atacama-Peruvian	360	140

Although there will be many aspects of the dry climates discussed in this section, perhaps the most characteristic feature of all, aside from the fact that yearly rainfall totals are relatively meager, is that the amount of precipitation received each year is very unreliable. Generally, the smaller the mean annual rainfall, the greater its variability (Figure 13-8). As a result, yearly averages are often very misleading. For example, during one 7-year period, Trujillo, Peru had an average rainfall of 6.1 centimeters per year. However, a closer look reveals that during the first 6 years and 11 months of the period the station received a scant 3.5 centimeters (an annual average of slightly more than 0.5 centimeters). Then during the twelfth month of the seventh

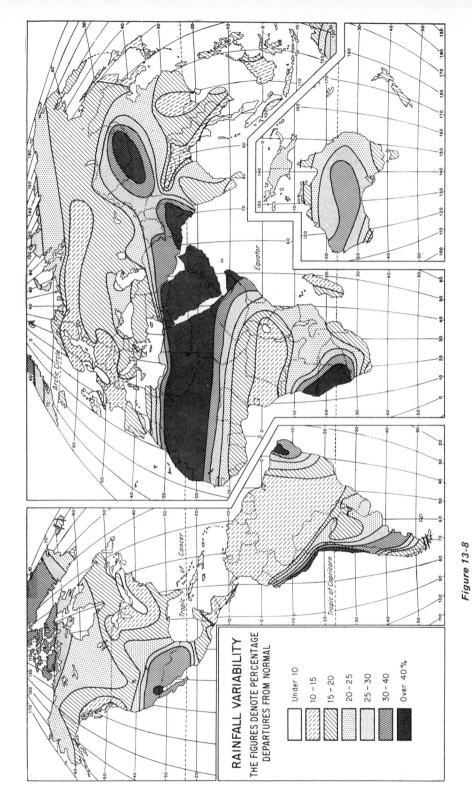

RAINFALL VARIABILITY

THE FIGURES DENOTE PERCENTAGE
DEPARTURES FROM NORMAL

	Under 10
	10 – 15
	15 – 20
	20 – 25
	25 – 30
	30 – 40
	Over 40%

Figure 13-8

One of the most characteristic features of dry regions is the high rainfall vari-
ability. Generally, the smaller the mean annual rainfall, the greater its variability.
(From C. E. Koeppe and G. C. DeLong, *Weather and Climate*, © 1958 by
McGraw-Hill Book Company, reprinted by permission of the publisher.)

year 39 centimeters of rain fell, 23 centimeters of it during a 3-day span. Although this may be an extreme case, it nevertheless illustrates very well a common feature of most dry regions—the irregularity of rainfall. It is also significant to note that there are usually more years when rainfall totals are below the average than above because, as the foregoing example showed, the occasional wet period tends to lift the average.

There are a number of common misconceptions about deserts that continue to persist in the minds of many people, especially those living in more humid regions. One common fallacy about deserts is that they are lifeless or almost lifeless. Although reduced in amount and different in character, plant and animal life are indeed present (Figure 13-9). Although desert plants differ widely from one part of the world to another, they all have one characteristic in common; they have developed adaptations that make them highly tolerant of drought. Many have waxy leaves, stems, or branches or a thickened cuticle (outermost protective layer) to reduce water loss. Others have very small leaves or no leaves at all. Further, the roots of some species often extend to great depths in order to tap the moisture found there, while others produce a shallow but widespread root system that enables them to quickly absorb great amounts of moisture from the infrequent desert down-

Figure 13-9

As this scene in the desert of the American Southwest illustrates, these environments are commonly far from lifeless. (Courtesy National Park Service, U.S. Department of the Interior.)

pours. Often the stems of these plants are thickened by a spongy tissue that can store enough water to sustain the plant until the next rainfall comes. Thus, although they are widely dispersed and provide little ground cover, plants of many kinds flourish in the desert.

A second widely held belief about the world's dry lands is that they are always hot. As we shall see, this is not true in every case. Unlike any other climatic group, the B climates are defined by precipitation criteria instead of temperature. Therefore, dry climates are found from the tropics poleward to the high middle latitudes. Consequently, although tropical deserts lack a cold season, deserts in the middle latitudes do experience seasonal temperature changes.

The last two commonly held misconceptions are more geologic than climatic. One mistaken assumption about the world's deserts is that they consist of mile after mile of drifting sand. It is true that sand accumulations do exist in some areas and may be striking features, but they represent only a small percentage of the total desert area. In the Sahara, the world's largest desert, accumulations of sand cover only one-tenth of its area. The sandiest of all deserts is the Arabian, one-third of which consists of sand. The final mistaken assumption is the seemingly logical idea that wind is the most important agent of erosion in deserts. Although wind is relatively more significant in dry areas than anywhere else, most desert landforms are created by running water. When the rains come, they usually take the form of thunderstorms. Since the heavy rain associated with these storms cannot all soak in, rapid runoff results. Without a thick vegetative cover to protect the ground, erosion is great.

what is meant by "dry"?

It is important to realize that the concept of dryness is a relative one and refers to any situation in which a water deficiency exists. Hence, climatologists define a *dry climate* as one in which the yearly precipitation is not as great as the potential loss of water by evaporation. Dryness then is not only related to annual rainfall totals, but it is also a function of evaporation, which, in turn, is closely dependent upon temperature. As temperatures climb, potential evaporation also increases. Twenty-five centimeters of precipitation may be sufficient to support forests in northern Scandinavia where evaporation into the cool, humid air is slight and a surplus of water remains in the soil. However, the same amount of rain falling on Nevada or Iran supports only a sparse vegetative cover because evaporation into the hot, dry air is great. Hence, it becomes clear that no specific amount of precipitation can serve as a universal boundary for dry climates.

To establish the boundary between dry and humid climates, the Koeppen classification uses formulas that involve three variables: (1) average annual precipitation, (2) average annual temperature, and (3) seasonal distribution of precipitation. The use of average annual temperature reflects its importance as an index of evaporation. The amount of rainfall defining the humid–dry boundary will be larger where mean annual temperatures are high and smaller where temperatures are low. The use of seasonal precipitation distribution as a variable is also related to this idea. If rain is concentrated in the warmest months, loss to evaporation is greater than if the precipitation were concentrated in the cooler months. Thus, there are considerable differences in the precipitation amounts received at various stations in the B climates. Table 13-6 summarizes these differences. Notice that if a station with an annual mean of 20°C and a summer rainfall maximum does not receive more than 680 millimeters of precipitaion per year, it is classified as dry. However, if the rain falls primarily in winter, the station must receive only 400 millimeters or more to be considered humid. If the precipitation is more evenly distributed, the figure defining the humid–dry boundary is between the other two.

Table 13-6

Average Annual Precipitation (mm) at BS–Humid Boundary

Ave. Ann. Temp. (°C)	Summer Dry Season	Even Distribution	Winter Dry Season
5	100	240	380
10	200	340	480
15	300	440	580
20	400	540	680
25	500	640	780
30	600	740	880

Within the regions defined by a general water deficiency there are two climatic types: *arid* or *desert* (BW) and *semiarid* or *steppe* (BS). These two groups have many features in common; their differences are primarily a matter of degree. The semiarid is a marginal and more humid variant of the arid and represents a transition zone that surrounds the desert and separates it from the bordering humid climates. The arid–semiarid boundary is commonly (and arbitrarily) set at one-half the annual precipitation separating dry regions from humid. Thus, if the humid–dry boundary happens to be 40 centimeters the steppe–desert boundary will be 20 centimeters.

tropical desert and steppe (BWh and BSh)

The heart of the low-latitude dry climates lies in the vicinity of the Tropic of Cancer and the Tropic of Capricorn. A glance at Figure 13-1 reveals a virtually unbroken desert environment stretching for more than 9300 kilometers, from the Atlantic coast of North Africa to the dry lands of northwestern India. In addition to this single great expanse, the Northern

Hemisphere contains another much smaller area of tropical desert and steppe in northern Mexico and the southwestern United States. In the Southern Hemisphere dry climates dominate Australia. Almost 40 percent of the continent is desert, while much of the remainder is steppe. In addition, arid and semiarid areas are found in southern Africa as well as making a rather limited appearance in coastal Chile and Peru. The existence and distribution of this dry tropical realm are primarily a consequence of the subsidence and marked stability of the subtropical anticyclones.

Within tropical deserts the scanty precipitation that falls is not only infrequent but also very erratic. Indeed, there is no well-defined seasonal precipitation regime that could be termed characteristic of the tropical deserts. These areas are simply located too far poleward to be influenced by the zone of intertropical convergence and too far equatorward to benefit from the frontal and cyclonic precipitation of the middle latitudes. Even during the summer months, when extensive daytime heating produces a steep lapse rate and considerable convective motion, clear skies are still the rule. The reason is that subsidence aloft prevents the lower air from rising high enough to penetrate the condensation level.

In the semiarid transitional belts surrounding the desert the situation is different, for here a seasonal pattern of rainfall becomes better defined. As exemplified by the data for Dakar in Table 13-7, stations located on the equatorward side of low-latitude deserts have a brief period of relatively heavy rainfall during the summer when the ITC is farthest poleward. The rainfall regime should look familiar since it is very similar to that found in the adjacent wet and dry tropics, except that the amount is less and the period of drought is longer. For steppe areas on the poleward margins of the tropical deserts, the precipitation regime is reversed. As the data for Marrakech illustrate (Table 13-7), the cool season is the period when nearly all precipitation falls. At this time of year middle-latitude cyclones often take more equatorward routes and hence bring occasional periods of rain.

Table 13-7

Data for Tropical Steppe and Desert Stations

	J	F	M	A	M	J	J	A	S	O	N	D	Yr.
				Marrakech, Morocco 31°37′N ; 458 m									
Temp. (°C)	11.5	13.4	16.1	18.6	21.3	24.8	28.7	28.7	25.4	21.2	16.5	12.5	19.9
Precip. (mm)	28	28	33	30	18	8	3	3	10	20	28	33	242
				Dakar, Senegal 14°44′N ; 23 m									
Temp. (°C)	21.1	20.4	20.9	21.7	23.0	26.0	27.3	27.3	27.5	27.5	26.0	25.2	24.49
Precip. (mm)	0	2	0	0	1	15	88	249	163	49	5	6	578
				Alice Springs, Australia 23°38′S ; 570 m									
Temp. (°C)	28.6	27.8	24.7	19.7	15.3	12.2	11.7	14.4	18.3	22.8	25.8	27.8	20.8
Precip. (mm)	43	33	28	10	15	13	8	8	8	18	30	38	252

To understand temperatures in the desert environment two factors, humidity and cloud cover, are important to consider. The cloudless sky and low humidity allow an abundance of solar radiation to reach the ground during the day and permit the rapid exit of terrestrial radiation at night. As one would expect, relative humidities are low throughout the year. Relative humidities of from 10 to 30 percent are typical at midday for interior location's. In regard to cloud cover, desert skies are almost always clear [Figure 13-10(a)]. In the Sonoran desert region of Mexico and the United States, for example, most stations receive nearly 85 percent of the possible sunshine [Figure 13-10(b)]. Yuma, Arizona averages 91 percent for the year, with a low of 83 percent in January and a high of 98 percent in June. The Sahara has an average winter cloud cover of about 10 percent, which in summer drops to a mere 3 percent.

During the summer season the desert surface heats rapidly after sunrise, for as we saw in the preceding examples, the clear skies permit almost all of the solar energy to reach the surface. By midafternoon ground-surface temperatures may approach 90°C. Under such circumstances, it is not surprising that the world's highest temperatures are recorded in the tropical deserts and that the daily maximums at many stations during the hot season are consistently close to the absolute maximum (the highest temperature ever recorded at a station). At Abadán, Iran the average daily maximum in July is a scorching 44.7°C, or only 8.3°C lower than the record high. Phoenix, Arizona is little better, recording an average July maximum of 40.5°C compared with an absolute maximum of 47.7°C. A contributing factor to the high ground and air temperatures is the fact that since very little insolation is used in evaporation, almost all of the energy goes to heating the surface. By contrast, humid regions are not as likely to have such extreme ground and air temperatures because more solar radiation is used to evaporate water and hence not as much remains to heat the ground.

At night temperatures typically drop very rapidly. In part this is because the water vapor content of the air is fairly low. However, the temperature of the ground surface is also a factor. Recall from the discussion of radiation in Chapter 2 that the higher the temperature of a radiating body, the more rapidly it loses heat. Thus, when applied to a desert setting, this means that not only do such environments heat up quickly by day, but at night they also cool very rapidly. Consequently, low-latitude deserts in the interior of continents have the greatest diurnal temperature ranges on earth. Daily variations of from 15°C to 25°C are common, and periodically they may reach even higher values. The highest daily temperature range ever recorded was at In Salah, Algeria in the Sahara. On October 13, 1927, this station experienced a 24-hour range of 55.5°C, from 52.2°C to −3.3°C.

Since most areas of BWh and BSh are located poleward of the A climates, annual temperature ranges are the highest among the tropical climates. During the low sun period averages are below those in other parts of the tropics, with monthly means of from 16°C to 24°C being typical. However,

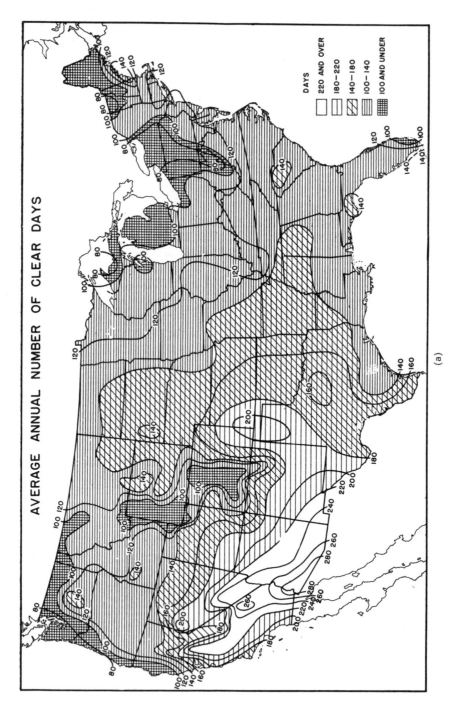

AVERAGE ANNUAL NUMBER OF CLEAR DAYS

DAYS

220 AND OVER
180—220
140—180
100—140
100 AND UNDER

(a)

Figure 13-10

With few exceptions, desert skies are typically cloudless and hence receive a very high percentage of the possible sunlight. These points are strikingly illustrated when the desert Southwest is examined in (a) and (b).

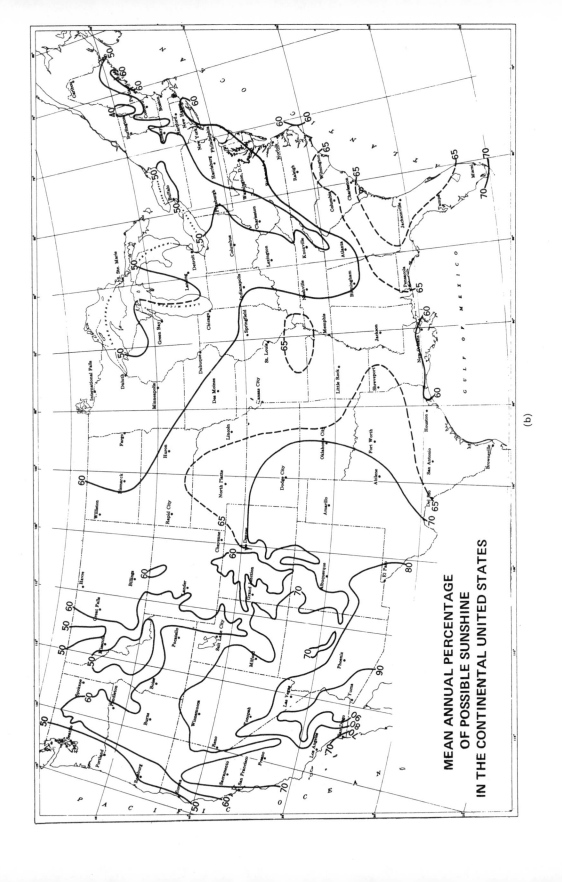

MEAN ANNUAL PERCENTAGE
OF POSSIBLE SUNSHINE
IN THE CONTINENTAL UNITED STATES

(b)

since temperatures during the summer are higher than those in the humid
tropics, annual means at many tropical desert and steppe stations are similar
to those in the A climates.

west coast tropical deserts

Where tropical deserts are found along the west coasts of continents, cold
ocean currents have a dramatic influence on the climate. The principal west
coast deserts are the Atacama in Peru and Chile and the Namib in South and
Southwest Africa. Other areas include portions of the Sonoran desert in
Baja California and coastal areas of the Sahara in northwestern Africa.
These areas deviate considerably from the general image we have of tropical
deserts. Among the most obvious effects of the cold currents are the reduced
temperatures, as exemplified by the data for Lima, Peru and Port Nolloth,
South Africa (Table 13-8). Compared to stations at similar latitudes, these
places have lower annual means and subdued annual and diurnal ranges.
Port Nolloth, for example, has an annual mean of only 14°C and an annual
range of just 4°C as compared with Durban, on the opposite side of South
Africa, which has a yearly mean of 20°C and an annual range that is twice
that at Port Nolloth.

Table 13-8

Data for West Coast Tropical Desert Stations

	J	F	M	A	M	J	J	A	S	O	N	D	Yr.
					Port Nolloth, South Africa 29°14′S; 7 m								
Temp. (°C)	15	16	15	14	14	13	12	12	13	13	15	15	14
Precip. (mm)	2.5	2.5	5.1	5.1	10.2	7.6	10.2	7.6	5.1	2.5	2.5	2.5	63.4
					Lima, Peru 12°02′; 155 m								
Temp. (°C)	22	23	23	21	19	17	16	16	16	17	19	21	19
Precip. (mm)	2.5	T	T	T	5.1	5.1	7.6	7.6	7.6	2.5	2.5	T	40.5

Although these stations are adjacent to the oceans, their yearly rainfall
totals are among the lowest in the world. The aridity along these coasts is
intensified because the lower air is chilled by the cold offshore waters and
hence further stabilized. In addition, the presence of the cold currents causes
temperatures to approach and often reach the dew point. As a result, these
areas are characterized by high relative humidities and much advection fog
and dense stratus cloud cover. Thus, it should be remembered that not all
tropical deserts are sunny and hot places and have low humidities and little
cloud cover. Indeed, the tropical west coast deserts are relatively cool and
humid and are often shrouded by low clouds or fog.

middle-latitude desert and steppe (BWk and BSk)

Unlike their low-latitude counterparts, middle-latitude deserts and steppes are not controlled by the subsiding air masses of the subtropical anticyclones. Instead, these dry lands exist principally because of their position in the deep interiors of large land masses far removed from the oceans. In addition, the presence of high mountains across the paths of the prevailing winds further acts to separate these areas from water-bearing maritime air masses. In North America the Coast Ranges, Sierra Nevada, and Cascades are the foremost barriers, while in Asia, the great Himalayan chain prevents the summertime monsoon flow of moist air from the Indian Ocean from reaching far into the interior. A glance at Figure 13-1 reveals that middle-latitude desert and steppe climates are most widespread in North America and Eurasia. Because of the lack of extensive land areas in the middle latitudes, the Southern Hemisphere has a much smaller area of BWk and BSk. It is found only at the southern tip of South America in the rain shadow of the towering Andes.

Like the tropical deserts and steppes, the dry regions of the middle latitudes have meager and unreliable precipitation. However, unlike the dry lands of the low latitudes, these more poleward regions have much lower winter temperatures and consequently lower annual means and higher annual ranges of temperature. The data in Table 13-9 illustrate this nicely. The data also reveal that rainfall is most abundant during the warm months. Although not all BWk and BSk stations have a summer precipitation maximum, most do because in winter high pressure and cold temperatures tend to dominate the continents. Both of these factors oppose precipitation. In summer, however, conditions are somewhat more conducive to cloud formation and precipitation because the anticyclone disappears over the heated continent and higher surface temperatures and greater specific humidities prevail.

Table 13-9

Data for Middle-Latitude Steppe and Desert Stations

	J	F	M	A	M	J	J	A	S	O	N	D	Yr.
			Ulan Bator, Mongolia 47°55'N ; 1311 m										
Temp. (°C)	−26	−21	−13	−1	6	14	16	14	9	−1	−13	−22	−3
Precip. (mm)	1	2	3	5	10	28	76	51	23	7	4	3	213
			Denver, Colorado 39°32'N ; 1588 m										
Temp. (°C)	0	1	4	9	14	20	24	23	18	12	5	2	11
Precip. (mm)	12	16	27	47	61	32	31	28	23	24	16	10	327

the humid subtropical climate (Cfa)

Located on the eastern sides of the continents, in the 25- to 40-degree latitude range, the *humid subtropical climate* dominates the southeastern United States, as well as other similarly situated areas around the world: all of

Uruguay and portions of Argentina and southern Brazil in South America, eastern China and southern Japan in Asia, and the eastern coast of Australia.

In the summer a visitor to the humid subtropics would experience hot, sultry weather of the type one expects to find in the rainy tropics. Daytime temperatures are generally in the lower thirties, but it is not uncommon for the thermometer to reach into the upper thirties or even forty (°C) on many afternoons. Since both specific and relative humidities are high, the night brings little relief. An afternoon or evening thunderstorm is also a possibility, for these areas experience such storms on an average of from 40 to 100 days each year, the majority during the summer months. The primary reason for the tropical summer weather in Cfa regions is the dominating influence of maritime tropical air masses. During the summer months this warm, moist, and unstable air moves inland from the western portions of the oceanic subtropical anticyclone. As the mT air passes over the heated continent, it is made increasingly unstable, giving rise to the common convectional showers and thunderstorms.

As summer turns to autumn the humid subtropics lose their similarity to the rainy tropics. Although winters are best described as mild, frosts are common in higher-latitude Cfa areas and occasionally plague the tropical margins as well. The winter precipitation is also different in character from summer. Some is in the form of snow, and most is generated along fronts of the frequent middle-latitude cyclones that sweep over these regions. Since the land surface is colder than the maritime air that now arrives less frequently from tropical source regions, the air is chilled in its lower layers as it moves poleward over the ground. Consequently, convectional showers are rare as the stabilized mT air masses produce clouds and precipitation only when forced to rise.

The data for two humid subtropical stations in Table 13-10 serve to summarize the general characteristics of the Cfa climate. Yearly precipitation totals are usually in excess of 100 centimeters and the rainfall is well distributed throughout the year. Although summer is ordinarily the season of greatest precipitation, there is considerable variation. For example, in the United States precipitation in the Gulf states is very evenly distributed. However, as one moves poleward or toward the drier western margins, a much higher percentage falls in summer. Some coastal stations, especially along the equatorward margins, have rainfall maximums in late summer or autumn as tropical cyclones or their remnants visit the area. In Asia the well-developed monsoon circulation favors a summer precipitation maximum. An examination of temperature figures shows that summer temperatures are comparable to the tropics and that winter values are markedly lower. This, of course, is to be expected since the higher latitudinal position of the subtropics results in a wider variation in sun angle and length of daylight as well as occasional or even frequent invasions of cP air masses during winter.

Table 13-10

Data for Humid Subtropical Stations

	J	F	M	A	M	J	J	A	S	O	N	D	Yr.
				New Orleans, Louisiana 29°59'N; 1 m									
Temp. (°C)	12	13	16	19	23	26	27	27	25	21	15	13	20
Precip. (mm)	98	101	136	116	111	113	171	136	128	72	85	104	1371
				Buenos Aires, Argentina 34°35'S; 27 m									
Temp. (°C)	24	23	21	17	14	11	10	12	14	16	20	22	17
Precip. (mm)	104	82	122	90	79	68	61	68	80	100	90	83	1027

the marine west coast climate (Cfb and Cfc)

Situated on the western (windward) side of continents from about 40 to 65 degrees north and south latitude is a climatic region dominated by the onshore flow of oceanic air. The prevalence of maritime air masses means that mild winters and cool summers are the rule, as is an ample amount of rainfall throughout the year. In North America this climatic zone begins near the United States–Canadian border and extends northward as a narrow belt into southern Alaska. A similar slender strip occurs in South America along the coast of Chile poleward of about 40 degrees south latitude. In both instances high mountains parallel the coast and prevent the marine climate from penetrating far inland. The largest area of Cfb climate is found in Europe, for here there is no mountain barrier blocking the movement of cool maritime air from the North Atlantic. Other locations include most of New Zealand as well as tiny slivers of South Africa and Australia.

The data for representative marine west coast stations in Table 13-11 reveal that although there is no pronounced dry period, there is a drop in monthly precipitation totals during the summer. The reason for the reduced summer rainfall is the poleward migration of the oceanic subtropical highs. Although the areas of marine west coast climate are situated too far poleward to be dominated by these dry anticyclones, their influence is sufficient to cause a decrease in warm season rainfall. A comparison of the precipitation

Table 13-11

Data for Marine West Coast Stations

	J	F	M	A	M	J	J	A	S	O	N	D	Yr.
				Vancouver, British Columbia 49°11'N; 0 m									
Temp. (°C)	2	4	6	9	13	15	18	17	14	10	6	4	10
Precip. (mm)	139	121	96	60	48	51	26	36	56	117	142	156	1048
				London, U.K. 51°28'N; 5 m									
Temp. (°C)	4	4	7	9	12	16	18	17	15	11	7	5	10
Precip. (mm)	54	40	37	38	46	46	56	59	50	57	64	48	595

data for London and Vancouver also demonstrate that the presence of coastal mountains has a considerable influence upon yearly rainfall totals. Vancouver's total is about two and one-half times that of London. In settings like Vancouver's the precipitation totals are not only higher because of orographic uplift, but the mountains also act to slow the passage of cyclonic storms, thus allowing them to drop a greater quantity of water than would otherwise be the case.

Because of the nearness to the ocean, winters are mild and summers are relatively cool. Therefore, a low annual temperature range is characteristic in the marine west coast climate. Since cP air masses generally drift eastward in the zone of the westerlies, periods of severe winter cold are uncommon. The western edge of North America is especially sheltered from incursions of frigid continental air by the high mountains that intervene between the coast and the source regions for cP air masses. Because of the lack of such a mountain barrier in Europe, cold waves in that region are somewhat more frequent.

The dominance of the ocean in controlling temperatures can be further demonstrated by a look at temperature gradients (changes in temperature per unit distance). Although this climate encompasses a wide latitudinal span, temperatures change much more abruptly moving inland from the coast than they do in a north-south direction. The transport of heat from the oceans more than offsets the latitudinal variation in the receipt of solar energy. For example, in both January and July the temperature change from coastal Seattle to more inland Spokane, a distance of about 375 kilometers, is equal to the variation between Seattle and Juneau, Alaska. Juneau is about 11 degrees of latitude, or roughly 1200 kilometers, north of Seattle.

the dry-summer subtropical climate (Csa and Csb)

The *dry-summer subtropical climate* is typically located along the west sides of continents between latitudes 30 and 45 degrees. Situated between the marine west coast climate on the poleward side and the tropical steppes on the equatorward side, this climate region is best described as transitional in character. It is unique for the fact that it is the only humid climate that has a strong winter rainfall maximum, a feature that reflects its intermediate position. In summer the region is dominated by the stable eastern side of the oceanic subtropical highs. In winter, as the wind and pressure systems follow the sun equatorward, it is within range of the cyclonic storms of the polar front. Thus, during the course of a year these areas alternate between being a part of the dry tropics and an extension of the humid middle latitudes. While middle-latitude changeability characterizes the winter, tropical constancy describes the summer.

As was the case for the marine west coast climate, mountain ranges limit the dry-summer subtropics to a relatively narrow coastal zone in both North and South America. Since Australia and southern Africa barely reach to the latitudes where dry summer climates exist, the development of this climatic type is limited on these continents as well. Consequently, because of the arrangement of the continents, and of their mountain ranges, inland development occurs only in the Mediterranean Basin. Here the zone of subsidence extends far to the east in summer; in winter the sea is a major route of cyclonic disturbances. Since the dry-summer climate is particularly extensive in this region, the name *Mediterranean climate* is often used as a synonym.

Two types of Mediterranean climate are recognized and are based primarily on summertime temperatures. The cool summer type (Csb), as exemplified by San Francisco, California and Santiago, Chile (Table 13-12), is limited to coastal areas. Here the cooler summer temperatures one expects on a windward coast are further intensified by cold ocean currents. The data for Izmir and Sacramento illustrate the features of the warm summer type (Csa). At both places winter temperatures are not very different from those in the Csb type. However, since Sacramento in the Central Valley of California is removed from the coast and Izmir is bordered by the warm waters of the Mediterranean, summer temperatures are noticeably higher. As a result, seasonal temperature ranges are also higher in Csa areas.

Table 13-12

Data for Dry-Summer Subtropical Stations

	J	F	M	A	M	J	J	A	S	O	N	D	Yr.
San Francisco, California 37°37'N; 5 m													
Temp. (°C)	9	11	12	13	15	16	17	17	18	16	13	10	14
Precip. (mm)	102	88	68	33	12	3	0	1	5	19	40	104	475
Sacramento, California 38°35'N; 13 m													
Temp. (°C)	8	10	12	16	19	22	25	24	23	18	12	9	17
Precip. (mm)	81	76	60	36	15	3	0	1	5	20	37	82	416
Izmir, Turkey 38°26'N; 25 m													
Temp. (°C)	9	9	11	15	20	25	28	27	23	19	14	10	18
Precip. (mm)	141	100	72	43	39	8	3	3	11	41	93	141	695
Santiago, Chile 33°27'S; 512 m													
Temp. (°C)	19	19	17	13	11	8	8	9	11	13	16	19	14
Precip. (mm)	3	3	5	13	64	84	76	56	30	13	8	5	360

Yearly precipitation totals within the dry-summer subtropics range between about 40 and 80 centimeters. In many areas such amounts mean that a station barely escapes being classified as semiarid. Consequently, some climatologists refer to the dry-summer climate as subhumid instead of humid. This is especially true along the equatorward margins because rainfall totals increase in the poleward direction. Los Angeles, for example, receives 38 centimeters of precipitation annually, while San Francisco, 400 kilometers to the north, receives 51 centimeters per year. Still farther north, at Portland, Oregon, the yearly rainfall average is over 90 centimeters.

the humid
continental
climate
(Dfa, Dfb,
Dwa, Dwb)

Situated in the zone dominated by the polar front, this climatic region is a battleground for tropical and polar air masses. There is no other climate where rapid nonperiodic changes in the weather are more pronounced. Cold waves, heat waves, blizzards, and heavy downpours are all yearly events in the humid continental realm.

The *humid continental climate*, as its name implies, is a land-controlled climate, the result of broad continents located in the middle latitudes. Because continentality is a basic feature this climate is not found in the Southern Hemisphere where the middle-latitude zone is dominated by the oceans. Instead, it is confined to the central and eastern portions of North America and Eurasia in the latitude range between approximately 40 and 50 degrees north latitude. It may at first seem unusual that a continental climate should extend eastward to the margins of the ocean. However, since the prevailing atmospheric circulation is from the west, deep and persistent incursions of maritime air from the east are not very likely.

Both winter and summer temperatures in the humid continental climate may be characterized as relatively severe. Consequently, annual temperature ranges are high throughout the climate. A comparison of the stations whose data are shown in Table 13-13 illustrates the temperature variations within the humid continental realm. July means are generally near and often above 20°C. Thus, summertime temperatures, although somewhat lower in the north than in the south, are not markedly different. This is also illustrated by the isothermal map of the eastern United States in Figure 13-11(*a*). Notice that there are only a few widely spaced isotherms, indicating a weak summer temperature gradient. In January, however, the monthly means, although consistently cold, are more variable. The decrease in midwinter values with increasing latitude is appreciable [Figure 13-11(*b*)]. A glance at Table 13-13 reveals that the temperature change between Omaha and Winnipeg is more than twice as great in winter as in summer. Because of the steeper winter

Table 13-13

Data for Humid Continental Stations

	J	F	M	A	M	J	J	A	S	O	N	D	Yr.
					Omaha, Nebraska 41°18′N; 330 m								
Temp. (°C)	−6	−4	3	11	17	22	25	24	19	12	4	−3	10
Precip. (mm)	20	23	30	51	76	102	79	81	86	48	33	23	652
					New York City 40°47′N; 40 m								
Temp. (°C)	−1	−1	3	9	15	21	23	22	19	13	7	1	11
Precip. (mm)	84	84	86	84	86	86	104	109	86	86	86	84	1065
					Winnipeg, Canada 49°54′N; 240 m								
Temp. (°C)	−18	−16	−8	3	11	17	20	19	13	6	−5	−13	3
Precip. (mm)	26	21	27	30	50	81	69	70	55	37	29	22	517
					Harbin, Manchuria 45°45′N; 143 m								
Temp. (°C)	−20	−16	−6	6	14	20	23	22	14	6	−7	−17	3
Precip. (mm)	4	6	17	23	44	92	167	119	52	36	12	5	557

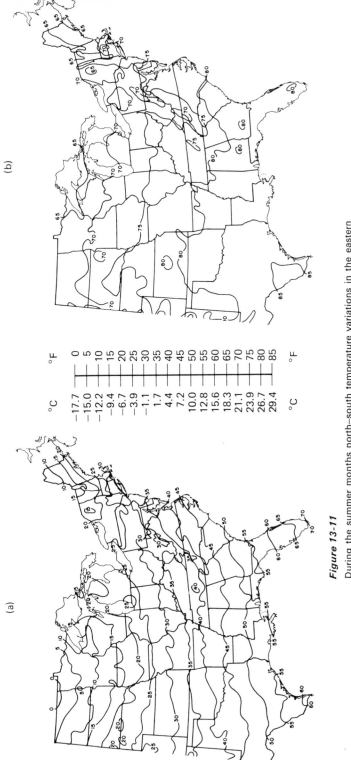

(a)

(b)

°F	°C
0	−17.7
5	−15.0
10	−12.2
15	−9.4
20	−6.7
25	−3.9
30	−1.1
35	1.7
40	4.4
45	7.2
50	10.0
55	12.8
60	15.6
65	18.3
70	21.1
75	23.9
80	26.7
85	29.4
°F	°C

Figure 13-11

During the summer months north–south temperature variations in the eastern United States are small, that is, the temperature gradient is weak (*a*). In winter, however, north–south temperature contrasts are sharp (*b*).

temperature gradient, shifts in wind direction during the cold season often result in sudden large temperature changes. This is not the case in summer, since temperatures throughout the region are more uniform.

Annual temperature ranges also vary within this climate, generally increasing from south to north as well as from the coast toward the interior. A comparison of the data for Omaha and Winnipeg illustrates the first situation, and a comparison of New York City and Omaha exemplifies the second.

The records of the four stations in Table 13-13 reveal the general pattern of precipitation within the regions having humid continental climates. A summer maximum occurs at each of the stations, but it is weakly defined at New York City because the east coast is more accessible to maritime air masses throughout the year. For the same reason, New York City also has the highest total of the four stations. Harbin, Manchuria, on the other hand, shows the most pronounced summer maximum of all, followed by a winter drought. This is characteristic of most east Asian stations in the middle latitudes and reflects the powerful control of the monsoon. Another pattern revealed by the data is that precipitation generally decreases toward the interior of the continents as well as from south to north. This is primarily a consequence of increasing distance from the sources of mT air. Further, the more northerly stations are also influenced for a greater part of the year by drier polar air masses.

Wintertime precipitation in these climates is associated primarily with the passage of fronts associated with traveling middle-latitude cyclones. A portion of this precipitation is in the form of snow, the proportion increasing with latitude. Although the precipitation is often considerably less during the cold season, it is usually much more conspicuous than the greater amounts of summer. An obvious reason is that snow remains on the ground, often for extended periods, and rain, of course, does not. Further, while summer rains are often in the form of relatively short convective showers, winter snows generally occur over a more prolonged period.

the subarctic climate (Dfc, Dfd, Dwc, Dwd)

Situated north of the humid continental climate and south of the polar tundra is an extensive *subarctic climate* region covering broad, uninterrupted expanses from western Alaska to Newfoundland in North America and from Norway to the Pacific coast of the Soviet Union in Eurasia. It is often referred to as the *taiga* climate since its extent closely corresponds to the northern coniferous forest region of the same name. Although they are scrawny, the spruce, fir, larch, and birch trees in the taiga represent the largest stretch of continuous forest on the surface of the earth.

The subarctic is well illustrated by the data for Yakutsk and Dawson (Table 13-14). Here, in the source regions of continental polar air masses, the outstanding feature is certainly the dominance of winter. Not only is it long, but temperatures are also bitterly cold to say the least. Winter minimum temperatures are among the lowest ever recorded outside of the ice caps of

Greenland and Antarctica. In fact, for many years the world's coldest temperature was attributed to Verkhoyansk in east-central Siberia, where the temperature dropped to −68°C on February 5 and 7, 1892. Over a 23-year period, this same station had an average monthly minimum of −62°C during January. Although these are exceptional temperatures, they illustrate the extreme cold that envelopes the taiga in winter.

By contrast, summers in the subarctic are remarkably warm, despite their short duration. However, when compared to regions farther south, this

Table 13-14

Data for Subarctic Stations

	J	F	M	A	M	J	J	A	S	O	N	D	Yr.
			Yakutsk, U.S.S.R. 62°05′N ; 103 m										
Temp. (°C)	−43	−37	−23	−7	7	16	20	16	6	−8	−28	−40	−10
Precip. (mm)	7	6	5	7	16	31	43	38	22	16	13	9	213
			Dawson, Canada 64°03′N ; 315 m										
Temp. (°C)	−30	−24	−16	−2	8	14	15	12	6	−4	−17	−25	−5
Precip. (mm)	20	20	13	18	23	33	41	41	43	33	33	28	346

short season must be characterized as cool; for despite the many hours of daylight, the sun never rises very high in the sky and hence solar radiation is not intense. The extremely cold winters and the relatively warm summers of the taiga combine to produce the highest annual temperature ranges on earth. Yakutsk holds the distinction of having the greatest average annual temperature range in the world–63°C. As the data for Dawson show, the North American subarctic is not so severe.

Since these far northerly continental interiors are the source regions of cP air masses, there is very limited moisture available throughout the year. Precipitation totals are therefore small, seldom exceeding 50 centimeters annually. By far, the greatest quantity of precipitation comes in the form of rain from scattered summer convectional showers. Snowfall is not as heavy as in the humid continental climate to the south. However, there is often the illusion of more. Since no melting occurs for months at a time, the entire winter accumulation (up to 1 meter) is visible all at once. Furthermore, during blizzards the dry powdery snow is easily picked up and whirled about by the high winds, creating high drifts and giving the impression more snow is falling than is actually the case. So although snowfall is not excessive, a visitor to this region could leave with that impression.

the polar (E) climates

According to the Koeppen classification, *polar climates* are those in which the mean temperature of the warmest month is below 10°C. Therefore, just as the tropics are defined by their year-round warmth, so the polar realm is known for its enduring cold. These regions hold the distinction of having the

347

lowest annual means for any part of the planet. Since winters are periods of perpetual night, or nearly so, temperatures at most polar locations are understandably bitter. During the summer months temperatures remain cool despite the long days, because the sun is so low in the sky that its oblique rays are not effective in bringing about a genuine warming. In addition, much solar radiation is reflected by the ice and snow or used in melting the snow cover. In either case, energy that could have warmed the land is lost or consumed. Although summer temperatures are cool, they are still much higher than those experienced during the extremely severe winter months. Consequently, annual temperature ranges are usually large.

Although polar climates are classified as humid, precipitation is generally meager, with many nonmarine stations receiving less than 25 centimeters annually. Evaporation, of course, is also limited. The scanty precipitation totals are easily understood in view of the temperature characteristics of the region. The amount of water vapor in the air is always small because low specific humidities must accompany low temperatures. In addition, steep lapse rates are not possible under polar temperature conditions. Usually precipitation is most abundant during the warmer summer months when the air's moisture content is highest.

the tundra climate (ET)

The *tundra climate* on land is found almost exclusively in the Northern Hemisphere. Here it occupies the coastal fringes of the Arctic Ocean, as well as many Arctic islands and the ice-free shores of northern Iceland and southern Greenland. In the Southern Hemisphere no extensive land areas exist in the latitudes where tundra climates prevail. Consequently, except for some small islands in the southern oceans, the ET climate occupies only the southwestern tip of South America and the northern portion of the Palmer Peninsula in Antarctica.

The 10°C summer isotherm that marks the equatorward limit of the tundra also marks the poleward limit of tree growth. Thus, this is essentially a treeless region of grasses, sedges, mosses, and lichens. During the long cold season plant life is dormant, but once the short cool summer commences, these plants mature and produce seeds with great rapidity. Since summers are cool and short, the frozen soils of the tundra generally thaw to depths of less than a meter and consequently the subsoil remains permanently frozen. This condition, known as *permafrost*, blocks the downward loss of water and results in poorly drained, boggy soils that make the summer tundra landscape a difficult one to traverse.

The data for Point Barrow, Alaska (Table 13-15), on the shores of the frozen Arctic Ocean, exemplifies the most common type of ET station, that is, one where continentality prevails. Because of the combination of high latitude and continentality, winters are severe, summers are cool, and annual temperature ranges are high. Further, yearly precipitation is small, with a modest summertime maximum.

Table 13-15

Data for Polar Stations

	J	F	M	A	M	J	J	A	S	O	N	D	Yr.
				Eismitte, Greenland 70°53′N; 2953 m									
Temp. (°C)	−42	−47	−40	−32	−24	−17	−12	−11	−11	−36	−43	−38	−29
Precip. (mm)	15	5	8	5	3	3	3	10	8	13	13	25	111
				Angmagssalik, Greenland 65°36′N; 29 m									
Temp. (°C)	−7	−7	−6	−3	2	6	7	7	4	0	−3	−5	0
Precip. (mm)	57	81	57	55	52	45	28	70	72	96	87	75	775
				Barrow, Alaska 71°18′N; 9 m									
Temp. (°C)	−28	−28	−26	−17	−8	0	4	3	−1	−9	−18	−24	−12
Precip. (mm)	5	5	3	3	3	10	20	23	15	13	5	5	110
				Cruz Loma, Ecuador 0°08′S; 3888 m									
Temp. (°C)	6.1	6.6	6.6	6.6	6.6	6.1	6.1	6.1	6.1	6.1	6.6	6.6	6.4
Precip. (mm)	198	185	241	236	221	122	36	23	86	147	124	160	1779

While Point Barrow represents the most common type of tundra setting, the data for Angmagssalik, Greenland (Table 13-15) reveals that some ET stations are different. Although summer temperatures at both stations are roughly equivalent, winters at Angmagssalik are much warmer and the annual precipitation is eight times greater than at Point Barrow. The reason for the contrast is Angmagssalik's location on the southeast coast of Greenland where there is considerable marine influence. The warm North Atlantic Drift keeps winter temperatures relatively high and mP air masses supply moisture throughout the year. Since winters are not so severe at stations like Angmagssalik, annual temperature ranges are much smaller than at ET stations like Point Barrow, where continentality is a major control.

Finally, it should be pointed out that tundra climates are not entirely confined to the high latitudes. The summer coolness of this climate is also found at increasingly higher elevations as one moves equatorward. Hence, even in the tropics you can find ET climates if you go high enough. When compared to the Arctic tundra, however, winter temperatures in these lower-latitude counterparts become milder and less distinct from summer, as the data for Cruz Loma, Equador (Table 13-15) illustrate.

ice cap climate (EF)

The *ice cap climate*, designated by Koeppen as EF, does not have a single monthly mean above 0°C. Consequently, since the average temperature for all months is below freezing, the growth of vegetation is prohibited and the landscape is one of permanent ice and snow. This climate of perpetual frost covers a surprisingly large area of more than 15.5 million square kilometers or about 9 percent of the earth's land area, and aside from scattered occurrences in high mountain areas, it is confined to the ice caps of Greenland and Antarctica.

Average annual temperatures are extremely low; for example, the annual mean at Eismitte, Greenland (Table 13-15) is —29°C; at Byrd Station, Antarctica, —21°C; and at Vostok, the U.S.S.R. Antarctic Meteorological Station, —57°C. Vostok has also experienced the lowest temperature ever recorded, —88.3°C on August 24, 1960. In addition to the effect of latitude, the primary cause for the very low temperatures is undoubtedly the presence of permanent ice. Ice has a very high albedo, reflecting up to 80 percent of the insolation striking it. Further, the energy that is not reflected is used largely to melt the ice and is therefore not available for raising the temperature of the air. An additional factor that contributes to the very low temperatures at many EF stations is elevation. Eismitte, at the center of the Greenland ice cap, rests at an elevation of almost 3000 meters, and much of Antarctica is even higher. Thus, because of the permanent ice and often high elevations, the already low temperatures of the polar realm are further reduced.

The intense chilling of the air close to the ice cap means that strong surface temperature inversions are common. Temperatures near the surface may be as much as 30°C colder than air just a few hundred meters above. Often, this cold dense air flows downslope because of the force of gravity and produces strong winds and blizzard conditions. Such air movements, called fall or katabatic winds, are an important aspect of ice cap weather at many locations. In some places, where the slope is sufficient, gravity winds are sometimes strong enough to flow in a direction that is opposite the pressure gradient.

REVIEW
1. Why is classification often a helpful or even a necessary task in science?
2. What climatic data are needed in order to classify a climate using the Koeppen scheme?
3. Should climatic boundaries, such as those shown on the world map in Figure 13-1, be regarded as fixed? Explain.
4. List the major climatic controls and briefly describe their influence.
5. How does the selva differ from a typical middle-latitude forest?
6. Distinguish between jungle and selva.
7. Explain each of the following characteristics of the wet tropics:
   a. This climate is restricted to elevations below 1000 meters.
   b. Mean monthly and annual temperatures are high and the annual temperature range is low.
   c. This climate is rainy throughout the year or nearly so.
8. Why are the wet tropics considered oppressive and monotonous?
9. What is the difference between Af areas and Am areas?
10. a. What primary factor distinguishes Aw climates from Af and Am?
    b. How is this difference reflected in the vegetation?

11. Describe the influence of the ITC and the subtropical high on the precipitation regime in the Aw climate.

12. In which of the following climates is the annual rainfall total likely to be most consistent from year to year: BSh, Aw, BWh, Af? In which of these climates is the annual rainfall total most variable from year to year? Explain your answers.

13. In the dry (B) climates there are usually more years when rainfall totals are below the average than above. Explain and give an example.

14. List and briefly discuss the four common misconceptions about deserts.

15. Why is the amount of precipitation that defines the humid–dry boundary variable?

16. What is the primary reason (control) for the existence of the dry tropical realm (BWh and BSh)?

17. a. Describe and explain the seasonal distribution of precipitation for a BSh station located on the poleward side of a tropical desert and a BSh station located on the equatorward side of the desert.
    b. If both stations barely meet the requirements for steppe climates (that is, with only a little more rainfall both stations would be considered humid), which station would likely have the lowest rainfall total? Explain.

18. Why do ground and air temperatures reach such high values in tropical deserts?

19. Tropical deserts such as the Atacama and Namib deviate considerably from the general image we have of tropical deserts. In what ways are these deserts not "typical" and why?

20. What is the primary cause for the existence of middle-latitude deserts and steppes?

21. Why are desert and steppe areas uncommon in the middle latitudes of the Southern Hemisphere?

22. Describe and explain the differences between summertime and wintertime precipitation in the humid subtropics (Cfa).

23. Why is the marine west coast climate (Cfb and Cfc) represented by only slender strips of land in North and South America and why is it very extensive in western Europe?

24. How do temperature gradients (north–south versus east–west) reveal the strong oceanic influence along the west coast of North America?

25. In this chapter the dry-summer subtropics were described as transitional. Explain why this statement is true.

26. What other name is given to the dry-summer subtropical climate?

27. Why are summer temperatures cooler at San Francisco than at Sacramento (Table 13-12)?

28. Why is the humid continental climate confined to the Northern Hemisphere?

29. Why do coastal stations such as New York City experience primarily continental climatic conditions?

30. Using the four stations whose data are shown in Table 13-13, describe the general pattern of precipitation in the humid continental climates.

351

31. In the dry-summer subtropics precipitation totals increase with an increase in latitude, but in the humid continental climates the reverse is true. Explain.

32. Although the subarctic and polar climates are generally characterized by small precipitation totals, they are considered humid. Explain.

33. Although snowfall in the subarctic climate is relatively small, a wintertime visitor to this region might leave with the impression that snowfall is great. Explain.

34. Describe and explain the annual temperature range one should expect in the realm of the taiga.

35. Although polar regions experience extended periods of almost perpetual sunlight in the summer, temperatures remain cool. Explain.

36. What is the significance of the 10°C summer isotherm?

37. Why is the tundra landscape characterized by poorly drained boggy soils?

38. Why are winter temperatures higher and the annual precipitation total much greater at Angmagssalik, Greenland than at Point Barrow, Alaska?

39. The tundra climate is not confined solely to the high latitudes. Under what circumstances might the ET climate be found in more equatorward locations?

40. Where are EF climates most extensively developed?

41. Aside from the effect of latitude, what other factor(s) contributes to the extremely low temperatures that characterize the EF climate? Explain.

VOCABULARY REVIEW

marine climate

continental climate

tropical rain forest (selva)

jungle

zone of intertropical convergence (ITC)

tropical wet and dry

savanna

monsoon

dry climate

desert (arid)

steppe (semiarid)

humid subtropical climate

marine west coast climate

dry-summer subtropical climate

Mediterranean climate

humid continental climate

subarctic climate

taiga

polar climates

tundra climate

permafrost

ice cap climate

Appendix A

metric units

I. Basic Units

Quantity	Unit	SI Symbol
Length	meter	m
Mass	kilogram	kg
Time	second	s
Electric current	ampere	A
Thermodynamic temperature	kelvin	K
Amount of substance	mole	mol
Luminous intensity	candela	cd

II. Prefixes

Prefix	Factor by Which Unit Is Multiplied	Symbol
tera	10^{12}	T
giga	10^{9}	G
mega	10^{6}	M
kilo	10^{3}	k
hecto	10^{2}	h
deka	10	da
deci	10^{-1}	d

Table A-1 continued

II. Prefixes

Prefix	Factor by Which Unit Is Multiplied	Symbol
centi	10^{-2}	c
milli	10^{-3}	m
micro	10^{-6}	μ
nano	10^{-9}	n
pico	10^{-12}	p
femto	10^{-15}	f
atto	10^{-18}	a

III. Derived Units

Quantity	Units	Expression
Area	square meter	m^2
Volume	cubic meter	m^3
Frequency	hertz (Hz)	s^{-1}
Density	kilogram per cubic meter	kg/m^3
Velocity	meter per second	m/s
Angular velocity	radian per second	rad/s
Acceleration	meter per second squared	m/s^2
Angular acceleration	radian per second squared	rad/s^2
Force	newton (N)	$kg \cdot m/s^2$
Pressure	newton per square meter	N/m^2
Work, energy, quantity of heat	joule (J)	$N \cdot m$
Power	watt (W)	J/s
Electric charge	coulomb (C)	$A \cdot s$
Voltage, potential difference, electromotive force	volt (V)	W/A
Luminance	candela per square meter	cd/m^2

Table A-2

Metric–English Conversion

When You Want to Convert:	Multiply by:	To Find:
Length		
inches	2.54	centimeters
centimeters	0.39	inches
feet	0.30	meters
meters	3.28	feet
yards	0.91	meters
meters	1.09	yards
miles	1.61	kilometers
kilometers	0.62	miles

354

When You Want to Convert:	Multiply by:	To Find:
Area		
square inches	6.45	square centimeters
square centimeters	0.15	square inches
square feet	0.09	square meters
square meters	10.76	square feet
square miles	2.59	square kilometers
square kilometers	0.39	square miles
Volume		
cubic inches	16.38	cubic centimeters
cubic centimeters	0.06	cubic inches
cubic feet	0.028	cubic meters
cubic meters	35.3	cubic feet
cubic miles	4.17	cubic kilometers
cubic kilometers	0.24	cubic miles
liters	1.06	quarts
liters	0.26	gallons
gallons	3.78	liters
Masses and Weights		
ounces	20.33	grams
grams	0.035	ounces
pounds	0.45	kilograms
kilograms	2.205	pounds

Temperature

When you want to convert degrees Fahrenheit (°F) to degrees Celsius (°C), subtract 32 degrees and divide by 1.8 (also see Table A-3).

When you want to convert degrees Celsius (°C) to degrees Fahrenheit (°F), multiply by 1.8 and add 32 degrees.

When you want to convert degrees Celsius (°C) to degrees Kelvin (K), delete the degree symbol and add 273.

When you want to convert degrees Kelvin (K) to degrees Celsius (°C), add the degree symbol and subtract 273.

Table A-3

Temperature Conversion Table. (To find either the Celsius or Fahrenheit equivalent, locate the known temperature in the center column. Then read the desired equivalent value from the appropriate column.)

°C		°F	°C		°F	°C		°F	°C		°F
−40.0	−40	−40	−17.2	+1	33.8	5.0	41	105.8	27.2	81	177.8
−39.4	−39	−38.2	−16.7	2	35.6	5.6	42	107.6	27.8	82	179.6
−38.9	−38	−36.4	−16.1	3	37.4	6.1	43	109.4	28.3	83	181.4
−38.3	−37	−34.6	−15.4	4	39.2	6.7	44	111.2	28.9	84	183.2
−37.8	−36	−32.8	−15.0	5	41.0	7.2	45	113.0	29.4	85	185.0
−37.2	−35	−31.0	−14.4	6	42.8	7.8	46	114.8	30.0	86	186.8
−36.7	−34	−29.2	−13.9	7	44.6	8.3	47	116.6	30.6	87	188.6
−36.1	−33	−27.4	−13.3	8	46.4	8.9	48	118.4	31.1	88	190.4
−35.6	−32	−25.6	−12.8	9	48.2	9.4	49	120.2	31.7	89	192.2
−35.0	−31	−23.8	−12.2	10	50.0	10.0	50	122.0	32.2	90	194.0
−34.4	−30	−22.0	−11.7	11	51.8	10.6	51	123.8	32.8	91	195.8
−33.9	−29	−20.2	−11.1	12	53.6	11.1	52	125.6	33.3	92	197.6
−33.3	−28	−18.4	−10.6	13	55.4	11.7	53	127.4	33.9	93	199.4
−32.8	−27	−16.6	−10.0	14	57.2	12.2	54	129.2	34.4	94	201.2
−32.2	−26	−14.8	−9.4	15	59.0	12.8	55	131.0	35.0	95	203.0
−31.7	−25	−13.0	−8.9	16	60.8	13.3	56	132.8	35.6	96	204.8
−31.1	−24	−11.2	−8.3	17	62.6	13.9	57	134.6	36.1	97	206.6
−30.6	−23	−9.4	−7.8	18	64.4	14.4	58	136.4	36.7	98	208.4
−30.0	−22	−7.6	−7.2	19	66.2	15.0	59	138.2	37.2	99	210.2
−29.4	−21	−5.8	−6.7	20	68.0	15.6	60	140.0	37.8	100	212.0
−28.9	−20	−4.0	−6.1	21	69.8	16.1	61	141.8	38.3	101	213.8
−28.3	−19	−2.2	−5.6	22	71.6	16.7	62	143.6	38.9	102	215.6
−27.8	−18	−0.4	−5.0	23	73.4	17.2	63	145.4	39.4	103	217.4
−27.2	−17	+1.4	−4.4	24	75.2	17.8	64	147.2	40.0	104	219.2
−26.7	−16	3.2	−3.9	25	77.0	18.3	65	149.0	40.6	105	221.0
−26.1	−15	5.0	−3.3	26	78.8	18.9	66	150.8	41.1	106	222.8
−25.6	−14	6.8	−2.8	27	80.6	19.4	67	152.6	41.7	107	224.6
−25.0	−13	8.6	−2.2	28	82.4	20.0	68	154.4	42.2	108	226.4
−24.4	−12	10.4	−1.7	29	84.2	20.6	69	156.2	42.8	109	228.2
−23.9	−11	12.2	−1.1	30	86.0	21.1	70	158.0	43.3	110	230.0
−23.3	−10	14.0	−0.6	31	87.8	21.7	71	159.8	43.9	111	231.8
−22.8	−9	15.8	0.0	32	89.6	22.2	72	161.6	44.4	112	233.6
−22.2	−8	17.6	+0.6	33	91.4	22.8	73	163.4	45.0	113	235.4
−21.7	−7	19.4	1.1	34	93.2	23.3	74	165.2	45.6	114	237.2
−21.1	−6	21.2	1.7	35	95.0	23.9	75	167.0	46.1	115	239.0
−20.6	−5	23.0	2.2	36	96.8	24.4	76	168.8	46.7	116	240.8
−20.0	−4	24.8	2.8	37	98.6	25.0	77	170.6	47.2	117	242.6
−19.4	−3	26.6	3.3	38	100.4	25.6	78	172.4	47.8	118	244.4
−18.9	−2	28.4	3.9	39	102.2	26.1	79	174.2	48.3	119	246.2
−18.3	−1	30.2	4.4	40	104.0	26.7	80	176.0	48.9	120	248.0
−17.8	0	32.0									

Appendix B

explanation and decoding of the Daily weather map

Weather maps showing the development and movement of weather systems are among the principal tools used by the weather forecaster. Of the several types of maps used, some portray conditions near the surface of the earth, while others depict conditions at various heights in the atmosphere. Some cover the entire Northern Hemisphere, while others cover only local areas as required for special purposes. The maps used for daily forecasting by the National Weather Service are similar in many respects to the printed Daily Weather Map. At Weather Service offices maps showing conditions at the earth's surface are drawn four times daily. Maps of upper-level temperature, pressure, and humidity are prepared twice each day.

principal surface weather map

To prepare the surface map and present the information quickly and pictorially, two actions are necessary: (*1*) Weather observers at many places must go to their posts at regular times each day to observe the weather and send the information by wire or radio to the offices where the maps are drawn; (*2*) The information must be quickly transcribed to the maps. In order that the necessary speed and economy of space and transmission time may be realized, codes have been devised for sending the information and for plotting it on the maps.

357

explanation of
the weather map

A great deal of information is contained in a brief coded weather message. If each item were named and described in plain language, a very lengthy message would be required, and it would be confusing to read and difficult to transfer to a map. A code permits the message to be condensed to a few five-figure numeral groups, each figure of which has a meaning, depending upon its position in the message. People trained in the use of the code can read the message as easily as plain language.

The location of the reporting station is printed on the map as a small circle (the station circle). A definite arrangement of the data around the station circle, called the *station model*, is used. When the report is plotted in these fixed positions around the station circle on the weather map, many of the code figures are transcribed exactly as sent. Entries in the station model that are not made in code figures or actual values found in the message are usually in the form of symbols that graphically represent the element concerned. In some cases, certain of the data may or may not be reported by the observer, depending upon local weather conditions. Precipitation and clouds are examples. In such cases, the absence of an entry on the map is interpreted as nonoccurrence or nonobservance of the phenomena. The letter "M" is entered where data are normally observed but not received.

Both the code and the station model are based on international agreements. These standardized numerals and symbols enable a meteorologist of one country to use the weather reports and weather maps of another country even though he does not understand the language. Weather codes are, in effect, an international language that make possible complete interchange and use of worldwide weather reports so essential in present-day activities.

The boundary between two different air masses is called a *front*. Important changes in weather, temperature, wind direction, and clouds often occur with the passage of a front. Half circles and/or triangular symbols are placed on the lines representing fronts to indicate the kind of front. The side on which the symbols are placed indicates the direction of frontal movement. The boundary of relatively cold air of polar origin advancing into an area occupied by warmer air, often of tropical origin, is called a *cold front*. The boundary of relatively warm air advancing into an area occupied by colder air is called a *warm front*. The line along which a cold front has overtaken a warm front at the ground is called an *occluded front*. A boundary between two air masses, which shows at the time of observation little tendency to advance into either the warm or cold areas, is called a *stationary front*. Airmass boundaries are known as *surface fronts* when they intersect the ground and as upper-air fronts when they do not. Surface fronts are drawn in solid black, fronts aloft are drawn in outline only. Front symbols are given in Table B-1.

A front that is disappearing or is weak and decreasing in intensity is labeled *frontolysis*. A front that is forming is labeled *frontogenesis*. A *squall line* is a line of thunderstorms or squalls usually accompanied by heavy showers and shifting winds (Table B-1).

Symbol	Explanation
▲▲▲	Cold front (surface)
●●●	Warm front (surface)
●▲●▲	Occluded front (surface)
●▼●▼	Stationary front (surface)
△△△	Cold front (aloft)
⌒⌒⌒	Warm front (aloft)
—··—··—··	Squall line
⟶ ⟶	Path of low-pressure center
⊠	Location of low pressure at 6-hour intervals
— — — —	32°F isotherm
—·—·—·—	0°F isotherm

The paths followed by individual disturbances are called *storm tracks* and are shown by arrows (Table B-1). A symbol (a box containing an "x") indicates past positions of a low-pressure center at 6-hour intervals. "HIGH" (H) and "LOW" (L) indicate the centers of high- and low-barometric pressure. Solid lines are isobars and connect points of equal sea-level barometric pressure. The spacing and orientation of these lines on weather maps are indications of speed and direction of wind flow. In general, wind direction is parallel to these lines with low pressure to the left of observer looking downwind. Speed is directly proportional to the closeness of the lines (termed *pressure gradient*). Isobars are labeled in millibars.

Isotherms are lines connecting points of equal temperature. Two isotherms are drawn on the large surface weather map when applicable. The freezing, or 32°F, isotherm is drawn as a dashed line and the 0°F isotherm is drawn as a dash–dot line (Table B-1). Areas where precipitation is occurring at the time of observation are shaded.

auxiliary maps   *500-millibar map*

Contour lines, isotherms, and wind arrows are shown on the insert map for the 500-millibar contour level. Solid lines are drawn to show height above sea level and are labeled in feet. Dashed lines are drawn at 5° intervals of temperature and are labeled in degrees Celsius. True wind direction is shown by "arrows" that are plotted as flying with the wind. The wind speed is shown by flags and feathers. Each flag represents 50 knots, each full feather represents 10 knots, and each half-feather represents 5 knots.

temperature map (highest and lowest)

Temperature data are entered from selected weather stations in the United States. The figures entered above the station dot denote maximum temperatures reported from these stations during the 24 hours ending 1:00 A.M. EST; the figures entered below the station dot denote minimum temperature during the 24 hours ending at 1:00 P.M. EST of the previous day. The letter "M" denotes missing data.

precipitation map

Precipitation data are entered from selected weather stations in the United States. When precipitation has occurred at any of these stations in the 24-hour period ending at 1:00 A.M. EST, the total amount, in inches and hundredths, is entered above the station dot. When the figures for total precipitation have been compiled from incomplete data and entered on the map, the amount is underlined. "T" indicates a trace of precipitation, and the letter "M" denotes missing data. The geographical areas where precipitation has fallen during the 24 hours ending at 1:00 A.M. EST are shaded. Dashed lines show depth of snow on ground in inches as of 7:00 A.M. EST of the previous day.

SYMBOLIC FORM OF MESSAGE ❶

iii Nddff VVwwW PPPTT N$C_h$$C_C$ $T_d$app 7RRR$_s$
              $_h$$_L$ $_M$$_H$ $_d$$_d$ $_t$

Note: This Abridged Code Shows Only Data Normally Plotted on Printed Maps

SAMPLE CODED MESSAGE

405 83220 12716 24731 67292 30228 74542

SYMBOLIC STATION MODEL	SAMPLE PLOTTED REPORT
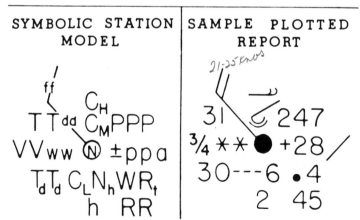	

N Sky coverage (total amount)—see block **6**

dd True direction *from* which wind is blowing

ff Wind speed in knots or miles per hour—see block **9**

VV Visibility in miles

ww Present weather—see block **8**

W Past weather—see block **11**

PPP Barometric pressure in millibars reduced to sea level. To decode barometric pressure, follow the steps listed below:

1. Place a decimal point to the left of the last number. For example, in
the sample plotted report above, 247 becomes 24.7.

2. Next, place either a 9 or 10 in front (to the left of the first digit). Thus,
24.7 becomes either 924.7 mb or 1024.7 mb.

3. To determine whether to place a 9 or 10 in front of the first digit,
follow the rule stated below:

If the number falls between 56.0 and 99.9, place a 9 before the first
digit.

If the number falls between 00.0 and 55.9, place a 10 before the first
digit.

Thus, 24.7 would be 1024.7 mb.

TT Current air temperature in Fahrenheit

N_h Fraction of sky covered by low or middle clouds—see block **7**

C_L Low clouds or clouds of vertical development—see block **3**

h Height in feet of the base of the lowest clouds—see block **5**

C_M Middle clouds—see block **3**

C_H High clouds—see block **3**

T_dT_d Dew-point temperature in Fahrenheit

a Pressure tendency—see block **10**

pp Pressure change (in millibars) in preceding 3 hours ($+28 = +2.8$
mb)

RR Amount of precipitation ($45 = .45$ inch)

R_t Time precipitation began or ended—see block **4**

Cloud Classification Chart

CLOUD ABBREVIATION	C_L		DESCRIPTION (Abridged From W M O Code)
St or Fs-Stratus or Fractostratus	1		Cu of fair weather, little vertical development and seemingly flattened
Ci-Cirrus	2		Cu of considerable development, generally towering, with or without other Cu or Sc bases all at same level
Cs-Cirrostratus	3		Cb with tops lacking clear-cut outlines, but distinctly not cirriform or anvil-shaped; with or without Cu, Sc, or St
Cc-Cirrocumulus	4		Sc formed by spreading out of Cu; Cu often present also
Ac-Altocumulus	5		Sc not formed by spreading out of Cu
As-Altostratus	6		St or Fs or both, but no Fs of bad weather
Sc-Stratocumulus	7		Fs and/or Fc of bad weather (scud)
Ns-Nimbostratus	8		Cu and Sc (not formed by spreading out of Cu) with bases at different levels
Cu or Fc-Cumulus or Fractocumulus	9		Cb having a clearly fibrous (cirriform) top, often anvil-shaped, with or without Cu, Sc, St, or scud

C_M		DESCRIPTION (Abridged From W M O Code)
1		Thin As (most of cloud layer semi-transparent)
2		Thick As, greater part sufficiently dense to hide sun (or moon), or Ns
3		Thin Ac, mostly semi-transparent; cloud elements not changing much and at a single level
4		Thin Ac in patches; cloud elements continually changing and/or occurring at more than one level
5		Thin Ac in bands or in a layer gradually spreading over sky and usually thickening as a whole
6		Ac formed by the spreading out of Cu
7		Double-layered Ac, or a thick layer of Ac, not increasing, or Ac with As and/or Ns
8		Ac in the form of Cu-shaped tufts or Ac with turrets
9		Ac of a chaotic sky, usually at different levels; patches of dense Ci are usually present also

C_H		DESCRIPTION (Abridged From W M O Code) ③
1		Filaments of Ci, or "mares tails," scattered and not increasing
2		Dense Ci in patches or twisted sheaves, usually not increasing, sometimes like remains of Cb; or towers or tufts
3		Dense Ci, often anvil-shaped, derived from or associated with Cb
4		Ci, often hook-shaped, gradually spreading over the sky and usually thickening as a whole
5		Ci and Cs, often in converging bands, or Cs alone; generally overspreading and growing denser; the continuous layer not reaching 45° altitude
6		Ci and Cs, often in converging bands, or Cs alone, generally overspreading and growing denser; the continuous layer exceeding 45° altitude
7		Veil of Cs covering the entire sky
8		Cs not increasing and not covering entire sky
9		Cc alone or Cc with some Ci or Cs, but the Cc being the main cirriform cloud

Rt	TIME OF PRECIPITATION (4)	h	HEIGHT IN FEET (Rounded Off)	HEIGHT IN METERS (Approximate)	N	SKY COVERAGE (Total Amount) (6)	Nh	SKY COVERAGE (Low And/Or Middle Clouds) (7)
0	No Precipitation	0	0 - 149	0 - 49	○	No clouds	0	No clouds
1	Less than 1 hour ago	1	150 - 299	50 - 99	⊖	Less than one-tenth or one-tenth	1	Less than one-tenth or one-tenth
2	1 to 2 hours ago	2	300 - 599	100 - 199	◔	Two-tenths or three-tenths	2	Two-tenths or three-tenths
3	2 to 3 hours ago	3	600 - 999	200 - 299	◑	Four-tenths	3	Four-tenths
4	3 to 4 hours ago	4	1,000 - 1,999	300 - 599	◕	Five-tenths	4	Five-tenths
5	4 to 5 hours ago	5	2,000 - 3,499	600 - 999	⊕	Six-tenths	5	Six-tenths
6	5 to 6 hours ago	6	3,500 - 4,999	1,000 - 1,499	◕	Seven-tenths or eight-tenths	6	Seven-tenths or eight-tenths
7	6 to 12 hours ago	7	5,000 - 6,499	1,500 - 1,999	⦸	Nine-tenths or over-cast with openings	7	Nine-tenths or over-cast with openings
8	More than 12 hours ago	8	6,500 - 7,999	2,000 - 2,499	●	Completely overcast	8	Completely overcast
9	Unknown	9	At or above 8,000, or no clouds	At or above 2,500, or no clouds	⊗	Sky obscured	9	Sky obscured

	0	1	2	3	4
00	Cloud development NOT observed or NOT observable during past hour	Clouds generally dissolving or becoming less developed during past hour	State of sky on the whole unchanged during past hour	Clouds generally forming or developing during past hour	Visibility reduced by smoke
10	Light fog	Patches of shallow fog at station, NOT deeper than 6 feet on land	More or less continuous shallow fog at station, NOT deeper than 6 feet on land	Lightning visible, no thunder heard	Precipitation within sight, but NOT reaching the ground
20	Drizzle (NOT freezing and NOT falling as showers) during past hour, but NOT at time of observation	Rain (NOT freezing and NOT falling as showers) during past hour, but NOT at time of observation	Snow (NOT falling as showers) during past hour, but NOT at time of observation	Rain and snow (NOT falling as showers) during past hour, but NOT at time of observation	Freezing drizzle or freezing rain (NOT falling as showers) during past hour, but NOT at time of observation
30	Slight or moderate dust storm or sand storm, has decreased during past hour	Slight or moderate dust storm or sand storm, no appreciable change during past hour	Slight or moderate dust storm or sand storm, has increased during past hour	Severe dust storm or sand storm, has decreased during past hour	Severe dust storm or sand storm, no appreciable change during past hour
40	Fog at distance at time of observation, but NOT at station during past hour	Fog in patches	Fog, sky discernible, has become thinner during past hour	Fog, sky NOT discernible, has become thinner during past hour	Fog, sky discernible, no appreciable change during past hour
50	Intermittent drizzle (NOT freezing) slight at time of observation	Continuous drizzle (NOT freezing) slight at time of observation	Intermittent drizzle (NOT freezing) moderate at time of observation	Continuous drizzle (NOT freezing), moderate at time of observation	Intermittent drizzle (NOT freezing), thick at time of observation
60	Intermittent rain (NOT freezing), slight at time of observation	Continuous rain (NOT freezing), slight at time of observation	Intermittent rain (NOT freezing) moderate at time of obs.	Continuous rain (NOT freezing), moderate at time of observation	Intermittent rain (NOT freezing), heavy at time of observation
70	Intermittent fall of snow flakes, slight at time of observation	Continuous fall of snowflakes, slight at time of observation	Intermittent fall of snowflakes, moderate at time of observation	Continuous fall of snowflakes, moderate at time of observation	Intermittent fall of snowflakes, heavy at time of observation
80	Slight rain shower(s)	Moderate or heavy rain shower(s)	Violent rain shower(s)	Slight shower(s) of rain and snow mixed	Moderate or heavy shower(s) of rain and snow mixed
90	Moderate or heavy shower(s) of hail, with or without rain or rain and snow mixed, not associated with thunder	Slight rain at time of observation; thunderstorm during past hour, but NOT at time of observation	Moderate or heavy rain at time of observation; thunderstorm during past hour, but NOT at time of observation	Slight snow or rain and snow mixed or hail at time of observation; thunderstorm during past hour, but not at time of observation	Moderate or heavy snow, or rain and snow mixed or hail at time of observation; thunderstorm during past hour, but NOT at time of obs.

5	6	7	8	9
Haze	Widespread dust in suspension in the air, NOT raised by wind, at time of observation	Dust or sand raised by wind, at time of observation	Well developed dust devil(s) within past hour	Dust storm or sand storm within sight of or at station during past hour
Precipitation within sight, reaching the ground, but distant from station	Precipitation within sight, reaching the ground, near to but NOT at station	Thunder heard, but no precipitation at the station	Squall(s) within sight during past hour	Funnel cloud(s) within sight during past hour
Showers of rain during past hour, but NOT at time of observation	Showers of snow, or of rain and snow, during past hour, but NOT at time of observation	Showers of hail, or of hail and rain, during past hour, but NOT at time of observation	Fog during past hour, but NOT at time of observation	Thunderstorm (with or without precipitation) during past hour, but NOT at time of obs.
Severe dust storm or sand storm, has increased during past hour	Slight or moderate drifting snow, generally low	Heavy drifting snow, generally low	Slight or moderate drifting snow, generally high	Heavy drifting snow, generally high
Fog, sky NOT discernible, no appreciable change during past hour	Fog, sky discernible, has begun or become thicker during past hour	Fog, sky NOT discernible, has begun or become thicker during past hour	Fog, depositing rime, sky discernible	Fog, depositing rime, sky NOT discernible
Continuous drizzle (NOT freezing), thick at time of observation	Slight freezing drizzle	Moderate or thick freezing drizzle	Drizzle and rain, slight	Drizzle and rain, moderate or heavy
Continuous rain (NOT freezing), heavy at time of observation	Slight freezing rain	Moderate or heavy freezing rain	Rain or drizzle and snow, slight	Rain or drizzle and snow; moderate or heavy
Continuous fall of snowflakes, heavy at time of observation	Ice needles (with or without fog)	Granular snow (with or without fog)	Isolated starlike snow crystals (with or without fog)	Ice pellets (sleet, U. S. definition)
Slight snow shower(s)	Moderate or heavy snow shower(s)	Slight shower(s) of soft or small hail with or without rain, or rain and snow mixed	Moderate or heavy shower(s) of soft or small hail with or without rain, or rain and snow mixed	Slight shower(s) of hail, with or without rain and snow mixed, not associated with thunder
Slight or moderate thunderstorm without hail, but with rain and/or snow at time of obs.	Slight or moderate thunderstorm, with hail at time of observation	Heavy thunderstorm, without hail, but with rain and/or snow at time of observation	Thunderstorm combined with dust storm or sand storm at time of obs.	Heavy thunderstorm with hail at time of observation

365

Wind Speed (ff)

ff	(MILES) (Statute) Per Hour	KNOTS
Calm	Calm	Calm
(symbol)	1 - 2	1 - 2
(symbol)	3 - 8	3 - 7
(symbol)	9 - 14	8 - 12
(symbol)	15 - 20	13 - 17
(symbol)	21 - 25	18 - 22
(symbol)	26 - 31	23 - 27
(symbol)	32 - 37	28 - 32
(symbol)	38 - 43	33 - 37
(symbol)	44 - 49	38 - 42
(symbol)	50 - 54	43 - 47
(symbol)	55 - 60	48 - 52
(symbol)	61 - 66	53 - 57
(symbol)	67 - 71	58 - 62
(symbol)	72 - 77	63 - 67
(symbol)	78 - 83	68 - 72
(symbol)	84 - 89	73 - 77
(symbol)	119 - 123	103 - 107

10 — BAROMETRIC TENDENCY (a)

Code Number	a	BAROMETRIC TENDENCY	
0	/\	Rising, then falling	
1	/	Rising, then steady; or rising, then rising more slowly	Barometer now higher than 3 hours ago
2	/	Rising steadily, or unsteadily	
3	✓	Falling or steady, then rising; or rising, then rising more quickly	
4	—	Steady, same as 3 hours ago	
5	\	Falling, then rising, same or lower than 3 hours ago	
6	\	Falling, then steady; or falling, then falling more slowly	Barometer now lower than 3 hours ago
7	\	Falling steadily, or unsteadily	
8	\/	Steady or rising, then falling; or falling, then falling more quickly	

11 — PAST WEATHER (W)

Code Number	W	PAST WEATHER	
0		Clear or few clouds	Not Plotted
1		Partly cloudy (scattered) or variable sky	
2		Cloudy (broken) or overcast	
3	(symbol)	Sandstorm or dust-storm, or drifting or blowing snow	
4	=	Fog, or smoke, or thick dust haze	
5	,	Drizzle	
6	•	Rain	
7	*	Snow, or rain and snow mixed, or ice pellets (sleet)	
8	▽	Shower(s)	
9	(symbol)	Thunderstorm, with or without precipitation	

Appendix C

correcting mercurial barometer readings

When barometric readings taken from separate locations are being compared, corrections must be made to insure that the true differences in atmospheric pressure are shown. In general, this is accomplished by reducing these barometric readings to normal sea-level conditions. Hence, a correction is made for temperature, elevation, and latitude (gravity correction).

temperature correction

This correction accounts for the expansion and contraction of the mercury column and brass scale caused by temperature variations.

elevation correction

Since pressure decreases with increases in elevation, barometers not located at sea level will show a reduced pressure. The elevation correction accounts for the position of the barometer as well as for the temperature of the assumed air column extending down to sea level.

latitude correction

The earth is not a perfect sphere; it has a shorter polar diameter than an equatorial diameter. Consequently, the pull of gravity on the mercury column is greater at the poles and lesser at the equator. Because the adjustment needed to correct this latitudinal variation in gravity is usually negligible compared to the other corrections, we have not included it in the following procedure for correcting the barometer.

To reduce a mercurial barometer reading to sea level, do the following:

1. Read the height of the mercury column on the millimeter scale.

2. Correct this reading for the temperature of the barometer by using Table C-1 and subtracting the obtained value from your reading.

3. Using the out-of-doors air temperature and the elevation (altitude) of the barometer, obtain the temperature–altitude factor from Table C-2.

4. Using the temperature–altitude factor from step 3, the temperature corrected barometric reading from step 2, and Table C-3, find the sea-level correction.

5. Add the sea-level correction to the temperature corrected barometer reading. This sum gives the desired barometric reading reduced to sea level.

Table C-1

Temperature Correction

Observed Height in Millimeters

Temp. °C	620	630	640	650	660	670	680	690	700	710	720	730	740	750	760	770	780	790
0	0.00	0.00	0.00	0.00	0.00	0.00	0.00	0.00	0.00	0.00	0.00	0.00	0.00	0.00	0.00	0.00	0.00	0.00
1	.10	.10	.10	.11	.11	.11	.11	.11	.11	.12	.12	.12	.12	.12	.12	.13	.13	.13
2	.20	.21	.21	.21	.22	.22	.22	.23	.23	.23	.24	.24	.24	.25	.25	.25	.25	.26
3	.30	.31	.31	.32	.32	.33	.33	.34	.34	.35	.35	.36	.36	.37	.37	.38	.38	.39
4	.40	.41	.42	.42	.43	.44	.44	.45	.46	.46	.47	.48	.48	.49	.50	.50	.51	.52
5	.51	.51	.52	.53	.54	.55	.56	.56	.57	.58	.59	.60	.60	.61	.62	.63	.64	.64
6	.61	.62	.63	.64	.65	.66	.67	.68	.69	.70	.71	.71	.72	.73	.74	.75	.76	.77
7	.71	.72	.73	.74	.75	.77	.78	.79	.80	.81	.82	.83	.85	.86	.87	.88	.89	.90
8	.81	.82	.84	.85	.86	.87	.89	.90	.91	.93	.94	.95	.97	.98	.99	1.01	1.02	1.03
9	.91	.92	.94	.95	.97	.98	1.00	1.01	1.03	1.04	1.06	1.07	1.09	1.10	1.12	1.13	1.15	1.16
10	1.01	1.03	1.04	1.06	1.08	1.09	1.11	1.13	1.14	1.16	1.17	1.19	1.21	1.22	1.24	1.26	1.27	1.29
11	1.11	1.13	1.15	1.17	1.18	1.20	1.22	1.24	1.26	1.27	1.29	1.31	1.33	1.35	1.36	1.38	1.40	1.42
12	1.21	1.23	1.25	1.27	1.29	1.31	1.33	1.35	1.37	1.39	1.41	1.43	1.45	1.47	1.49	1.51	1.53	1.55
13	1.31	1.34	1.36	1.38	1.40	1.42	1.44	1.46	1.48	1.50	1.53	1.55	1.57	1.59	1.61	1.63	1.65	1.67
14	1.41	1.44	1.46	1.48	1.51	1.53	1.55	1.57	1.60	1.62	1.64	1.67	1.69	1.71	1.73	1.76	1.78	1.80
15	1.52	1.54	1.56	1.59	1.61	1.64	1.66	1.69	1.71	1.74	1.76	1.78	1.81	1.83	1.86	1.88	1.91	1.93
16	1.62	1.64	1.67	1.69	1.72	1.75	1.77	1.80	1.82	1.85	1.88	1.90	1.93	1.96	1.98	2.01	2.03	2.06
17	1.72	1.74	1.77	1.80	1.83	1.86	1.88	1.91	1.94	1.97	1.99	2.02	2.05	2.08	2.10	2.13	2.16	2.19
18	1.82	1.85	1.88	1.91	1.93	1.96	1.99	2.02	2.05	2.08	2.11	2.14	2.17	2.20	2.23	2.26	2.29	2.32
19	1.92	1.95	1.98	2.01	2.04	2.07	2.10	2.13	2.17	2.20	2.23	2.26	2.29	2.32	2.35	2.38	2.41	2.44
20	2.02	2.05	2.08	2.12	2.15	2.18	2.21	2.25	2.28	2.31	2.34	2.38	2.41	2.44	2.47	2.51	2.54	2.57
21	2.12	2.15	2.19	2.22	2.26	2.29	2.32	2.36	2.39	2.43	2.46	2.50	2.53	2.56	2.60	2.63	2.67	2.70
22	2.22	2.26	2.29	2.33	2.36	2.40	2.43	2.47	2.51	2.54	2.58	2.61	2.65	2.69	2.72	2.76	2.79	2.83
23	2.32	2.36	2.40	2.43	2.47	2.51	2.54	2.58	2.62	2.66	2.69	2.73	2.77	2.81	2.84	2.88	2.92	2.96
24	2.42	2.46	2.50	2.54	2.58	2.62	2.66	2.69	2.73	2.77	2.81	2.85	2.89	2.93	2.97	3.01	3.05	3.08
25	2.52	2.56	2.60	2.64	2.68	2.72	2.77	2.81	2.85	2.89	2.93	2.97	3.01	3.05	3.09	3.13	3.17	3.21
26	2.62	2.66	2.71	2.75	2.79	2.83	2.88	2.92	2.96	3.00	3.04	3.09	3.13	3.17	3.21	3.26	3.30	3.34
27	2.72	2.77	2.81	2.85	2.90	2.94	2.99	3.03	3.07	3.12	3.16	3.20	3.25	3.29	3.34	3.38	3.42	3.47
28	2.82	2.87	2.91	2.96	3.00	3.05	3.10	3.14	3.19	3.23	3.28	3.32	3.37	3.41	3.46	3.51	3.55	3.60
29	2.92	2.97	3.02	3.06	3.11	3.16	3.21	3.25	3.30	3.35	3.39	3.44	3.49	3.54	3.58	3.63	3.68	3.72
30	3.02	3.07	3.12	3.17	3.22	3.27	3.32	3.36	3.41	3.46	3.51	3.56	3.61	3.66	3.71	3.75	3.80	3.85
31	3.12	3.17	3.22	3.27	3.32	3.37	3.43	3.48	3.53	3.58	3.63	3.68	3.73	3.78	3.83	3.88	3.93	3.98
32	3.22	3.28	3.33	3.38	3.43	3.48	3.54	3.59	3.64	3.69	3.74	3.79	3.85	3.90	3.95	4.00	4.05	4.11
33	3.32	3.38	3.43	3.48	3.54	3.59	3.64	3.70	3.75	3.81	3.86	3.91	3.97	4.02	4.07	4.13	4.18	4.23
34	3.42	3.48	3.53	3.59	3.64	3.70	3.75	3.81	3.87	3.92	3.98	4.03	4.09	4.14	4.20	4.25	4.31	4.36
35	3.52	3.58	3.64	3.69	3.75	3.81	3.86	3.92	3.98	4.03	4.09	4.15	4.21	4.26	4.32	4.38	4.43	4.49

Table C-2

Temperature–Altitude Factor Values
(enter in Table C-3)

Altitude in Meters	Assumed Temperature of Air Column °C									
	−16°	−8°	0°	+4°	+8°	+12°	+16°	+20°	+24°	+28°
10	1.2	1.1	1.1	1.1	1.0	1.0	1.0	1.0	1.0	1.0
50	5.8	5.6	5.4	5.3	5.2	5.2	5.1	5.0	4.9	4.9
100	11.5	11.2	10.8	10.7	10.5	10.3	10.2	10.0	9.9	9.7
150	17.3	16.7	16.2	16.0	15.7	15.5	15.3	15.0	14.8	14.6
200	23.0	22.3	21.6	21.3	21.0	20.7	20.3	20.0	19.7	19.5
250	28.8	27.9	27.0	26.6	26.2	25.8	25.4	25.0	24.7	24.3
300	34.5	33.5	32.5	32.0	31.5	31.0	30.5	30.1	29.6	29.2
350	40.3	39.0	37.9	37.3	36.7	36.2	35.6	35.1	34.6	34.0
400	46.0	44.6	43.3	42.6	42.0	41.3	40.7	40.1	39.5	38.9
450	51.8	50.2	48.7	47.9	47.2	46.5	45.8	45.1	44.4	43.8
500	57.5	55.8	54.1	53.3	52.4	51.6	50.9	50.1	49.4	48.6
550	63.3	61.4	59.5	58.6	57.7	56.8	55.9	55.1	54.3	53.5
600	69.0	66.9	64.9	63.9	62.9	62.0	61.0	60.1	59.2	58.3
650	74.8	72.5	70.3	69.2	68.2	67.1	66.1	65.1	64.2	63.2
700	80.6	78.1	75.7	74.6	73.4	72.3	71.2	70.1	69.1	68.1
750	86.3	83.7	81.1	79.9	78.7	77.5	76.3	75.1	74.0	72.9
800	92.1	89.2	86.5	85.2	83.9	82.6	81.4	80.1	79.0	77.8
850	97.8	94.8	92.0	90.5	89.2	87.8	86.4	85.2	83.9	82.7
900	103.6	100.4	97.4	95.9	94.4	93.0	91.5	90.2	88.8	87.5
950	109.3	106.0	102.8	101.2	99.6	98.1	96.6	95.2	93.8	92.4
1000	115.1	111.5	108.2	106.5	104.9	103.3	101.7	100.2	98.7	97.3
1050	120.8	117.1	113.6	111.8	110.1	108.4	106.8	105.2	103.6	102.1
1100	126.6	122.7	119.0	117.2	115.4	113.6	111.9	110.2	108.6	107.0
1150	132.3	128.3	124.4	122.5	120.6	118.8	117.0	115.2	113.5	111.8
1200	138.1	133.8	129.8	127.8	125.9	123.9	122.0	120.2	118.4	116.7
1250	143.8	139.4	135.2	133.1	131.1	129.1	127.1	125.2	123.4	121.6
1300	149.6	145.0	140.6	138.5	136.3	134.3	132.2	130.2	128.3	126.4
1350	155.3	150.6	146.0	143.8	141.6	139.4	137.3	135.2	133.2	131.3
1400	161.1	156.2	151.4	149.1	146.8	144.6	142.4	140.2	138.2	136.2
1450	166.8	161.7	156.8	154.5	152.1	149.7	147.5	145.3	143.1	141.0
1500	172.6	167.3	162.3	159.8	157.3	154.9	152.5	150.3	148.0	145.9
1550	178.3	172.9	167.7	165.1	162.6	160.1	157.6	155.3	153.0	150.7
1600	184.1	178.5	173.1	170.4	167.8	165.2	162.7	160.3	157.9	155.6
1650	189.8	184.0	178.5	175.7	173.0	170.4	167.8	165.3	162.8	160.5
1700	195.6	189.6	183.9	181.1	178.3	175.6	172.9	170.3	167.8	165.3
1750	201.4	195.2	189.3	186.4	183.5	180.7	178.0	175.3	172.7	170.2
1800	207.1	200.8	194.7	191.7	188.8	185.9	183.1	180.3	177.6	175.0
1850	212.9	206.3	200.1	197.0	194.0	191.0	188.1	185.3	182.6	179.9
1900	218.6	211.9	205.5	202.4	199.3	196.2	193.2	190.3	187.5	184.8
1950	224.4	217.5	210.9	207.7	204.5	201.4	198.3	195.3	192.4	189.6
2000	230.1	223.0	216.3	213.0	209.7	206.5	203.4	200.3	197.4	194.5
2050	235.9	228.6	221.7	218.3	215.0	211.7	208.5	205.3	202.3	199.3
2100	241.6	234.2	227.1	223.7	220.2	216.8	213.5	210.4	207.2	204.2
2150	247.4	239.8	232.5	229.0	225.5	222.0	218.6	215.4	212.2	209.1
2200	253.1	245.4	237.9	234.3	230.7	227.2	223.7	220.4	217.1	213.9
2250	258.9	250.9	243.4	239.6	235.9	232.3	228.8	225.4	222.0	218.8

Table C-2 continued

Altitude in Meters	Assumed Temperature of Air Column °C									
	−16°	−8°	0°	+4°	+8°	+12°	+16°	+20°	+24°	+28°
2300	264.6	256.5	248.8	245.0	241.2	237.5	233.9	230.4	227.0	223.6
2350	270.4	262.1	254.2	250.3	246.4	242.7	239.0	235.4	231.9	228.5
2400	276.1	267.7	259.6	255.6	251.7	247.8	244.0	240.4	236.8	233.4
2450	281.9	273.2	265.0	260.9	256.9 ·	253.0	249.1	245.4	241.8	238.2
2500	287.6	278.8	270.4	266.2	262.2	258.1	254.2	250.4	246.7	243.1
2550	293.4	284.4	275.8	271.6	267.4	263.3	259.3	255.4	251.6	247.9
2600	299.1	290.0	281.2	276.9	272.6	268.5	264.4	260.4	256.6	252.8
2650	304.9	295.5	286.6	282.2	277.9	273.6	269.5	265.4	261.5	257.7
2700	310.6	301.1	292.0	287.5	283.1	278.8	274.5	270.4	266.4	262.5
2750	316.4	306.7	297.4	292.9	288.4	283.9	279.6	275.4	271.4	267.4
2800	322.1	312.3	302.8	298.2	293.6	289.1	284.7	280.4	276.3	272.2
2850	327.9	317.8	308.2	303.5	298.8	294.3	289.8	285.4	281.2	277.1
2900	333.6	323.4	313.6	308.8	304.1	299.4	294.9	290.4	286.2	282.0
2950	339.4	329.0	319.0	314.2	309.3	304.6	299.9	295.5	291.1	286.8
3000	345.1	334.5	324.4	319.5	314.6	309.7	305.0	300.5	296.0	291.7

Table C-3
Barometric Correction Values

Temp.–Alt. Factor	Barometer Reading				
	780 mm	760 mm	740 mm	720 mm	700 mm
1	0.9	0.9	0.9	0.8	0.8
5	4.5	4.4	4.3	4.2	4.0
10	9.0	8.8	8.6	8.3	8.1
15	13.6	13.2	12.9	12.5	12.2
20	18.2	17.7	17.2	16.8	16.3
25	22.8	22.2	21.6	21.0	20.4
30	27.4	26.7	26.0	25.3	24.6
35	......	31.2	30.4	29.6	28.8

Temp.–Alt. Factor	760 mm	740 mm	720 mm	700 mm	680 mm	660 mm
40	35.8	34.9	33.9	33.0	32.0	31.1
45	40.4	39.3	38.3	37.2	36.2	35.1
50	45.0	43.8	42.7	41.5	40.3	39.1
55	49.7	48.4	47.1	45.8	44.5	43.1
60	......	52.9	51.5	50.1	48.6	47.2
65	......	57.5	55.9	54.4	52.8	51.3
70	......	62.1	60.4	58.7	57.1	55.4
75	......	66.7	64.9	63.1	61.3	59.5

Temp.–Alt. Factor	720 mm	700 mm	680 mm	660 mm	640 mm
80	69.5	67.5	65.6	63.7	61.7
85	74.0	72.0	69.9	67.9	65.8
90	78.6	76.4	74.2	72.1	69.9
95	83.2	80.9	78.6	76.3	74.0
100	87.9	85.4	83.0	80.5	78.1
105	......	89.9	87.4	84.8	82.2
110	......	94.5	91.8	89.1	86.4
115	......	99.1	96.3	93.4	90.6
120	......	103.7	100.7	97.8	94.8
125	......	108.3	105.3	102.2	99.1

Temp.–Alt. Factor	680 mm	660 mm	640 mm	620 mm	600 mm
125	105.3	102.2	99.1	96.0	92.9
130	109.8	106.6	103.3	100.1	96.9
135	114.3	111.0	107.6	104.3	100.9
140	118.9	115.4	111.9	108.4	104.9
145	123.5	119.9	116.3	112.6	109.0
150	128.2	124.4	120.6	116.9	113.1
155	......	128.9	125.0	121.1	117.2
160	......	133.5	129.4	125.4	121.4
165	,.....	138.1	133.9	129.7	125.5
170	......	142.7	138.4	134.0	129.7

Temp.–Alt. Factor	Barometer Reading				
	640 mm	620 mm	600 mm	580 mm	560 mm
170	138.4	134.0	129.7	125.4	121.1
175	142.9	138.4	133.9	129.5	125.0
180	147.4	142.8	138.2	133.6	129.0
185	151.9	147.2	142.4	137.7	132.9
190	156.5	151.6	146.7	141.8	136.9
195	161.1	156.1	151.0	146.0	141.0
200	165.7	160.5	155.4	150.2	145.0
205	170.4	165.0	159.7	154.4	149.1
210	......	169.6	164.1	158.6	153.2
215	......	174.1	168.5	162.9	157.3

Temp.–Alt. Factor	620 mm	600 mm	580 mm	560 mm	540 mm
215	174.1	168.5	162.9	157.3	151.7
220	178.7	172.9	167.2	161.4	155.7
225	183.3	177.4	171.5	165.6	159.7
230	188.0	181.9	175.8	169.8	163.7
235	192.6	186.4	180.2	174.0	167.8
240	......	191.0	184.6	178.2	171.9
245	......	195.5	189.0	182.5	176.0
250	......	200.1	193.4	186.8	180.1
255	......	204.7	197.9	191.1	184.3
260	......	209.4	202.4	195.4	188.4

Temp.–Alt. Factor	580 mm	560 mm	540 mm	520 mm
260	202.4	195.4	188.4	181.5
265	206.9	199.8	192.6	185.5
270	211.5	204.2	196.9	189.6
275	216.0	208.6	201.1	193.7
280	220.6	213.0	205.4	197.8
285	225.2	217.5	209.7	201.9
290	229.9	222.0	214.0	206.1
295	......	226.5	218.4	210.3
300	......	231.0	222.8	214.5

Temp.–Alt. Factor	560 mm	540 mm	520 mm	500 mm	480 mm
305	235.6	227.2	218.8	210.3	201.9
310	240.2	231.6	223.0	214.4	205.9
315	244.8	236.0	227.3	218.6	209.8
320	249.4	240.5	231.6	222.7	213.8
325	254.1	245.0	236.0	226.9	217.8
330	......	249.6	240.3	231.1	221.8
335	......	254.1	244.7	235.3	225.9
340	......	258.7	249.1	239.6	230.0
345	......	263.3	253.6	243.8	234.1

Appendix D

Forces and air motions

Newton's second law

When motions in the atmosphere are being considered, Newton's laws relate the forces that are involved to acceleration. In particular, Newton's second law states: The acceleration of a body is directly proportional to the net force acting on the body and is inversely proportional to the mass of the body. Stated mathematically,

$$A = \frac{F}{M}$$

where A is the acceleration that a force F will produce on mass M. In the metric system force is measured in newtons, where 1 newton of force will accelerate 1 kilogram of matter 1 meter per second squared. Two forces frequently encountered in basic meteorology are the Coriolis force and the pressure gradient force.

Coriolis force

The magnitude of the Coriolis force can be expressed as:

$$Fc = 2\mu\Omega \sin \phi$$

where Fc is the Coriolis force per unit mass of air, μ is the wind speed, Ω is the earth's rate of rotation (angular velocity) which is 7.29×10^{-5} radians per second, and ϕ is the latitude. Note that the sin ϕ is a trigonometric function that equals zero for an angle of 0 degrees (equator), 0.64 when $\phi = 40$ degrees, and one when $\phi = 90$ degrees (poles). Since the earth's rate of rotation remains rather constant, we can see from the preceding expression that the magnitude of the Corilois force is dependent upon wind speed and latitude. When this formula is used to determine the Coriolis force for a specific latitude and wind speed, it is common practice to use the letter f to represent $2\Omega \sin \phi$, which has the following values at various latitudes:

ϕ	0	10	20	30	40	50	60	70	80	90
$f \cdot 10^5 \text{ sec}^{-1}$	0	2.5	5.0	7.3	9.4	11.2	12.6	13.7	14.4	14.6

Using the above information, we can calculate the following: At 30 degrees latitude with a wind speed of 10 meters per second the Coriolis force would equal 0.073 centimeter per second squared, while with the same wind speed at 60 degrees latitude the Coriolis force would equal 0.126 centimeter per second squared.

pressure gradient force

The magnitude of the pressure gradient force can be expressed as:

$$F_p = \frac{1}{d} \frac{\Delta p}{\Delta n}$$

where F_p is the pressure gradient force per unit mass, d is the density of air, Δp is the pressure difference obtained from the isobar spacing, and Δn is the distance between isobars. Using this formula and the standard sea-level density of $1.293 \cdot 10^{-3}$ grams per cubic centimeter, we can calculate the pressure gradient force for any isobaric surface. For example, suppose 4-millibar isobars are spaced 200 kilometers apart. Then,

$$F_p = \frac{1 \times 4 \times 10^3}{1.293 \times 10^{-3} \times 1 \times 10^7} = 0.155 \text{ centimeter per second squared}$$

Comparing this value with that obtained for the Coriolis force calculated earlier suggests that a balance between these forces is very possible at reasonable wind speeds.

Appendix E

Laws relating to Gases

kinetic energy

All moving objects, by virtue of their motion, are capable of doing work. We call this energy of motion, or *kinetic energy*. The kinetic energy of a moving object is equal to one-half its mass multiplied by its velocity squared. Stated mathematically,

$$\text{kinetic energy} = \frac{1}{2}M\mu^2$$

We can see from this that by doubling the velocity of a moving object, the object's kinetic energy will increase four times.

first law of thermodynamics

According to the kinetic theory, the temperature of a gas is proportional to the kinetic energy of the moving molecules. When a gas is heated, its kinetic energy increases because of an increase in molecular motion. Further, when a gas is compressed, the kinetic energy will also be increased, and the temperature of the gas will rise. These relationships are expressed in the first law of thermodynamics, which states: The temperature of a gas may be changed by the addition or subtraction of heat, or by changing the pressure (compression or expansion), or by a combination of both. It is easy to understand how the atmosphere is heated or cooled by the gain or loss of heat. However, when we consider rising and sinking air, the relationships between temperature and pressure become more important. Here an increase in temperature is brought about by performing work on the gas and not by the addition of heat. This phenomenon is called the *adiabatic form* of the first law of thermodynamics.

Boyle's law

About 1660 the Englishman, Robert Boyle, showed that if the temperature is kept constant when the pressure exerted on a gas is increased, the volume decreases. This principle, called Boyle's law, states: At a constant temperature the volume of a given mass of gas varies inversely with the pressure. Stated mathematically,

$$P_1 V_1 = P_2 V_2$$

The symbols P_1 and V_1 refer to the original pressure and volume, while P_2 and V_2 indicate the new pressure and volume after a change occurred. Boyle's law shows that if a given volume of gas is compressed so that the volume is reduced by one-half, the pressure exerted by the gas is doubled. This increase in pressure can be explained by the kinetic theory that predicts that when the volume of the gas is reduced by one-half the molecules collide with the walls of the container twice as often. Since density is defined as the mass per unit volume, an increase in pressure results in increased density.

Charles' law

The relationships between temperature and volume (therefore density) of a gas were recognized about 1787 by the French scientist, Jacques Charles, and were stated formally by J. Gay-Lussac in 1802. Charles' law states: At a constant pressure the volume of a given mass is directly proportional to the absolute temperature. In other words, when a quantity of gas is kept at a constant pressure, an increase in temperature results in an increase in volume, and vice versa. Stated mathematically,

$$\frac{V_1}{V_2} = \frac{T_1}{T_2}$$

where V_1 and T_1 represent the original volume and temperature, respectively, and V_2 and T_2 represent the final volume and temperature, respectively. This law explains the fact that a gas expands when it is heated. According to the kinetic theory, when the particles are heated, they move more rapidly and therefore collide more often.

the ideal gas law or equation of state

In describing the atmosphere three variable quantities must be considered; pressure, temperature, and density (mass per unit volume). The relationships between these variables can be found by combining in a single statement the laws of Boyle and Charles as follows:

$$PV = RT \quad \text{or} \quad P = \rho RT$$

where P is the pressure, V is the volume, R is the constant of proportionality, T is the absolute temperature, and ρ is the density. This law, called the ideal gas law, states:

1. When the volume is kept constant, the pressure of a gas is directly proportional to its absolute temperature.
2. When the temperature is kept constant, the pressure of the gas is proportional to its density and inversely proportional to its volume.
3. When the pressure is kept constant, the absolute temperature of a gas is proportional to its volume and inversely proportional to its density.

Appendix F

worldwide extremes of temperature and precipitation recorded by continental area

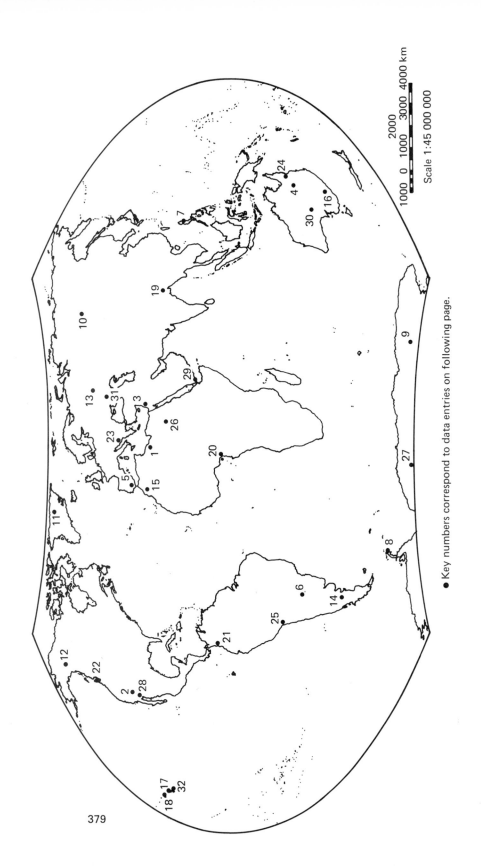

Scale 1:45 000 000

1000 0 1000 2000 3000 4000 km

● Key numbers correspond to data entries on following page.

379

Table F-1

Temperature Extremes

Key No.	Area	Highest (°C)	Place	Elevation (ms)	Date
1	Africa	57.8	Azizia, Libya	114	Sept. 13, 1922
2	North America	56.7	Death Valley, Calif.	−53	July 10, 1913
3	Asia	53.9	Tirat Tsvi, Israel	−217	June 21, 1942
4	Australia	53.3	Cloncurry, Queensland	187	Jan. 16, 1889
5	Europe	50.0	Seville, Spain	8	Aug. 4, 1881
6	South America	48.9	Rivadavia, Argentina	203	Dec. 11, 1905
7	Oceania	42.2	Tuguegarao, Philippines	22	Apr. 29, 1912
8	Antarctica	14.4	Esperanza, Palmer Peninsula	8	Oct. 20, 1956

Key No.	Area	Lowest (°C)	Place	Elevation (ms)	Date
9	Antarctica	−88.3	Vostok	3366	Aug. 24, 1960
10	Asia	−67.8	Oymykon, U.S.S.R.	788	Feb. 6, 1933
11	Greenland	−66.1	Northice	2307	Jan. 9, 1954
12	North America	−62.8	Snag, Yukon, Canada	578	Feb. 3, 1947
13	Europe	−55.0	Ust'Shchugor, U.S.S.R.	84	January*
14	South America	−32.8	Sarmiento, Argentina	264	June 1, 1907
15	Africa	−23.9	Ifrane, Morocco	1609	Feb. 11, 1935
16	Australia	−22.2	Charlotte Pass, N.S.W.	‡	July 22, 1947†
17	Oceania	−10.0	Haleakala Summit, Hawaii	293	Jan. 2, 1961

*Exact date unknown; lowest in 15-year period.
†And earlier date.
‡Elevation unknown.

Table F-2

Extremes of Average Annual Precipitation

Key No.	Area	Greatest Amount (mm)	Place	Elevation (ms)	Years of Record
18	Oceania	11,684	Mt. Waialeale, Kauai, Hawaii	1,523	32
19	Asia	11,430	Cherrapunji, India	1,313	74
20	Africa	10,277	Debundscha, Cameroon	9	32
21	South America	8,989	Quibdo, Colombia	72	10–16
22	North America	6,657	Henderson Lake, B.C., Canada	4	14
23	Europe	4,643	Crkvica, Yugoslavia	1,001	22
24	Australia	4,554	Tully, Queensland	—	31

Key No.	Area	Least Amount (mm)	Place	Elevation (ms)	Years of Record
25	South America	0.76	Arica, Chile	29	59
26	Africa	< 2.54	Wadi Halfa, Sudan	123	39
27	Antarctica	20.32*	South Pole Station	2,756	10
28	North America	30.48	Batagues, Mexico	5	14
29	Asia	45.72	Aden, Arabia	7	50
30	Australia	102.87	Mulka, South Australia	—	34
31	Europe	162.56	Astrakhan, U.S.S.R.	14	25
32	Oceania	226.82	Puako, Hawaii	2	13

*The value given is the average amount of solid snow accumulating in 1 year as indicated by snow markers. The liquid content of the snow is undetermined.

381

Appendix G

CLIMATIC DATA

The climatic data that follow are for 51 representative stations from around the world. Temperatures are given in degrees Celsius and precipitation is given in millimeters. Names and locations are given at the end of this appendix along with the elevation (in meters) of each station and its Koeppen classification. This format was selected so that you can use the data in exercises to reinforce your understanding of climatic controls and classification. For example, after classifying a station using Table 13-1, write out a likely location based upon such things as mean annual temperature, annual temperature range, total precipitation, and seasonal precipitation distribution. Your location need not be a specific city; it could be a description of the station's setting, such as "middle-latitude continental" or "subtropical with a strong monsoon influence." It would also be a good idea to list the reasons for your selection. You may then check your answer by examining the list of stations at the end of this appendix.

If you simply wish to examine the data for a specific place or data for a specific climatic type, consult the list at the end of this appendix.

	J	F	M	A	M	J	J	A	S	O	N	D	Yr.
1.	1.7	4.4	7.9	13.2	18.4	23.8	25.8	24.8	21.4	14.7	6.7	2.8	13.8
	10	10	13	13	20	15	30	32	23	18	10	13	207
2.	−10.4	−8.3	−4.6	3.4	9.4	12.8	16.6	14.9	10.8	5.5	−2.3	−6.4	3.5
	18	25	25	30	51	89	64	71	33	20	18	15	459
3.	10.2	10.8	13.7	17.9	22.2	25.7	26.7	26.5	24.2	19.0	13.3	10.0	18.4
	66	84	99	74	91	127	196	168	147	71	53	71	1247
4.	−23.9	−17.5	−12.5	−2.7	8.4	14.8	15.6	12.8	6.4	−3.1	−15.8	−21.9	−3.3
	23	13	18	8	15	33	48	53	33	20	18	15	297
5.	−17.8	−15.3	−9.2	−4.4	4.7	10.9	16.4	14.4	10.3	3.3	−3.6	−12.8	−0.3
	58	58	61	48	53	61	81	71	58	61	64	64	738
6.	−8.2	−7.1	−2.4	5.4	11.2	14.7	18.9	17.4	12.7	7.4	−0.4	−4.6	5.4
	21	23	29	35	52	74	39	40	35	29	26	21	424
7.	12.8	13.9	15.0	15.0	17.8	20.0	21.1	22.8	22.2	18.3	17.2	15.0	17.6
	69	74	46	28	3	3	0	0	5	10	28	61	327
8.	18.9	20.0	21.1	22.8	25.0	26.7	27.2	27.8	27.2	25.0	21.1	20.0	23.6
	51	48	58	99	163	188	172	178	241	208	71	43	1520
9.	−4.4	−2.2	4.4	10.6	16.7	21.7	23.9	22.7	18.3	11.7	3.8	−2.2	10.4
	46	51	69	84	99	97	97	81	97	61	61	51	894
10.	−5.6	−4.4	0.0	6.1	11.7	16.7	20.0	18.9	15.5	10.0	3.3	−2.2	7.5
	112	96	109	94	86	81	74	61	89	81	107	99	1089
11.	−2.1	0.9	4.7	9.9	14.7	19.4	24.7	23.6	18.3	11.5	3.4	−0.2	10.7
	34	30	40	45	36	25	15	22	13	29	33	31	353
12.	12.8	13.9	15.0	16.1	17.2	18.8	19.4	22.2	21.1	18.8	16.1	13.9	15.9
	53	56	41	20	5	0	0	2	5	13	23	51	269
13.	−0.1	1.8	6.2	13.0	18.7	24.2	26.4	25.4	21.1	14.9	6.7	1.6	13.3
	50	52	78	94	95	109	84	77	70	73	65	50	897
14.	2.7	3.2	7.1	13.2	18.8	23.4	25.7	24.7	20.9	15.0	8.7	3.4	13.9
	77	63	82	80	105	82	105	124	97	78	72	71	1036
15.	12.8	15.0	18.9	21.1	26.1	31.1	32.7	33.9	31.1	22.2	17.7	13.9	23.0
	10	9	6	2	0	0	6	13	10	10	3	8	77
16.	25.6	25.6	24.4	25.0	24.4	23.3	23.3	24.4	24.4	25.0	25.6	25.6	24.7
	259	249	310	165	254	188	168	117	221	183	213	292	2619
17.	25.9	25.8	25.8	25.9	26.4	26.6	26.9	27.5	27.9	27.7	27.3	26.7	26.7
	365	326	383	404	185	132	68	43	96	99	189	143	2433
18.	13.3	13.3	13.3	13.3	13.9	13.3	13.3	13.3	13.9	13.3	13.3	13.9	13.5
	99	112	142	175	137	43	20	30	69	112	97	79	1115
19.	25.9	26.1	25.2	23.9	22.3	21.3	20.8	21.1	21.5	22.3	23.1	24.4	23.2
	137	137	143	116	73	43	43	43	53	74	97	127	1086
20.	13.8	13.5	11.4	8.0	3.7	1.2	1.4	2.9	5.5	9.2	11.4	12.9	7.9
	21	16	18	13	25	15	15	17	12	7	15	18	171
21.	1.5	1.3	3.1	5.8	10.2	12.6	15.0	14.7	12.0	8.3	5.5	3.3	7.8
	179	139	109	140	83	126	141	167	228	236	207	203	1958
22.	−0.5	0.2	3.9	9.0	14.3	17.7	19.4	18.8	15.0	9.6	4.7	1.2	9.5
	41	37	30	39	44	60	67	65	45	45	44	39	556
23.	6.1	5.8	7.8	9.2	11.6	14.4	15.6	16.0	14.7	12.0	9.0	7.0	10.8
	133	96	83	69	68	56	62	80	87	104	138	150	1126

	J	F	M	A	M	J	J	A	S	O	N	D	Yr.
24.	10.8	11.6	13.6	15.6	17.2	20.1	22.2	22.5	21.2	18.2	14.4	11.5	16.6
	111	76	109	54	44	16	3	4	33	62	93	103	708
25.	−9.9	− 9.5	−4.2	4.7	11.9	16.8	19.0	17.1	11.2	4.5	−1.9	−6.8	4.4
	31	28	33	35	52	67	74	74	58	51	36	36	575
26.	8.0	9.0	10.9	13.7	17.5	21.6	24.4	24.2	21.5	17.2	12.7	9.5	15.9
	83	73	52	50	48	18	9	18	70	110	113	105	749
27.	−9.0	−9.0	−6.6	−4.1	0.4	3.6	5.6	5.5	3.5	−0.6	−4.5	−7.6	−1.9
	202	180	164	166	197	249	302	278	208	183	190	169	2488
28.	−2.9	−3.1	−0.7	4.4	10.1	14.9	17.8	16.6	12.2	7.1	2.8	0.1	6.6
	43	30	26	31	34	45	61	76	60	48	53	48	555
29.	12.8	13.9	17.2	18.9	22.2	23.9	25.5	26.1	25.5	23.9	18.9	15.0	20.3
	66	41	20	5	3	0	0	0	3	18	46	66	268
30.	24.6	24.9	25.0	24.9	25.0	24.2	23.7	23.8	23.9	24.2	24.2	24.7	24.4
	81	102	155	140	133	119	99	109	206	213	196	122	1675
31.	21.1	20.4	20.9	21.7	23.0	26.0	27.3	27.3	27.5	27.5	26.0	25.2	24.3
	0	2	0	0	1	15	88	249	163	49	5	6	578
32.	20.4	22.7	27.0	30.6	33.8	34.2	33.6	32.7	32.6	30.5	25.5	21.3	28.7
	0	0	0	0	0	2	1	11	2	0	0	0	16
33.	17.8	18.1	18.8	18.8	17.8	16.2	14.9	15.5	16.8	18.6	18.3	17.8	17.5
	46	51	102	206	160	46	18	25	25	53	109	81	922
34.	20.6	20.7	19.9	19.2	16.7	13.9	13.9	16.3	19.1	21.8	21.4	20.9	18.7
	236	168	86	46	13	8	0	3	8	38	94	201	901
35.	11.7	13.3	16.7	18.6	19.2	20.0	20.3	20.5	20.5	19.1	15.9	12.9	17.4
	20	41	179	605	1705	2875	2455	1827	1231	447	47	5	11437
36.	26.2	26.3	27.1	27.2	27.3	27.0	26.7	27.0	27.4	27.4	26.9	26.6	26.9
	335	241	201	141	116	97	61	50	78	91	151	193	1755
37.	−0.8	2.6	5.3	8.5	13.1	17.0	17.2	17.3	15.3	11.5	5.7	0.3	9.4
	0	0	1	1	18	72	157	151	68	4	1	0	473
38.	24.5	25.8	27.9	30.5	32.7	32.5	30.7	30.1	29.7	28.1	25.9	24.6	28.6
	24	7	15	25	52	53	83	124	118	267	308	157	1233
39.	−18.7	−18.1	−16.7	−11.7	−5.0	0.6	5.3	5.8	1.4	−4.2	−12.3	−15.8	−7.5
	8	8	8	8	15	20	36	43	43	33	13	12	247
40.	−21.9	−18.6	−12.5	−5.0	9.7	15.6	18.3	16.1	10.3	0.8	−10.6	−18.4	−1.4
	15	8	8	13	30	51	51	51	28	25	18	20	318
41.	−4.7	−1.9	4.8	13.7	20.1	24.7	26.1	24.9	19.9	12.8	3.8	−2.7	11.8
	4	5	8	17	35	78	243	141	58	16	10	3	623
42.	24.3	25.2	27.2	29.8	29.5	27.8	27.6	27.1	27.6	28.3	27.7	25.0	27.3
	8	5	6	17	260	524	492	574	398	208	34	3	2530
43.	25.8	26.3	27.8	28.8	28.2	27.4	27.1	27.1	26.7	26.5	26.1	25.7	27.0
	6	13	12	65	196	285	242	277	292	259	122	37	1808
44.	3.7	4.3	7.6	13.1	17.6	21.1	25.1	26.4	22.8	16.7	11.3	6.1	14.7
	48	73	101	135	131	182	146	147	217	220	101	60	1563
45.	−15.8	−13.6	−4.0	8.5	17.7	21.5	23.9	21.9	16.7	6.1	−6.2	−13.0	5.3
	8	15	15	33	25	33	16	35	15	47	22	11	276
46.	−46.8	−43.1	−30.2	−13.5	2.7	12.9	15.7	11.4	2.7	−14.3	−35.7	−44.5	−15.2
	7	5	5	4	5	25	33	30	13	11	10	7	155

	J	F	M	A	M	J	J	A	S	O	N	D	Yr.
47.	19.2	19.6	18.4	16.4	13.8	11.8	10.8	11.3	12.6	16.3	15.9	17.7	15.2
	84	104	71	109	122	140	140	109	97	106	81	79	1242
48.	28.2	27.9	28.3	28.2	26.8	25.4	25.1	25.8	27.7	29.1	29.2	28.7	27.6
	341	338	274	121	9	1	2	5	17	66	156	233	1562
49.	21.9	21.9	21.2	18.3	15.7	13.1	12.3	13.4	15.3	17.6	19.4	21.0	17.6
	104	125	129	101	115	141	94	83	72	80	77	86	1205
50.	−7.2	−7.2	−4.4	−0.6	4.4	8.3	10.0	8.3	5.0	1.1	−3.3	−6.1	0.7
	84	66	86	62	89	81	79	94	150	145	117	79	1132
51.	−4.4	−8.9	−15.5	−22.8	−23.9	−24.4	−26.1	−26.1	−24.4	−18.8	−10.0	−3.9	−17.4
	13	18	10	10	10	8	5	8	10	5	5	8	110

Station No.	City	Location	Elevation	Koeppen Classification
		North America		
1.	Albuquerque, N.M.	lat. 35°05′N long. 106°40′W	1593 m	BWk
2.	Calgary, Canada	lat. 51°03′N long. 114°05′W	1062 m	Dfb
3.	Charleston, S.C.	lat. 32°47′N long. 79°56′W	18 m	Cfa
4.	Fairbanks, Alaska	lat. 64°50′N long. 147°48′W	134 m	Dfc
5.	Goose Bay, Canada	lat. 53°19′N long. 60°33′W	45 m	Dfb
6.	Lethbridge, Canada	lat. 49°40′N long. 112°39′W	920 m	Dfb
7.	Los Angeles, Calif.	lat. 34°00′N long. 118°15′W	29 m	BSk
8.	Miami, Fla.	lat. 25°45′N long. 80°11′W	2 m	Am
9.	Peoria, Ill.	lat. 40°45′N long. 89°35′W	180 m	Dfa
10.	Portland, Me.	lat. 43°40′N long. 70°16′W	14 m	Dfb
11.	Salt Lake City, Utah	lat. 40°46′N long. 111°52′W	1288 m	BSk
12.	San Diego, Calif.	lat. 32°43′N long. 117°10′W	26 m	BSk
13.	St. Louis, Mo.	lat. 38°39′N long. 90°15′W	172 m	Cfa
14.	Washington, D.C.	lat. 38°50′N long. 77°00′W	20 m	Cfa
15.	Yuma, Ariz.	lat. 32°40′N long. 114°40′W	62 m	BWh

Station No.	City	Location	Elevation	Koeppen Classification
		South America		
16.	Iquitos, Peru	lat. 3°39'S long. 73°18'W	115 m	Af
17.	Manaus, Brazil	lat. 3°01'S long. 60°00'W	60 m	Am
18.	Quito, Ecuador	lat. 0°17'S long. 78°32'W	2766 m	Cfb
19.	Rio de Janeiro, Brazil	lat. 22°50'S long. 43°20'W	26 m	Aw
20.	Santa Cruz, Argentina	lat. 50°01'S long. 60°30'W	111 m	BSk
		Europe		
21.	Bergen, Norway	lat. 60°24'N long. 5°20'E	44 m	Cfb
22.	Berlin, Germany	lat. 52°28'N long. 13°26'E	50 m	Cfb
23.	Brest, France	lat. 48°24'N long. 4°30'W	103 m	Cfb
24.	Lisbon, Portugal	lat. 38°43'N long. 9°05'W	93 m	Csa
25.	Moscow, U.S.S.R.	lat. 55°45'N long. 37°37'E	156 m	Dfb
26.	Rome, Italy	lat. 41°52'N long. 12°37'E	3 m	Csa
27.	Santis, Switzerland	lat. 47°15'N long. 9°21'E	2496 m	ET
28.	Stockholm, Sweden	lat. 59°21'N long. 18°00'E	52 m	Dfb
		Africa		
29.	Benghazi, Libya	lat. 32°06'N long. 20°06'E	25 m	BSh
30.	Coquilhatville, Zaire	lat. 0°01'N long. 18°17'E	21 m	Af
31.	Dakar, Senegal	lat. 14°40'N long. 17°28'W	23 m	BSh
32.	Faya, Chad	lat. 18°00'N long. 21°18'E	251 m	BWh
33.	Nairobi, Kenya	lat. 1°16'S long. 36°47'E	1791 m	Csb
34.	Salisbury, Rhodesia	lat. 17°50'S long. 30°52'E	1449 m	Cwb
		Asia		
35.	Cherrapunji, India	lat. 25°15'N long. 91°44'E	1313 m	Cwb
36.	Djakarta, Indonesia	lat. 6°11'S long. 106°45'E	8 m	Am

Station No.	City	Location	Elevation	Koeppen Classification
37.	Lhasa, Tibet	lat. 29°40'N long. 91°07'E	3685 m	Cwb
38.	Madras, India	lat. 13°00'N long. 80°11'E	16 m	Aw
39.	Novaya Zemlya, U.S.S.R.	lat. 72°23'N long. 54°46'E	15 m	ET
40.	Omsk, U.S.S.R.	lat. 54°48'N long. 73°19'E	85 m	Dfb
41.	Peking, China	lat. 39°57'N long. 116°23'E	52 m	Dwa
42.	Rangoon, Burma	lat. 16°46'N long. 96°10'E	23 m	Am
43.	Saigon, Viet Nam	lat. 10°49'N long. 106°40'E	10 m	Aw
44.	Tokyo, Japan	lat. 35°41'N long. 139°46'E	6 m	Cfa
45.	Urumchi, China	lat. 43°47'N long. 87°43'E	912 m	Dfa
46.	Verkhoyansk, U.S.S.R.	lat. 67°33'N long. 133°23'E	137 m	Dfd
	Australia and New Zealand			
47.	Auckland, New Zealand	lat. 37°43'S long. 174°53'E	49 m	Csb
48.	Darwin, Australia	lat. 12°26'S long. 131°00'E	27 m	Aw
49.	Sydney, Australia	lat. 33°52'S long. 151°17'E	42 m	Cfb
	Greenland			
50.	Ivigtut, Greenland	lat. 61°12'N long. 48°10'W	29 m	ET
	Antarctica			
51.	McMurdo Station, Antarctica	lat. 77°53'S long. 167°00'E	2 m	EF

Glossary

Absolute Humidity   The weight of water vapor per volume of air (usually expressed as grams of water vapor per cubic meter of air).

Absolute Instability   Air that has a lapse rate that is greater than the dry adiabatic rate (1°C per 100 meters).

Absolute Stability   Air with a lapse rate less than the wet adiabatic rate.

Absolute Zero   The zero point on the Kelvin temperature scale representing the temperature at which all molecular motion is presumed to cease.

Adiabatic Temperature Change   Cooling or warming of air caused when air is allowed to expand or is compressed, not because heat is added or subtracted.

Advection   Horizontal convective motion, such as wind.

Advection Fog   Fog formed when warm, moist air is blown over a cool surface and chilled below the dew point.

Air   A mixture of many discrete gases, of which nitrogen and oxygen are most abundant, in which varying quantities of tiny solid and liquid particles are suspended.

Air Mass   A large body of air, usually 1600 kilometers or more across, that is characterized by homogeneous physical properties at any given altitude.

Air-Mass Weather   Conditions experienced in an area as an air mass passes over it. Since air masses are large and relatively homogeneous, air-mass weather will be very constant and may last for several days.

Air Pressure   The pressure exerted by the weight of a column of air above a given point.

Albedo The reflectivity of a substance, usually expressed as a percentage of the incident radiation reflected.

Aleutian Low A large cell of low pressure centered over the Aleutian Islands of the North Pacific during the winter.

Altimeter An aneroid barometer calibrated to indicate altitude instead of pressure.

Altitude (of the sun) The angle of the sun above the horizon.

Analog Method A statistical approach to weather forecasting in which current conditions are matched with records of similar past weather events with the idea that the succession of events in the past will be paralleled by current conditions.

Anemometer An instrument used to determine wind speed.

Aneroid Barometer An instrument for measuring air pressure that consists of evacuated metal chambers that are very sensitive to variations in air pressure.

Annual March of Temperature The annual cycle of air temperature changes. Usually, the lowest monthly mean occurs a month, or on occasion two months, after the winter solstice while the highest monthly mean commonly occurs a month or two after the summer solstice. January and July generally represent the extreme months.

Annual Mean An average of the 12 monthly means.

Annual Temperature Range The difference between the warmest and coldest monthly means.

Anticyclone An area of high atmospheric pressure characterized by diverging and rotating winds and subsiding air aloft.

Anticyclonic Flow Winds blow out and clockwise about an anticyclone (high) in the Northern Hemisphere, and they blow out and counterclockwise about an anticyclone in the Southern Hemisphere.

Aphelion The point in the orbit of a planet that is farthest from the sun.

Arctic (A) Air Mass A bitterly cold air mass that formed over the frozen Arctic Ocean.

Arctic Sea Smoke A dense and often extensive steam fog that occurs over high-latitude ocean areas in winter.

Arid See Desert.

Astronomical Theory A theory of climatic change first developed by the Yugoslavian astronomer Milankovitch. It is based upon changes in the shape of the earth's orbit, variations in the obliquity of the earth's axis, and the wobbling of the earth's axis.

Atmosphere The gaseous portion of a planet; the planet's envelope of air; one of the traditional subdivisions of the earth's physical environment.

Aurora A bright display of ever-changing light caused by solar radiation interacting with the upper atmosphere in the region of the poles. Termed aurora borealis in the Northern Hemisphere and aurora australis in the Southern Hemisphere.

Backing Wind Shift A wind shift in a counterclockwise direction, such as a shift from east to north.

Barograph A recording barometer.

Barometric Tendency See Pressure Tendency.

Beaufort Scale A scale that may be used for estimating wind speed when an anemometer is not available.

Bergeron Process A theory that relates the formation of precipitation to super-cooled clouds, freezing nuclei, and the different saturation levels of ice and liquid water.

Bimetal Strip A thermometer consisting of two thin strips of metal that are welded together and have widely different coefficients of thermal expansion. When temperature changes, the two metals expand or contract unequally and cause changes in the curvature of the element. Commonly used in thermographs.

Biosphere The totality of life forms on earth.

Blackbody A material that is able to absorb 100 percent of the radiation it receives.

Blizzard A violent and extremely cold wind laden with dry snow picked up from the ground.

Bora In the region of the eastern shore of the Adriatic Sea, a cold, dry north-easterly wind that blows down from the mountains.

Buys Ballott's Rule With your back to the wind in the Northern Hemisphere low pressure will be to your left and high pressure to your right. The reverse is true in the Southern Hemisphere.

Calorie The amount of heat required to raise the temperature of 1 gram of water 1°C.

Ceiling The height ascribed to the lowest layer of clouds or obscuring phenomena when the sky is reported as broken, overcast, or obscured and the clouds are not classified "thin" or "partial." The ceiling is termed unlimited when the foregoing conditions are not present.

Celsius Scale A temperature scale (at one time called the centigrade scale) devised by Anders Celsius in 1742 and used where the metric system is used. For water at sea level, 0° is designated the ice point and 100° the steam point.

Centrifugal Force The tendency of a particle to move in a straight line when rotated creates this imaginary outward force.

Centripetal Acceleration The inward turning of air resulting from a force acting to keep it in circular motion about a fixed center.

Chinook The name applied to a foehn wind in the Rocky Mountains.

Circle of Illumination The line (great circle) separating daylight from darkness on the earth.

Cirrus One of three basic cloud forms; also one of the three high cloud types. They are thin, delicate ice crystal clouds often appearing as veil-like patches or thin, wispy fibers.

Climate A description of aggregate weather conditions; the sum of all statistical weather information that helps describe a place or region.

Climatic Change A study dealing with variations in climate on many different time scales, from decades to millions of years, and the possible causes of such variations.

391

Climatic Feedback Mechanisms Since the atmosphere is a very complex interactive physical system, when one of the system's elements is altered, several different possible outcomes may result. These various possibilities are climatic feedback mechanisms.

Cloud A form of condensation best described as dense concentrations of suspended water droplets or tiny ice crystals.

Cloud Seeding The introduction of freezing nuclei (often silver-iodide crystals) into a cloud in order to increase or induce rainfall.

Clouds of Vertical Development Clouds that have their bases in the low height range but extend upward into the middle or high altitudes.

Cold Front The discontinuity at the forward edge of an advancing cold air mass that is displacing warmer air in its path.

Cold-Type Occluded Front A front that forms when the air behind the cold front is colder than the air underlying the warm front it is overtaking.

Cold Wave A rapid and marked fall of temperature. The Weather Service applies this term to a fall of temperature in 24 hours equaling or exceeding a specified number of degrees and reaching a specified minimum temperature or lower. These specifications vary for different parts of the country and for different periods of the year.

Collision–Coalescence Process A theory of raindrop formation in warm clouds (above 0°C) in which large cloud droplets ("giants") collide and join together with smaller droplets to form a raindrop. Opposite electrical charges may bind the cloud droplets together.

Condensation The change of state from a gas to a liquid.

Condensation Nuclei Microscopic particles that serve as surfaces upon which water vapor condenses.

Conditional Instability Moist air with a lapse rate between the dry and wet adiabatic rates.

Conduction The transfer of heat through matter by molecular activity. Energy is transferred through collisions from one molecule to another.

Constant Pressure Surface A surface along which the atmospheric pressure is everywhere equal at any given moment.

Continental (c) Air Mass An air mass that formed over land; it is normally relatively dry.

Continental Climate A climate lacking marine influence and that is characterized by more extreme temperatures than in marine climates; hence, it has a relatively high annual temperature range for its latitude.

Contrail A cloudlike streamer frequently observed behind aircraft flying in clear, cold, humid air caused by the addition to the atmosphere of water vapor from engine exhaust gases.

Controls of Temperature Those factors that cause variations in temperature from place to place, such as latitude and altitude.

Convection The transfer of heat by the movement of a mass or substance. It can only take place in fluids.

Convection Cell Circulation that results from the uneven heating of a fluid; the warmer parts of the fluid expand and rise because of their buoyancy and the cooler parts sink.

Convergence The condition that exists when the distribution of winds within a given area results in a net horizontal inflow of air into the area. In convergence at lower levels the removal of the resulting excess is accomplished by an upward movement of air; consequently, areas of convergent winds are regions favorable to cloud formation and precipitation.

Coriolis Effect (Force) The deflective force of the earth's rotation on all free-moving objects, including the atmosphere and oceans. Deflection is to the right in the Northern Hemisphere and to the left in the Southern Hemisphere.

Country Breeze A circulation pattern characterized by a light wind blowing into a city from the surrounding countryside. It is best developed on clear and otherwise calm nights when the urban heat island is most pronounced.

Cumulus One of three basic cloud forms; also the name given one of the clouds of vertical development. Cumulus are billowy individual cloud masses that often have flat bases.

Cumulus Stage The initial stage in thunderstorm development in which the growing cumulonimbus is dominated by strong updrafts.

Cup Anemometer See Anemometer.

Cyclogenesis The process that creates or develops a new cyclone; also the process that produces an intensification of a preexisting cyclone.

Cyclone An area of low atmospheric pressure characterized by rotating and converging winds and ascending air.

Cyclonic Flow Winds blow in and counterclockwise about a cyclone (low) in the Northern Hemisphere and in and clockwise about a cyclone in the Southern Hemisphere.

Daily March of Temperature The daily cycle of air temperature changes. Commonly the minimum temperature occurs near sunrise and the maximum temperature occurs during midafternoon.

Daily Mean The mean temperature for a day that is determined by averaging the 24 hourly readings or, more commonly, by averaging the maximum and minimum temperatures for a day.

Daily Range The difference between the maximum and minimum temperatures for a day.

Degassing The release of gases dissolved in molten rock.

Desert One of the two types of dry climate; the driest of the dry climates.

Dew Point The temperature to which air has to be cooled in order to reach saturation.

Diffused Daylight Solar energy scattered and reflected in the atmosphere that reaches the earth's surface in the form of diffuse blue light from the sky.

Discontinuity A zone characterized by a comparatively rapid transition of meteorological elements.

Dissipating Stage The final stage of a thunderstorm that is dominated by downdrafts and entrainment that leads to the evaporation of the cloud structure.

Diurnal Daily, especially pertaining to actions that are completed within 24 hours and that recur every 24 hours.

Divergence The condition that exists when the distribution of winds within a given area results in a net horizontal outflow of air from the region. In divergence at lower levels the resulting deficit is compensated for by a downward movement of air from aloft; hence, areas of divergent winds are unfavorable to cloud formation and precipitation.

Doldrums The equatorial belt of calms or light variable winds lying between the two trade-wind belts.

Drizzle Precipitation from stratus clouds consisting of numerous tiny droplets.

Dry Adiabatic Rate The rate of adiabatic cooling or warming in unsaturated air. The rate of temperature change is 1°C per 100 meters.

Dry Climate A climate in which yearly precipitation is not as great as the potential loss of water by evaporation.

Dry-Summer Subtropical Climate A climate located on the west sides of continents between latitudes 30 degrees and 45 degrees. It is the only humid climate with a strong winter precipitation maximum.

Eccentricity The variation of an ellipse from a circle.

Electromagnetic Radiation See Radiation.

Elements (atmospheric) Those quantities or properties of the atmosphere that are measured regularly and that are used to express the nature of weather and climate.

Entrainment The influx of cool, dry air from the surrounding environment into the downdraft of a cumulonimbus cloud; a process that acts to intensify the downdraft.

Equatorial (E) Air Mass A warm to hot air mass that formed in the tropics near the equator.

Equatorial Low A quasicontinuous belt of low pressure lying near the equator and between the subtropical highs.

Equinox The point in time when the vertical rays of the sun are striking the equator. March 21 is the vernal or spring equinox in the Northern Hemisphere and September 22 or 23 is the autumnal equinox in the Northern Hemisphere. Length of daylight and darkness is equal at all latitudes at equinox.

Evaporation Process by which a liquid is transformed into a gas.

Eye A zone of calm and scattered clouds at the center of a hurricane; usually averages 20 kilometers in diameter.

Fahrenheit Scale A temperature scale devised by Gabriel Daniel Fahrenheit in 1714 and used in the English System. For water at sea level, 32° is designated the ice point and 212° the steam point.

Fall Wind See Katabatic Wind.

Fixed Points   Reference points, such as the steam point and the ice point, used in the construction of temperature scales.

Foehn   A warm, dry wind on the lee side of a mountain range that owes its relatively high temperature largely to adiabatic heating during descent down mountain slopes.

Fog   A cloud with its base at or very near the earth's surface.

Freezing   The change of state from a liquid to a solid.

Freezing Nuclei   Solid particles that have a crystal form resembling that of ice; they serve as cores for the formation of ice crystals.

Front   A boundary (discontinuity) separating air masses of different densities, one warmer and often higher in moisture content than the other.

Frontal Fog   Fog formed when rain evaporates as it falls through a layer of cool air.

Frontal Wedging   Lifting of air resulting when cool air acts as a barrier over which warmer, lighter air will rise.

Frontogenesis   The beginning or creation of a front.

Frontolysis   The destruction and dying of a front.

Gas Law   The pressure exerted by a gas is proportional to its density and absolute temperature.

Geostationary Satellite   A satellite that remains over a fixed point because its rate of travel corresponds to the earth's rate of rotation. Because the satellite must orbit at distances of about 35,000 kilometers, photographs from this type of satellite are not as detailed as those from polar satellites.

Geostrophic Wind   A wind, usually above a height of 600 meters, that blows parallel to the isobars.

Glaze   A coating of ice on objects formed when supercooled rain freezes on contact. A storm that produces glaze is called an icing storm.

Global Circulation   The general circulation of the atmosphere; the average flow of air over the entire globe.

Greenhouse Effect   The transmission of shortwave solar radiation by the atmosphere coupled with the selective absorption of longer wavelength terrestrial radiation, especially by water vapor and carbon dioxide.

Hadley Cell   The thermally driven circulation system of equatorial and tropical latitudes consisting of two convection cells, one in each hemisphere. The existence of this circulation system was first proposed by George Hadley in 1735 as an explanation for the trade winds.

Hail   Precipitation in the form of hard, round pellets or irregular lumps of ice that may have concentric shells formed by the successive freezing of layers of water.

Heat   The kinetic energy of random molecular motion.

Heterosphere   A zone of the atmosphere beyond about 80 kilometers where the gases are arranged into four roughly spherical shells, each with a distinctive composition.

395

High Clouds Clouds that normally have their bases above 6000 meters; bases may be lower in winter and at high latitude locations.

Homosphere A zone of the atmosphere extending from the surface to about 80 kilometers that is uniform in terms of the proportions of its component gases.

Horse Latitudes A belt of calms or light variable winds and subsiding air located near the center of the subtropical high.

Humid Continental Climate A relatively severe climate characteristic of broad continents in the middle latitudes between approximately 40 degrees and 50 degrees north latitude. This climate is not found in the Southern Hemisphere where the middle latitudes are dominated by the oceans.

Humidity A general term referring to water vapor in the air but not to liquid droplets of fog, cloud, or rain.

Humid Subtropical Climate A climate generally located on the eastern side of a continent and characterized by hot, sultry summers and cool winters.

Hurricane A tropical cyclonic storm having winds in excess of 119 kilometers per hour; also known as typhoons (western Pacific), willy-willys (Australia), and cyclones (Indian Ocean).

Hydrologic Cycle The continuous movement of water from the oceans to the atmosphere (by evaporation), from the atmosphere to the land (by condensation and precipitation), and from the land back to the sea (via stream flow).

Hydrosphere The water portion of our planet; one of the traditional subdivisions of the earth's physical environment.

Hygrometer An instrument designed to measure relative humidity. Unlike a psychrometer, the relative humidity can be read directly without the use of tables.

Hygroscopic Nuclei Condensation nuclei having a high affinity for water, such as salt particles.

Ice Cap Climate A climate that has no monthly means above freezing and supports no vegetative cover except for a few scattered high mountain areas. This climate, with its perpetual ice and snow, is confined to the ice caps of Greenland and Antarctica.

Icelandic Low A large cell of low pressure centered over Iceland and Southern Greenland in the North Atlantic during the winter.

Ice Point The temperature at which ice melts.

Inclination of the Axis The tilt of the earth's axis from the perpendicular to the plane of the earth's orbit (plane of the ecliptic). Currently, the inclination is about 23.5 degrees away from the perpendicular.

Infrared Radiation with a wavelength from 0.7 to 200 micrometers.

Ionosphere A complex atmospheric zone of ionized gases that extends between 80 and 400 kilometers, thus coinciding with the lower thermosphere and heterosphere.

Isobar A line drawn on a map connecting points of equal barometric pressure, usually corrected to sea level.

Isohyet A line connecting places having equal rainfall.

Isotherm A line connecting points of equal air temperature.

ITC See Zone of Intertropical Convergence.

Jet Stream Swift geostrophic air streams in the upper troposphere that meander in relatively narrow belts.

Jungle An almost impenetrable growth of tangled vines, shrubs, and short trees that characterizes areas where the tropical rain forest has been cleared.

Katabatic Wind The flow of cold dense air downslope under the influence of gravity; the direction of flow is controlled largely by topography.

Kelvin Scale A temperature scale (also called the absolute scale) used primarily for scientific purposes and having intervals equivalent to those on the Celsius scale but beginning at absolute zero.

Kennelly–Heaviside Layer Layer in the lower portion of the ionosphere in which most of the significant reflection of long radio waves occurs.

Kinetic Energy The energy of motion.

Lake-Effect Snow Snow showers associated with a cP air mass to which moisture and heat are added from below as it traverses a large and relatively warm lake (such as one of the Great Lakes), rendering the air mass very moist and unstable.

Land and Sea Breeze Local winds in coastal areas in which flow is from the water by day and from the land by night.

Latent Heat The energy absorbed or released during a change of state.

Lightning A sudden flash of light generated by the flow of electrons between oppositely charged parts of a cumulonimbus cloud or between the cloud and the ground.

Lithosphere The solid portion of the earth; one of the traditional subdivisions of the earth's physical environment.

Long-Range Forecasts An estimate of rainfall and temperatures for a period beyond 3 to 5 days, usually for 7-to 30-day period. Such forecasts are not as detailed or reliable as those for shorter periods.

Low Clouds Clouds that form below a height of 2000 meters.

Macroscale Wind Systems Phenomena such as cyclones and anticyclones that persist for days or weeks and have a horizontal dimension of hundreds to several thousands of kilometers; also features of the atmospheric circulation that persist for weeks or months and have horizontal dimensions of up to 10,000 kilometers.

Marine Climate A climate dominated by the ocean; because of the moderating effect of water, places having this climate are considered relatively mild.

Marine West Coast Climate A climate found on windward coasts from latitudes 40 degrees to 65 degrees and dominated by maritime air masses. Winters are mild and summers are cool.

Maritime (m) Air Mass An air mass that originated over an oceanic source region. These air masses are relatively humid.

Mature Stage The second of the three stages of a thunderstorm. This stage is characterized by violent weather as downdrafts exist side by side with updrafts.

Maximum Thermometer   A thermometer that measures the maximum temperature for a given period of time, usually 24 hours. A constriction in the base of the glass tube allows mercury to rise but prevents it from returning to the bulb until the thermometer is shaken or whirled.

Mediterranean Climate   A common name applied to the dry-summer subtropical climate.

Melting   The change of stage from a solid to a liquid.

Mercurial Barometer   A mercury-filled glass tube in which the height of the mercury column is a measure of air pressure.

Mesopause   The boundary between the mesosphere and the thermosphere.

Mesoscale Wind Systems   Small convective cells that exist for minutes or hours, such as thunderstorms, tornadoes, and the land and sea breeze. Typical horizontal dimensions range from 1 to 100 kilometers.

Mesosphere   The layer of the atmosphere above the stratosphere where temperatures drop fairly rapidly with increasing height.

Microscale Wind Systems   Phenomena, such as turbulence, with life spans of less than a few minutes that affect small areas and are strongly influenced by local conditions of temperature and terrain.

Middle Clouds   Clouds occupying the height range from 2000 to 6000 meters.

Middle-Latitude Cyclone   See Wave Cyclone.

Millibar   The standard unit of pressure measurement used by the National Weather Service. One millibar (mb) equals 100 newtons per square meter.

Minimum Thermometer   A thermometer that measures the minimum temperature for a given period of time, usually 24 hours. By checking the small dumbbell-shaped index, the minimum temperature can be read.

Mistral   A cold northwest wind that blows into the western Mediterranean basin from higher elevations to the north.

Monsoon   Seasonal reversal of wind direction associated with large continents, especially Asia. In winter the wind blows from land to sea; in summer it blows from sea to land.

Monthly Mean   The mean temperature for a month that is calculated by averaging the daily means.

Mountain and Valley Breeze   Daily upvalley winds and nightly downvalley winds commonly encountered in mountain valleys.

Newton   A unit of force used in physics. One newton is the force that is necessary to accelerate 1 kilogram of mass 1 meter per second per second.

Normal   The average value of a meteorological element over any fixed period of years that is recognized as standard for the country and for the element of concern.

Normal Lapse Rate   The average drop in temperature with increasing height in the troposphere; about $6.5°C$ per kilometer.

Northeaster   The term applied to describe the weather associated with an incursion of mP air into the Northeast from the North Atlantic; strong northeast winds, freezing or near-freezing temperatures, and the possiblity of precipitation make this an unwelcome weather event.

Numerical Weather Prediction (NWP) Forecasting the behavior of atmospheric disturbances based upon the solution of the governing fundamental equations of hydrodynamics, subject to the observed initial conditions. Because of the vast number of calculations involved, high-speed computers are always used.

Obliquity The angle between the planes of the earth's equator and orbit.

Occluded Front A front formed when a cold front overtakes a warm front.

Occlusion The overtaking of one front by another.

Orographic Lifting Mountains or highlands acting as barriers to the flow of air force the air to ascend. The air cools adiabatically, and clouds and precipitation may result.

Overrunning Warm air gliding up a retreating cold air mass.

Oxygen-Isotope Analysis A method of deciphering past temperatures based upon the precise measurement of the ratio between two isotopes of oxygen, ^{16}O and ^{18}O. Analysis is commonly made of sea-floor sediments and cores from ice sheets.

Ozone Molecule of oxygen containing three oxygen atoms.

Paleosol An old buried soil that may furnish some evidence of the nature of past climates, because climate is the most important factor in soil formation.

Perihelion The point in the orbit of a planet closest to the sun.

Persistent Forecasts Forecasts that assume that the weather occurring upstream will persist and move on and will affect the areas in its path in much the same way. Persistent forecasts do not account for changes that might occur in the weather system.

Photochemical Reaction A chemical reaction in the atmosphere that is triggered by sunlight, often yielding a secondary pollutant.

Photosynthesis The production of sugars and starches by plants using air, water, sunlight, and chlorophyll. In the process atmospheric carbon dioxide is changed to organic matter and oxygen is released.

Plate Tectonic Theory A theory that states that the outer portion of the earth is made up of several individual pieces, called plates, which move in relation to one another upon a partially molten zone below. As plates move, so do continents, which might explain some climatic changes in the geologic past.

Polar (P) Air Mass A cold air mass that forms in a high-latitude source region.

Polar Climates Climates in which the mean temperature of the warmest month is below 10°C; climates that are too cold to support the growth of trees.

Polar Easterlies In the global pattern of prevailing winds, winds that blow from the ploar high toward the subpolar low. These winds, however, should not be thought of as persistent winds, such as the trade winds.

Polar Front The stormy frontal zone separating air masses of polar origin from air masses of tropical origin.

Polar Front Theory A theory developed by J. Bjerknes and other Scandinavian meteorologists in which the polar front, separating polar and tropical air masses, gives rise to cyclonic disturbances that intensify and move along the front and pass through a succession of stages.

Polar High Anticyclones that are assumed to occupy the inner polar regions and are believed to be thermally induced, at least in part.

Polar Satellite Satellites that orbit the poles at rather low altitudes of a few hundred kilometers and require only 100 minutes per orbit.

Potential Energy The energy that exists by virtue of its position with respect to gravitation.

Precession The slow migration of the earth's axis that traces out a cone over a period of 26,000 years.

Precipitation Fog See Frontal Fog.

Pressure Gradient The amount of pressure change occurring over a given distance.

Pressure Tendency The nature of the change in atmospheric pressure over the past several hours. It can be a useful aid in short-range weather prediction.

Prevailing Westerlies The dominant west-to-east motion of the atmosphere that characterizes the regions on the poleward side of the subtropical highs.

Prevailing Wind A wind that consistently blows from one direction more than from any other.

Primary Pollutant A pollutant emitted directly from an identifiable source.

Psychrometer Device consisting of two thermometers (wet bulb and dry bulb) that is rapidly whirled and, with the use of tables, yields the relative humidity.

Radiation Wavelike energy emitted by any substance that possesses heat. This energy travels through space at 300,000 kilometers per second (the speed of light).

Radiation Fog Fog resulting from radiation cooling of the ground and adjacent air; primarily a nighttime and early morning phenomenon.

Radiosonde A lightweight package of weather instruments fitted with a radio transmitted and carried aloft by a balloon.

Rain shadow A dry area on the lee side of a mountain range.

Rawinsonde A radiosonde that is tracked by radar in order to obtain data on upper-air winds.

Relative Humidity Ratio of the air's water vapor content to its water vapor capacity.

Revolution The motion of one body about another, as the earth about the sun.

Ridge An elongate region of high atmospheric pressure.

Rossby Waves Upper-air waves in the middle and upper troposphere of the middle-latitudes with wavelengths of from 4000 to 6000 kilometers; named for C. G. Rossby, the meteorologist who developed the equations for parameters governing the waves.

Rotation   The spinning of a body, such as the earth, about its axis.

Santa Ana   Local name given a foehn wind in Southern California.

Saturation   The maximum possible quantity of water vapor the air can hold at any given temperature and pressure.

Savanna   A tropical grassland usually with scattered trees and shrubs.

Sea Breeze   A local wind blowing from the sea during the afternoon in coastal areas.

Secondary Pollutant   Pollutant that is produced in the atmosphere by chemical reactions that occur among primary pollutants.

Selva   See Tropical Rain Forest.

Sensible Temperature   The sensation of temperature the human body feels in contrast to the actual temperature of the air as recorded by a thermometer.

Sleet   Frozen or semifrozen rain formed when raindrops pass through a subfreezing layer of air.

Snow   Precipitation in the form of white or translucent ice crystals, chiefly in complex branched hexagonal form and often clustered into snowflakes.

Solar Constant   The rate at which solar radiation is received outside the earth's atmosphere on a surface perpendicular to the sun's rays when the earth is at an average distance from the sun.

Solstice   The point in time when the vertical rays of the sun are striking either the Tropic of Cancer (summer solstice in Northern Hemisphere) or the Tropic of Capricorn (winter solstice in Northern Hemisphere). Solstice represents the longest or shortest day (length of daylight) of the year.

Source Region   The area where an air mass acquires its characteristic properties of temperature and moisture.

Specific Heat   The amount of heat needed to raise 1 gram of a substance 1°C at sea-level atmospheric pressure.

Specific Humidity   The weight of water vapor per weight of a chosen mass of air including the water vapor (usually expressed as grams of water vapor per kilogram of air).

Squall Line   Any nonfrontal line or narrow band of active thunderstorms.

Stable Air   Air that resists vertical displacement. If it is lifted, adiabatic cooling will cause its temperature to be lower than the surrounding environment, and, if it is allowed, it will sink to its original position.

Standard Rain Gauge   Having a diameter of about 20 centimeters, this gauge funnels rain into a cylinder that magnifies precipitation amounts by a factor of 10, allowing for accurate measurement of small amounts.

Stationary Front   A situation in which the surface position of a front does not move; the flow on either side of such a boundary is nearly parallel to the position of the front.

Statistical Methods (in forecasting)   Methods in which tables or graphs are prepared from a long series of observations to show the probability of certain weather events under certain conditions of pressure, temperature, or wind direction.

Steam Fog   Fog having the appearance of steam; produced by evaporation from a warm water surface into the cool air above.

Steam Point   The temperature at which water boils.

Steppe   One of the two types of dry climate. A marginal and more humid variant of the desert that separates it from bordering humid climates. Steppe also refers to the short-grass vegetation associated with this semiarid climate.

Stratopause   The boundary between the stratosphere and the mesosphere.

Stratosphere   The zone of the atmosphere above the troposphere that is characterized at first by isothermal conditions and then a gradual temperature increase. The earth's ozone is concentrated here.

Stratus   One of three basic cloud forms; also the name given one of the low clouds. They are sheets or layers that cover much or all of the sky.

Subarctic Climate   A climate found north of the humid continental climate and south of the polar climate and characterized by bitterly cold winters and short cool summers. Places within this climatic realm experience the highest annual temperature ranges on earth.

Sublimation   The conversion of a solid directly to a gas without passing through the liquid state, or vice versa.

Subpolar Low   Low pressure located at about the latitudes of the Arctic and Antarctic circles. In the Northern Hemisphere the low takes the form of individual oceanic cells; in the Southern Hemisphere there is a deep and continuous trough of low pressure.

Subsidence   An extensive sinking motion of air, most frequently occurring in anticyclones. The subsiding air is warmed by compression and becomes more stable.

Sunspot   A dark area on the sun associated with powerful magnetic storms that extend from the sun's surface deep into the interior.

Subtropical High   Not a continuous belt of high pressure but rather several semipermanent anticyclonic centers characterized by subsidence and divergence located roughly between latitudes 25 degrees and 35 degrees.

Supercooled Water   Water droplets that remain in the liquid state at temperatures well below freezing.

Synoptic Weather Charts   Weather charts describing the state of the atmosphere over a large area at a given moment.

Synoptic Weather Forecasting   A system of forecasting based upon careful studies of synoptic weather charts over a period of years; from these studies a set of empirical rules was established to aid the forecaster in estimating the rate and direction of weather system movements.

Taiga   The northern coniferous forest; also a name applied to the subarctic climate.

Temperature   A measure of the degree of hotness or coldness of a substance.

Temperature–Humidity Index (THI) A well-known and often used guide to human comfort or discomfort based upon the conditions of temperature and relative humidity.

Temperature Inversion A layer in the atmosphere of limited depth where the temperature increases rather than decreases with height.

Thermal Low An area of low atmospheric pressure created by abnormal surface heating.

Thermistor An electric thermometer consisting of a conductor whose resistance to the flow of current is temperature-dependent; commonly used in radiosondes.

Thermocouple An electric thermometer that operates on the principle that differences in temperature between the junction of two unlike metal wires in a circuit will induce a current to flow.

Thermograph An instrument that continuously records temperature.

Thermometer An instrument for measuring temperature; in meteorology, generally used to measure the temperature of the air.

Thermosphere The zone of the atmosphere beyond the mesosphere in which there is a rapid rise in temperature with height.

Thunder The sound emitted by rapidly expanding gases along the channel of a lightning discharge.

Thunderstorm A storm produced by a cumulonimbus cloud and always accompanied by lightning and thunder. It is of relatively short duration and usually accompanied by strong wind gusts, heavy rain, and sometimes hail.

Tipping-Bucket Gauge A recording rain gauge consisting of two compartments ("buckets") each capable of holding 0.025 centimeter of water. When one compartment fills, it tips and the other compartment takes its place.

TIROS 1 The first weather satellite; launched by the United States on April 1, 1960.

Tornado A violently rotating column of air attended by a funnel-shaped or tubular cloud extending downward from a cumulonimbus cloud.

Tornado Warning A warning issued when a tornado has actually been sighted in an area or is indicated by radar.

Tornado Watch A forecast issued for areas of about 65,000 square kilometers indicating that conditions are such that tornadoes may develop; they are intended to alert people to the possibility of tornadoes.

Trace of Precipitation An amount less than 0.025 centimeter.

Trade Winds Two belts of winds that blow almost constantly from easterly directions and are located on the equatorward sides of the subtropical highs.

Transpiration The release of water vapor to the atmosphere by plants.

Tropic of Cancer The parallel of latitude, $23\frac{1}{2}$ degrees north latitude, marking the northern limit of the sun's vertical rays.

Tropic of Capricorn The parallel of latitude, $23\frac{1}{2}$ degrees south latitude, marking the southern limit of the sun's vertical rays.

Tropical (T) Air Mass A warm to hot air mass that formed in the subtropics.

403

Tropical Rain Forest   A luxuriant broadleaf evergreen forest; also the name given the climate associated with this vegetation.

Tropical Wet and Dry   A climate that is transitional between the wet tropics and the tropical steppes.

Tropopause   The boundary between the troposphere and the stratosphere.

Troposphere   The lowermost layer of the atmosphere marked by considerable turbulence and generally a decrease in temperature with increasing height.

Trough   An elongate region of low atmospheric pressure.

Tundra Climate   Found almost exclusively in the Northern Hemisphere or at high altitudes in many mountainous regions. A treeless climatic realm of sedges, grasses, mosses, and lichens that is dominated by a long, bitterly cold winter.

Ultraviolet   Radiation with a wavelength from 0.2 to 0.4 micrometer.

Unstable Air   Air that does not resist vertical displacement. If it is lifted, its temperature will not cool as rapidly as the surrounding environment, and, hence, it will continue to rise on its own.

Upslope Fog   Fog created when air moves up a slope and cools adiabatically.

Urban Heat Island   Refers to the fact that temperatures within a city are generally higher than in surrounding rural areas.

Vapor Pressure   That part of the total atmospheric pressure attributable to its water vapor content.

Variable Sun Theory   One of the most persistent theories of climatic change that is based upon the idea that the sun is a variable star and that its output of energy varies through time.

Veering Wind Shift   A wind shift in a clockwise direction, such as a shift from east to south.

Virga   Wisps or streaks of water or ice particles falling out of a cloud but evaporating before reaching the earth's surface.

Visibility   The greatest distance that prominent objects can be seen and identified by unaided, normal eyes.

Visible Light   Radiation with a wavelength from 0.4 to 0.7 micrometer.

Volcanic Dust Theory   A theory of climatic change based upon the premise that suspended volcanic material in the atmosphere filters out a portion of the incoming solar radiation, which, in turn, leads to lower air temperatures. Thus, atmospheric temperature fluctuations may be linked to the amount of explosive volcanism.

Warm Front   A gently sloping frontal surface in which there is active movement of warm air over cold air.

Warm-Type Occluded Front   A front that forms when the air behind the cold front is warmer than the air underlying the warm front it is overtaking.

Water Hemisphere   A term used to refer to the Southern Hemisphere where the oceans cover 81 percent of the surface (compared to 61 percent in the Northern Hemisphere).

Wave Cyclone   A cyclone that forms and moves along a front. The circulation around the cyclone tends to produce the wavelike deformation of the front.

Weather   The state of the atmosphere at any given time.

Weather Analysis   The stage prior to developing a weather forecast. This stage involves collecting, compiling, and transmitting observational data.

Weather Forecasting   Predicting the future state of the atmosphere.

Weighing Gauge   A recording precipitation gauge consisting of a cylinder that rests on a spring balance.

Westerlies   See Prevailing Westerlies.

Wet Adiabatic Rate   The rate of adiabatic cooling or warming in saturated air. The rate of temperature change is variable, but it is always less than the dry adiabatic rate.

White Frost   Ice crystals that form on surfaces instead of dew when the dew point is below freezing.

Wind   Air flowing horizontal with respect to the earth's surface.

Wind-Chill   A measure of sensible temperature that uses the effects of wind and temperature on the cooling rate of the human body. The wind-chill chart translates the cooling power of the atmosphere with the wind to a temperature under calm conditions.

Wind Vane   An instrument used to determine wind direction.

World Meteorological Organization (WMO)   Established by the United Nations, the WMO consists of more than 130 nations. The organization is responsible for gathering needed observational data and compiling some general prognostic charts.

Zone of Intertropical Convergence (ITC)   The zone of general convergence between the Northern and Southern Hemisphere trade winds.

Index

T

Trough, 136, 160, 208
Tundra climate, 348–349
Typhoon, 232

U

Urban climate, 272–283
Urban heat island, 273–278

V

Valley breeze, 163
Vapor pressure, 68–69
Variable sun theory, 298–301
Veering winds, 205
Volcanic dust theory, 292, 294–295

W

Warm front, 133–195, 358
Water balance (*see* Hydrologic cycle)
Water vapor, 4
Wave cyclone, 191–192, 200–208
   life cycle, 200–202
   weather, 202–206, 247–257
Weather, 2
   elements, 3
Weather analysis, 243–246
Weather forecasting (*see* Forecasting)
Weather maps, 243–247
   decoding, 357–366
Weather modification, 114–117
Weather prediction (*see* Forecasting)
Westerlies, 151, 155–161
Willy-willys, 232

Wind, 126–168
   anticyclonic, 136–137
   backing, 205
   Beaufort scale, 143–144
   bora, 164
   Buys Ballott's law, 135
   centripetal acceleration, 138
   chinook, 163–164
   convergence, 139–142
   curved flow, 136–138
   cyclonic, 136–137
   divergence, 139–142
   foehn, 163–164
   formation, 139–142
   friction layer, 138–139
   geostrophic, 133–135
   Hadley cell, 150
   intertropical convergence zone, 152, 154
   katabatic, 164
   land breeze, 162
   local, 161–164
   macroscale, 146–147
   measurement, 143–144
   microscale, 147
   mistral, 164
   mountain breeze, 163
   norther, 161
   polar easterlies, 152
   prevailing, 143
   Rossby waves, 159
   Santa Ana, 164
   sea breeze, 162
   synoptic scale, 146–147
   trades, 151–152
   valley breeze, 163
   veering, 205
Wind chill, 41–43
Wind vane, 143
World Meteorological Organization, 243